FLAVONOIDS AS NUTRACEUTICALS

AAP Advances in Nutraceuticals

FLAVONOIDS AS NUTRACEUTICALS

Edited by

Rajesh K. Kesharwani, PhD

Deepika Saini, PhD

Raj K. Keservani, PhD, MPharm

Anil Kumar Sharma, PhD, MPharm

AAP | APPLE ACADEMIC PRESS

First edition published 2024

Apple Academic Press Inc.
1265 Goldenrod Circle, NE,
Palm Bay, FL 32905 USA

760 Laurentian Drive, Unit 19,
Burlington, ON L7N 0A4, CANADA

CRC Press
2385 NW Executive Center Drive,
Suite 320, Boca Raton FL 33431

4 Park Square, Milton Park,
Abingdon, Oxon, OX14 4RN UK

Library and Archives Canada Cataloguing in Publication

Title: Flavonoids as nutraceuticals / edited by Rajesh K. Kesharwani, PhD, Deepika Saini, PhD, Raj K. Keservani, PhD, Anil Kumar Sharma, PhD.
Names: Kesharwani, Rajesh K., 1978- editor. | Saini, Deepika, editor. | Keservani, Raj K., 1981-editor. | Sharma, Anil K., 1980- editor.
Series: AAP advances in nutraceuticals.
Description: First edition. | Series statement: AAP advances in nutraceuticals | Includes bibliographical references and index.
Identifiers: Canadiana (print) 20230555004 | Canadiana (ebook) 20230555047 | ISBN 9781774913826 (hardcover) | ISBN 9781774913833 (softcover) | ISBN 9781003412441 (ebook)
Subjects: Flavonoids—Therapeutic use.
Classification: LCC QP671.F52 F53 2024 | DDC 572/.2—dc23

ISBN: 978-1-77491-382-6 (hbk)
ISBN: 978-1-77491-383-3 (pbk)
ISBN: 978-1-00341-244-1 (ebk)

ABOUT THE BOOK SERIES: AAP ADVANCES IN NUTRACEUTICALS

SERIES EDITORS:

Raj K. Keservani, PhD, MPharm
Faculty of B. Pharmacy, CSM Group of Institutions, Allahabad, India
Telephone: +91-7897803904
email: rajksops@gmail.com

Rajesh K. Kesharwani, PhD
Nehru Gram Bharati (Deemed to be University), Prayagraj, India

Anil K. Sharma, PhD, MPharm
Department of Pharmacy, School of Medical and Allied Sciences,
GD Goenka University, Gurugram, India

In the modern era, mankind has witnessed a paradigm shift with respect to fundamental eating behavior. The lack of physical workouts and busy schedules at offices and in households have promoted the consumption of junk foods, which eventually results in numerous diseases and disorders since the nutritional content from fast food is inadequate. Plumpness and obesity have become a global health threat. The leading causes of death from most developing nations are noncommunicable diseases such as cardiovascular diseases, cancer, arthritis, osteoporosis, and liver toxicity. Patients suffering from such lifestyle ailments are a bit apprehensive towards prolonged use of costly modern therapeutics, which encourages instead the use of alternative approaches for management of such diseases and disorders.

The emerging sector of the nutraceutical industry encompasses products derived from nature, dietary supplements, and functional foods. Nutraceuticals are used for treatment and prevention of a broad range of diseases, such as the common cold, arthritis, sleep-related disorders, cancers, cardiovascular complications, metabolic disorders, and others.

The research on nutraceuticals is increasing day by day, considering the beneficial effects of food or food supplements in the management of diverse diseases. The issue of paramount concern is standardization and establishment of clinical efficacy of nutraceuticals, which is in fact a challenge for researchers around the globe.

This book series aims at realizing the significance of the variety of nutraceuticals for human well-being. The books in the series will also emphasize the role of dieticians and nutritionists for the prescription of judicious eating habits. The key food components such as carbohydrates, proteins, and lipids as well as micronutrients (vitamins, minerals) are demonstrated to maintain good health, obviating the need for medicines. Thus, nutraceuticals that are indeed derived from food ingredients are believed to be potential alternative therapeutics.

Asymptomatic diseases/disorders necessitate proper diagnosis. Stressed circumstances have been reported to cause weight loss. Workplace stress is a key variable in the genesis and propagation of metabolic disorders. The nuances of the underlying mechanisms are still to be deciphered. Poor lifestyle trends have been attributed along with the consumption of junk foods in several such instances. Sugar-rich carbonated beverages, the use of contraband drugs, and the consumption of liquor in excess are assigned as risk factors. This book series will cover advances and applications in addition to providing the basics of nutraceuticals.

The books in this series may be valuable resources for industry professionals to design and develop quality products for end users. In addition, the book series will be fortified by ideas of innovation, concept building, manufacturing aspects, quality control, and regulatory status of nutraceuticals.

Ostensibly, business has seen a rise in the nutraceutical consumption in recent years. Pure, safe, efficacious products are the need of the hour globally. There are a number of books on nutraceuticals in the market, yet these are only a few books presenting certain aspects of nutraceuticals therein. This book series aims to offer a holistic view of this promising strategy, which is picking up pace with time. This book series will be a unique endeavor to bring manufacturing, research and development, and marketing strategies under a single umbrella.

CURRENT BOOKS IN THE SERIES

- **Micronutrients and Macronutrients as Nutraceuticals**
  Editors: Prakash Chandra Gupta, PhD, Sayan Bhattacharyya, MBBS, MD, Nisha Sharma, PhD, MPharm, Rajesh K. Kesharwani, PhD, and Raj K. Keservani, PhD

- **Formulations, Regulations, and Challenges of Nutraceuticals**
  Editors: Tingirikari Jagan Mohan Rao, PhD, Rajesh K. Kesharwani, PhD, Raj. K. Keservani, PhD, and Anil K. Sharma, PhD
- **Food Supplements and Dietary Fiber in Health and Disease**
  Editors: Bhushan R. Rane, PhD, MPharm, Raj K. Keservani, PhD, Durgesh Singh, DPhil, Nayan A. Gujarathi, PhD, and Ashish S. Jain, PhD
- **Nutraceuticals in Cancer Prevention, Management, and Treatment**
  Editors: Raj K. Keservani, PhD, Bui Thanh Tung, PhD, Sippy Singh, DPhil, and Rajesh K. Kesharwani, PhD
- **Advances in Flavonoids for Human Health and Prevention of Diseases**
  Editors: Nisha Sharma, PhD, Deepika Saini, PhD, Rajesh K. Kesharwani, PhD, Prakash Chandra Gupta, and Raj. K. Keservani, PhD
- **The Flavonoids: Extraction and Applications**
  Editors: Deepika Saini, PhD, Rajesh K. Kesharwani, PhD, and Raj K. Keservani, PhD
- **Immune-Boosting Nutraceuticals for Better Human Health: Novel Applications**
  Editors: Urmila Jarouliya, PhD, Raj K. Keservani, PhD, Rajesh K. Kesharwani, PhD, Virendra K. Patel, PhD, and Adi D. Bharti
- **Applications of Functional Foods in Disease Prevention**
  Editors: Raj K. Keservani, PhD, and Eknath D. Ahire, MSPharm
- **Plant Metabolites and Vegetables as Nutraceuticals**
  Editors: Raj K. Keservani, PhD, Bui Thanh Tung, PhD, Rajesh K. Kesharwani, PhD, PhD, and Eknath D. Ahire, MSPharm
- **Preventive and Therapeutic Role of Vitamins as Nutraceuticals**
  Editors: Khemchand R. Surana, Eknath D. Ahire, MSPharm, Raj K. Keservani, PhD, and Rajesh K. Kesharwani, PhD
- **Nutraceuticals and Bone Health**
  Editors: Deepak Sharma, PhD, Madan Mohan Gupta, PhD, Anil K. Sharma, PhD, MPharm, Raj K. Keservani, PhD, and Rajesh K. Kesharwani, PhD
- **Flavonoids as Nutraceuticals**
  Editors: Rajesh K. Kesharwani, PhD, Deepika Saini, PhD, Raj K. Keservani, PhD, and Anil Kumar Sharma, PhD, MPharm

- **Herbals as Nutraceuticals: Their Role in Healthcare**
  Editors: Raj K. Keservani, PhD, Rajeshwar Kamal Kant Arya, PhD, MPharm, and Rajesh K. Kesharwani, PhD, MTech
- **Nutraceutical Fruits: Overview and Disease Prevention**
  Editor: Raj K. Keservani, PhD
- **Nutrigenomics and Nutraceuticals**
  Editors: Raj K. Keservani, PhD, Eknath D. Ahire, MSPharm, Shubham J. Khainar, PhD, Sanjay J. Kshirsagar, PhD, and Rajesh K. Kesharwani, PhD
- **The Nature of Nutraceuticals: History, Properties, Sources, and Nanotechnology**
  Editors: Rajesh K. Kesharwani, PhD, Prashant Kumar, PhD, and Raj K. Keservani, PhD
- **Nutraceuticals for the Treatment and Prevention of Sexual Disorders**
  Editors: Raj K. Keservani, PhD, Sharangouda J. Patil, PhD, and Ivan Aranha, PhD
- **Antioxidants as Nutraceuticals**
  Editors: Nayan A. Gujaratih, PhD, Raj K. Keservani, PhD, Rajesh K. Kesharwani, PhD, Bhushan R. Rane, PhD, and Yogeeta Sameer Goyal, PhD
- **Nutraceuticals in Insomnia and Sleep Problems**
  Editors: Raj K. Keservani, PhD, Sayan Bhattacharyya, PhD, and Rajesh K. Kesharwani, PhD
- **Nutraceuticals in Respiratory and Pulmonary Diseases**
  Editors: Deepika Saini, PhD, Rajesh K. Kesharwani, PhD, Raj K. Keservani, PhD
- **Nutraceuticals in Arthritis and Psoriasis**
  Editors: Editors: Meenakshi Jaiswal, PhD, Raj K. Keservani, PhD, Rajesh K. Kesharwani, PhD, and Swati G. Talele, PhD

ABOUT THE EDITORS

Rajesh K. Kesharwani, PhD
Associate Professor, Department of Computer Application, Nehru Gram Bharati (Deemed to be University), Prayagraj, India

Rajesh K. Kesharwani, PhD, MTech, is working as Associate Professor, Department of Computer Application, Nehru Gram Bharati (Deemed to be University), Prayagraj, India. He has more than 11 years of research and nine years of teaching experience at various institutes of India, imparting bioinformatics and biotechnology education. He has received several awards, including the NASI-Swarna Jayanti Puruskar by The National Academy of Sciences of India. He has supervised one PhD and more than 20 undergraduate and graduate students for their research work. Dr. Kesharwani has authored over 49 peer-reviewed articles, 20 book chapters, and 14 edited books with international publishers. He has been a member of many scientific communities as well as a reviewer for many international journals. He has presented many papers at various national and international conferences. He has been a recipient of a Ministry of Human Resource Development (India) Fellowship and a Senior Research Fellowship from the Indian Council of Medical Research, India. His research fields of interest are medical informatics, protein structure and function prediction, computer-aided drug designing, structural biology, drug delivery, cancer biology, nanobiotechnology, and biomedical sciences. Dr. Kesharwani received his PhD from the Indian Institute of Information Technology, Allahabad, and worked at NIT Warangal for two semesters.

Deepika Saini, PhD
Assistant Professor, Department of Zoology, Chamanlal Mahavidyalaya, Haridwar, Uttarakhand, India

Deepika Saini, PhD, is currently working as Assistant Professor in the Department of Zoology at Chamanlal Mahavidyalaya, Haridwar, Uttarakhand, India. She has teaching experience of about five years. She has also worked as an adjunct faculty member in the Natural Science Department of the University of Maryland Eastern Shore, Salisbury, Maryland, USA. She has published over 14 papers in national and international seminars and conferences and has also attended many workshops. She has also been awarded a Best Teacher Award 2020 by the Society of Research in Biological Studies. Dr. Saini serves as a reviewer and editorial board member for several international and national journals, including *American Journal of Life Sciences, International Journal of Zoology Studies, International Journal of Biological Studies,* among others. She has successfully edited four books. During the pandemic in 2020, she has organized webinars with reputed institutes, such as the Zoological Survey of India and the Botanical Survey of India. She has organized two national conferences funded by the Uttarakhand Council of Science and Technology, Dehradun and Uttarakhand Council for Biotechnology, Dehradun, India.

Raj K. Keservani, PhD, MPharm
Associate Professor, Faculty of B. Pharmacy, CSM Group of Institutions, Allahabad, India

Raj K. Keservani, PhD, MPharm, is a Faculty of B. Pharmacy, CSM Group of Institutions, Allahabad, India. He has more than 12 years of academic (teaching) experience from various institutes of India in pharmaceutical education. He has published over 30 peer-reviewed papers in the field of pharmaceutical sciences in national and international journals, more than 40 book chapters, three co-authored books, and 19 edited books, with several in the works now. He is also active as a reviewer for several international scientific journals. Dr. Keservani graduated with a pharmacy degree from the Department of Pharmacy, Kumaun University, Nainital (Uttarakhand), India. He received his Master of Pharmacy (MPharm) (specialization in pharmaceutics) from the School of Pharmaceutical Sciences, Rajiv Gandhi Proudyogiki Vishwavidyalaya, Bhopal, India. Dr. Keservani is a life member of Society of Pharmaceutical Education and Research (SPER). His research interests include nutraceutical and functional foods, novel drug delivery systems (NDDS), transdermal drug delivery/drug delivery, health science, cancer biology, and neurobiology.

Anil Kumar Sharma, PhD, MPharm
Assistant Profesoor, Department of Pharmacy, School of Medical and Allied Sciences, GD Goenka University, Gurugram, India

Anil Kumar Sharma, PhD, MPharm, is an expert in the area of pharmaceutics with a background in drug delivery. He has taught these subjects for nearly 10 years at universities such as the Delhi Institute of Pharmaceutical Sciences and Research, the University of Delhi, and the School of Medical and Allied Sciences, G. D. Goenka University, India. Prior to taking up his current role in 2018, Dr. Sharma served in academic positions such as Lecturer (Pharmaceutics) at the Delhi Institute of Pharmaceutical Sciences and Research, University of Delhi, India. He has published over 28 peer-reviewed papers in the field of pharmaceutical sciences in both national and international journals as well as many book chapters and several edited books. He holds a PhD (Pharmaceutical Sciences) from the University of Delhi, an MPharmacy (Pharmaceutics) from the Rajiv Gandhi Proudyogiki Vishwavidyalaya and a BPharmacy from the University of Rajasthan.

CONTENTS

CONTRIBUTORS

V. Aswathy
Ethnomedicine and Ethnopharmacology Division, KSCSTE–Jawaharlal Nehru Tropical Botanic Garden and Research Institute, Palode, Thiruvananthapuram, Kerala, India

Ittishree Bhardwaj
G.V.M. College of Pharmacy, Sonipat, Haryana, India

Priyanka Bhardwaj
MD Scholar, Department of Sharir Rachana, CBPACS, New Delhi, India

Mandar Bhutkar
Department of Bioscience and Bioengineering, Indian Institute of Technology, Roorkee, Uttarakhand, India

B. S. Bijukumar
Post-Graduate, Department of Zoology and Research Center, Mahatma Gandhi College, Thiruvananthapuram, Kerala, India

Diwakar Chauhan
Department of Chemistry, School of Basic and Applied Sciences, Galgotias University, Greater Noida, Uttar Pradesh, India

Monika Chauhan
Department of Forensic Science, School of Basic and Applied Sciences, Galgotias University, Greater Noida, Uttar Pradesh, India

Niharika Dewangan
Kalinga University, Naya Raipur; Swami Shri Swaroopanand Saraswati Mahavidyalaya, Bhilai, Chhattisgarh, India

Nancy Gautam
Research Scholar, Department of Chemistry, Meerut College, Meerut, Uttar Pradesh, India

Ritu Kataria
G.V.M. College of Pharmacy, Sonipat, Haryana, India

Inderjeet Kaur
PG Department of Biotechnology, Lyallpur Khalsa College, Jalandhar, Punjab, India

N. M. Krishnakumar
Department of Biosciences, Rajagiri College of Social Sciences, Kalamassery, Kochi, Ernakulam, Kerala, India

Ajay Kumar
Department of Life Science, School of Basic and Applied Sciences, Galgotias University, Greater Noida, Uttar Pradesh, India

Ashwani Kumar
Department of Pharmaceutical Sciences (FAMS), Gurukul Kangri Deemed University, Haridwar, Uttarakhand, India

Sunil Kumar
Gurukul Kangri Deemed University, Haridwar, Uttarakhand, India

Alka Mishra
Government VYTPG Autonomous College, Durg, Chhattisgarh, India

Harsh Mohan
Department of Forensic Science, School of Basic and Applied Sciences, Galgotias University, Greater Noida, Uttar Pradesh, India

Chandrashekhar Murumkar
Post-Graduate Research Center, Department of Botany, Tuljaram Chaturchand College of Arts, Science, and Commerce, Baramati (Autonomous), Maharashtra, India

Keshari Nandan
Department of Chemistry, Gurukula Kangri (Deemed to be University), Haridwar, Uttarakhand, India

Preeti Panchal
G.V.M. College of Pharmacy, Sonipat, Haryana, India

Madhuri Patil
Post-Graduate Research Center, Department of Botany, Tuljaram Chaturchand College of Arts, Science, and Commerce, Baramati (Autonomous), Maharashtra, India

R. Prakashkumar
Ethnomedicine and Ethnopharmacology Division, KSCSTE–Jawaharlal Nehru Tropical Botanic Garden and Research Institute, Palode, Thiruvananthapuram, Kerala, India

Deepanshu Rana
Assistant Professor, Department of Microbiology, Sardar Bhagwan Singh University, Balawala, Dehradun, Uttarakhand, India

Ruchi Rani
Department of Bioscience and Bioengineering, Indian Institute of Technology, Roorkee, Uttarakhand, India

Pooja Sabharwal
Assistant Professor, PG Department of Sharir Rachana, CBPACS, Affiliated to GGSIP University New Delhi, India

Neha Saini
G.V.M. College of Pharmacy, Sonipat, Haryana, India

Parul Saini
Alumni, John Curtin School of Medical Research, Australian National University, Canberra, Australia

Pragati Saini
Department of Life Science, School of Basic and Applied Sciences, Galgotias University, Greater Noida, Uttar Pradesh, India

Arun Dev Sharma
PG Department of Biotechnology, Lyallpur Khalsa College, Jalandhar, Punjab, India

Ankit Singh
Research Scholar, Department of Chemical Engineering, Rajiv Gandhi Institute of Petroleum Technology, Jais, Amethi, Uttar Pradesh, India

Shyam Vir Singh
Department of Chemistry, Shri Guru Ram Rai (PG) College, Pathribagh, Dehradun, Uttarakhand, India

Peeush Singhal
Department of Pharmaceutical Sciences (FAMS), Gurukul Kangri Deemed University, Haridwar, Uttarakhand, India

Ritu Vishnoi Singhal
Department of Botany, Chinmaya Degree College, Haridwar, Uttarakhand, India

S. R. Suja
Ethnomedicine and Ethnopharmacology Division, KSCSTE–Jawaharlal Nehru Tropical Botanic Garden and Research Institute, Palode, Thiruvananthapuram, Kerala, India

Shailly Tomar
Department of Bioscience and Bioengineering, Indian Institute of Technology, Roorkee, Uttarakhand, India

ABBREVIATIONS

ACE2	angiotensin-converting enzyme 2
ACGs	apigenin, C-glycosides
AChE	acetylcholinesterase
AD	Alzheimer's disease
AGEs	advanced glycation end products
AIDS	acquired immune deficiency syndrome
AMPK	AMP-activated protein kinase
AP-1	activating protein-1
AQP4	aquaporin 4
AR	analytical reagents
ARDS	acute respiratory distress syndrome
ARE	antioxidant response element
AS	aureusidin synthase
ASFV	African swine fever virus
ATL	adult T-cell leukemia
ATP	adenosine triphosphate
BChE	butyrylcholinesterase
CAD	coronary artery disease
cAMP	cyclic adenosine monophosphate
CDKs	cyclin-dependent kinases
CE	capillary electrophoresis
CGH	comparative genomic hybridization
CH_3	methyl group
CHI	chalcone isomerase
CHIKV	chikungunya virus
CHS	chalcone synthase
CoQ10	coenzyme Q10
CoV	coronavirus
COVID-19	coronavirus disease
COX	cyclo-oxygenase
CRP	reactive C-protein
CSI	chronic systemic inflammation
CV-B3	coxsackievirus B3
CVD	cardiovascular diseases
DCs	dendritic cells
DENV	dengue virus
DHAP	3-deoxy-D-arabino-heptulosonic acid-7-Phosphate

dsDNA	deoxyribonucleic acid
EAC	Ehrlich ascites carcinoma
EBV	Epstein-Barr virus
EGCG	epigallocatechin gallate
EGF	epidermal growth factor
EGFR	EGF receptor
ER	estrogen receptors
ERK1/2	extracellular signal-regulated kinases 1/2
EUFRIN	European Fruit Research Institutes Network
FISH	fluorescence in situ hybridization
GCG	gallocatechin gallate
GNP	gold nanoparticle-immobilized
GSH-Px	glutathione peroxidase
H2L	hit-to-lead
HBV	hepatitis B virus
HCV	hepatitis C virus
HD	Huntington's disease
HIV	human immunodeficiency virus
HPLC	high-performance liquid chromatography
HPV	human papillomavirus
HRV	human rhinovirus
HSD	3β-hydroxysteroid dehydrogenase
HSV-1	herpes simplex virus type 1
HTLV	human T-cell lymphotropic virus type
IE	immediate-early
IFN-γ	interferon-γ
IgE	immunoglobulin E
IL	interleukin
iNOS	inducible nitric oxide synthase
IOM	Institute of Medicine
IRFs	interferon regulatory factors
IκB	inhibitor of kappa B
JEV	Japanese Encephalitis Virus
KSHV	Kaposi's sarcoma-associated herpesvirus
LCMV	lymphocytic choriomeningitis virus
LPS	lipopolysaccharide
MAD	malondialdehyde
MAPK	mitogen-activated protein kinase
MAYV	Mayaro virus
MERS	middle east respiratory syndrome
MIP-2	macrophage inflammatory protein-2
Mpro	main protease
mRNA	messenger RNA

MS	multiple sclerosis
NAD+	nicotinamide adenine dinucleotide
NAFLD	non-alcoholic fatty liver disease
NASH	non-alcoholic steatohepatitis
NF-κB	nuclear factor kappa B
NMN	nicotinamide mononucleotide
NMR	nuclear magnetic resonance
NO	nitric oxide
NR	nicotinamide-riboside
nsP1	non-structural protein 1
PAC	proanthocyanidin
PAH	pulmonary arterial hypertension
PAL	phenylalanine lyase
PBMC	peripheral blood mononuclear cell
PD	Parkinson's disease
PD-ACE-2	peptidase domain of ACE-2
PEP	phosphoenol pyruvate
PPARα	peroxisome proliferator-activated receptor
PRRs	pattern recognition receptors
Rb	retinoblastoma
RBD	receptor binding domain
RNAs	ribonucleic acid
RNS	reactive nitrogen species
ROS	reactive oxygen species
rRNA	ribosomal RNA
RSV	respiratory syncytial virus
SARS	severe acute respiratory syndrome
SARS-CoV-2	severe acute respiratory disease syndrome coronavirus-2
SFV	Semliki-forest virus
SOCS	suppressor of cytokine signaling
SOD	superoxide dismutase
STAT	signal transducer and activator of transcription
TBEV	tick-borne encephalitis virus
TF	transcription factors
TGFB	transforming growth factor B
Th2	T-helper type 2
TLR9	toll-like receptor 9
TMC	traditional Chinese medicine
TNF-α	tumor necrosis factor-alpha
tRNA	transfer RNA
UV	ultraviolet
VEGF	vascular endothelial growth factor
ViPR	virus pathogen database and analysis resource

VZV	varicella-zoster virus
WNV	West Nile virus
XO	xanthine oxidase
XRE	xenobiotic response element
YFV	yellow fever virus
ZIKV	ZIKA virus

PREFACE

People's lives have become so busy today, that everyone is rushing around without taking care of their health. This has given rise to many people experiencing health issues early on. In seeking treatment though, surveys have shown that people today are increasingly seeking ayurvedic and plant-derived remedies in comparison to allopathy due to their potential for reduced side effects, easy availability, and a range of health benefits.

Flavonoids are well-known plant metabolites which have come to researchers' attention due to their extraordinary properties in treating health issues. The pharmaceutical importance of flavonoids can be summarized in the fact that they are widely accepted as having anti-depressant and antioxidant properties, are antiviral as well as anti-inflammatory. They are also of great benefit when it comes to neuroprotection and cardiovascular disease. Nowadays, oncology and flavonoids have proven to be a great combination for many types of cancers.

This book, *Flavonoids as Nutraceuticals*, comprises 13 chapters that describe the use of flavonoids for the prevention and treatment of diseases and their mechanism.

Chapter 1, *Nutraceuticals: The New Anti-Aging Method,* written by Peeush Singhal et al., emphasizes the importance of anti-aging nutritional foods and discusses modern aging theories. In addition, it also describes the role of anti-aging nutrients, how dietary needs change with age and diet, food quality problems for the aging population, and the diet patterns of the elderly.

Chapter 2, *Introduction to Flavonoids*, written by Deepanshu Rana and his associates discusses about the flavonoid's classification and its significance. Neha Saini and her associates outline the uses of flavonoids and demonstrate that metabolomics is a useful means for studying flavonoid metabolism in various agricultural crops in Chapter 3, *Importance of Flavonoids in Agriculture*.

Despite many efforts, very few flavonoids are in clinical trials for antiviral treatment against various virus infections. Flavonoids have been explored as evidence-based natural sources of antivirals against a variety of virus classes in Chapter 4, *Therapeutic Antiviral Potential of Flavonoids*, written by Shally Tomar and her colleagues.

Priyanka Bhardwaj and Pooja Sabharwal describe the naturally occurring polyphenolic compounds – flavonoids as well-known for their ability to aid in the treatment of a variety of ailments as well as to mitigate the adverse effects of various treatment regimens in Chapter 5, *Regulation of Gene Expression by Flavonoids.*

Chapter 6, *Flavonoids of Asteraceae-Promising Anti-Inflammatory Agents*, written by S. R. Suja and colleagues, explores the role of flavonoids of plants of the Asteraceae family in combating several inflammatory processes underlying chronic disease conditions. S. R. Suja and associates focus on the immuno-enhancing potential of various bioactive flavonoids isolated from natural sources and their mechanism of action in Chapter 7, *Bioactive Flavonoids From Natural Sources: Potential Immune-Boosters.*

Chapter 8, *Role of Flavonoids as Anti-Inflammatory Agents*, incorporates current knowledge of the mechanisms involved in flavonoids' anti-inflammatory properties and the implications of these effects on protection against various chronic inflammatory diseases written by Parul Saini.

Monika Chauhan and her associates summarize how flavonoids have become an essential factor for a wide range of nutraceutical, pharmacological, cosmetic, therapeutic uses. However, due to the intricacy of flavonoids' existence in diverse food sources, the diversity of dietary culture, and the incidence of a vast quantity of flavonoids in nature, precisely quantifying daily flavonoids consumption remains a challenge in Chapter 9, *Current Trends in the Health Benefits of Flavonoids.*

The fastness of washing, rubbing, light, and perspiration of the fastness values for colored samples range from fair to outstanding, and this evaluation is also useful in the textile industry described in Chapter 10, *Analysis of Color Fastness Properties of Natural Dye Extracted From Rhus parviflora (TUNG) on Wool Fibers Using a Combination of Natural and Synthetic Mordants* by Shyam Vir Singh.

Chapter 11, *Flavonoids in Treating Pregnancy-Induced Disorders*, written by Niharika Dewangan and Alka Mishra, gives details about the effects of flavonoids on pregnancy disorders.

Chapter 12, *Flavonoids: Their Classes and Biosynthesis*, deals with the flavonoid's function and classification along with their biosynthesis, discussed by Madhuri Patil and Chandrashekhar Murumkar.

Flavonoids and their antiviral perspective against COVID-19 have been covered in Chapter 13, *Plant-Based Flavonoids as a Promising Tool to Combat COVID-19 Infection*, written by Arun Dev Sharma and Inderjeet Kaur.

CHAPTER 1

NUTRACEUTICALS: THE NEW ANTI-AGING METHOD

PEEUSH SINGHAL,[1] ASHWANI KUMAR,[1] RITU VISHNOI SINGHAL,[2] and SUNIL KUMAR[3]

[1]*Department of Pharmaceutical Sciences (FAMS), Gurukul Kangri Deemed University, Haridwar, Uttarakhand, India*

[2]*Department of Botany, Chinmaya Degree College, Haridwar, Uttarakhand, India*

[3]*Gurukul Kangri Deemed University, Haridwar, Uttarakhand, India*

ABSTRACT

At different times in life, people have different views on aging. As a process, aging not only affects the medical and economic levels at the individual level but also at the social and national levels. Aging is a natural process, but its standard definition in the healthcare field is unclear. Delaying the Aging process and maintaining a high quality of life until old age are the two most important goals. Various healthcare methods are being considered and tested to best regard aging as a disease. Nutritious food is a value-added dietary supplement product which has great potential to change the key structure and function of aging. Nutritional health products can be the key to changing the physiological and metabolic system abnormalities caused by aging. Nutraceuticals for Aging and anti-aging: Basic understanding and clinical evidence are based on 10 main challenges to address aging and anti-aging nutritional drugs, such as cognitive health, malnutrition, drug abuse, bladder control, and oral health. It explores how to supplement these challenges with nutritious foods and connects the application to the traditional

Flavonoids as Nutraceuticals. Rajesh K. Kesharwani, Deepika Saini, Raj K. Keservani, and Anil Kumar Sharma (Eds.)

wisdom of the aging process. The purpose of this chapter is to elucidate the significance of anti-aging dietary components and go over current views on ageing. Additionally, it discusses the function of anti-aging nutrients, how dietary requirements change as people age, issues with diet quality among the ageing population, and senior people's eating habits.

1.1 INTRODUCTION

The nutrients obtained from oral dietary ingredients are related to medical and health benefits (Dumoulin et al., 2016). In the past few decades, these ingredients have shown great potential in anti-aging effects by preventing or delaying the degeneration of skin cells (Bolognia et al., 2012). For example, these factors can cause collagen and elastic fibers to degrade, leading to Aging. Digital internal and external factors cause degradation: biological progression of cells, tobacco, nutritional deficiencies, hormonal imbalances, and ultraviolet (UV) radiation (Ganceviciene et al., 2012; Parrado et al., 2019). The use of nutrients helps prevent skin roughness, skin elasticity and wrinkles, and pigmentation changes, all of which can lead to aging (Shamloul et al., 2019). Advances in skincare have greatly expanded the market for anti-aging nutrients. Some of the key nutrients that have been widely studied are amino acids, carotenoids (β-carotene, lutein, and zeaxanthin, and lycopene); fatty acids, minerals, polyphenols, and epigallocatechin gallate (EGCG) (Shamloul et al., 2019; Finch et al., 2010).

The connection between nutrition and aging has been extensively studied in animals and humans (Finch et al., 2010). Nutritional products are foods that have medicinal properties, which is why it's called a nutraceutical (Dhanjal et al., 2020). According to the definition in The Foundation for Medical Innovation, a nutraceutical is a meal that has medicinal price and offers fitness advantages, especially with inside the prevention and remedy of age-associated diseases (Bhowmik et al., 2013; Keservani et al., 2010a). That product includes realistic foods, dietary supplements (Keservani & Sharma, 2014), and herbal extracts, which give fitness advantages while fed overtime when fed as dietary supplements inside a diet plan (Himalian et al., 2021). Even researchers have encouraged that antioxidants have propitious consequences on each continual in addition to age-associated sicknesses, specifically neurodegenerative illnesses and maximum cancers (Hajhashemi et al., 2010). Various meals and dietary supplements (Keservani et al., 2020) that display an antioxidant capacity, which encompasses carotenoids, flavonoids, and nutrients, prevent and address ROS-related continual situations, which end up in more healthy and longer lifespans (Hajhashemi et al., 2010). Food dietary supplements produce antagonistic outcomes in opposition to the degenerative and inflammatory strategies inside the frame and feature beneficial results on the immune and digestive system, therefore enhancing the fine of life (Rinninella et al., 2019). The present-day evaluation focuses on highlighting the manifestations of developing vintage and theories associated with developing older. Additionally, it additionally discusses the importance of weight loss plan control in aging and practical meals, in addition to nutraceuticals with anti-aging potential (Carocho & Ferreira, 2013; Keservani et al., 2010b).

1.2 WHY DO YOU AGE?

Gerontology, the look at growing old, can be an exceedingly new technological know-how that has made splendid development over the last 30 years. In the past, scientists searched for one theory that defined getting old but has found out that getting older may be a complex interplay of genetics, chemistry, physiology, and conduct. There at the moment are dozens of theories of aging to explain this inevitable truth of being human (Kirkwood & Austad, 2000).

1.2.1 PROGRAMMED THEORIES OF AGING

Programmed theories state that the human body is intended to age, and there is a sure organic course of events that bodies follow. These hypotheses share the possibility that maturing is normal and "modified" into the body (Goldsmith, 2016; Sikora, 2014; Laughrea, 1982).

There are a couple of various programmed theories of maturing:

1. **Programmed Longevity Theory:** It is the idea that maturing is brought about by specific genes turning on and off after some time.
2. **Endocrine Theory:** It is the possibility that standard changes in hormones control maturing.
3. **Immunological Theory:** Expresses that the safe framework is modified to decay after some time, leaving individuals more vulnerable to sicknesses.

1.2.2 ERROR THEORIES OF AGING

Error theories attest that maturing is brought about by ecological harm to the body's frameworks, which aggregates over time (Curtis, 1971).

There are a few error theories of maturing:

1. **Wear and Tear Theory:** This asserts that cells and tissues basically wear out.
2. **Pace of Living Theory:** It is the possibility that the quicker a living being utilizes oxygen, the more limited it lives.
3. **Cross-Linking Theory:** It states that cross-connected proteins gather and dial back the body's processes.
4. **Free Radicals Theory:** It asserts that free radicals in the climate cause harm to cells, which in the end impedes their function.
5. **Somatic DNA Damage Theory:** It is the possibility that hereditary transformations cause cells to break down.

1.2.3 GENETIC THEORY OF AGING

Studies have exhibited that hereditary qualities can assume a significant part in maturing. In one review, when scientists eliminated cells containing certain qualities from the organs of mice, they had the option to broaden the life expectancy of the creatures by as much as 35%. The significance of these analyses for people isn't known. However, analysts believe that hereditary qualities represent a large part of the variety in maturing among individuals (Chipalkatti et al., 1983).

Some vital ideas in hereditary qualities and maturing include:

1. **Longevity Genes:** These are explicit qualities that help an individual live more.
2. **Cell Senescence:** It is the interaction by which cells disintegrate over time.
3. **Telomeres:** These are structures on the finish of DNA that, in the long run, are drained, bringing about cells stopping to recreate.
4. **Stem Cells:** These are cells that can turn out to be any kind of cell in the body and hold a guarantee to fix harm brought about by maturing.

1.2.4 BIOCHEMICAL THEORY OF AGING

Regardless qualities you have acquired, your body is persistently going through complex biochemical responses. A portion of these responses cause harm and, eventually, maturing in the body. Concentrating on these perplexing responses is assisting scientists with seeing how the body changes as it ages (Li et al., 2005).

Significant ideas in the biochemistry of maturing include:

1. **Free Radicals:** These are unsteady oxygen molecules that can harm cells.
2. **Protein Cross-Linking:** It means that abundance of sugars in the circulation system can make protein particles, in a real sense, stay together.
3. **DNA Repair:** It is the idea that, for obscure reasons, the frameworks in the body that maintain DNA appears to turn out to be less powerful in older people.
4. **Heatshock Proteins:** These are proteins that assist cells with enduring pressure and are available in less numbers in older individuals.
5. **Hormones:** These change as we age, causing many changes in organ frameworks and different capacities.

1.3 ANTI-AGING BEHAVIORS

Fortunately, a significant number of the reasons for maturing that might be going on rashly can be altered through your behaviors (Wengreen et al., 2009).

The following are a couple of approaches to keep your body feeling as youthful as could really be expected:

- Eat food varieties stacked with antioxidants to limit harm brought about by free radicals.
- Exercise consistently to restrict bone and muscle loss.
- Keep your cholesterol low so you can slow the solidifying of your arteries and ensure your heart.
- Practice mental wellness to keep your mind sharp.
- Eventually, maturing is unavoidable. Deal with your body and brain and embrace the progressions.

1.4 DIET QUALITY ISSUES FOR AGING POPULATION

Tucker remarked that the focus of her Research would be on how dietary needs change with Aging, which nutrients in particular are important for Aging populations, and the challenge of achieving access to and consumption of a high-quality diet given the obstacles already discussed by other speakers (e.g., loss of appetite, oral health decline, mobility constraints) (Drewnowski & Warren-Mears, 2001).

1.5 HOW DIETARY NEEDS CHANGE WITH AGING?

Dietary requirements change with maturing in more ways than one (Lyubomirsky et al., 1999):

- People become less dynamic, their digestion eases back, their energy necessity diminishes, all of which imply that they need to eat less.
- Recent research shows that on the grounds that more established grown-ups' capacities to ingest and use numerous supplements become less proficient, their supplement prerequisites (especially as a component of weight) really increment. Tucker referenced that the last arrangement of nourishment suggestions given by the Institute of Medicine (IOM) incorporated separate proposals for individuals aged 70 or more consequently (IOM, 2006).
- Tucker noticed that, as a portion of the past researcher had examined, ongoing conditions and drugs can influence sustenance prerequisites. For instance, notwithstanding drug-supplement associations influencing drug digestion, some medication supplement connections are additionally supplementing squandering. This is particularly valid for the B nutrients (Wysocki & Pelchat, 1993).
- Keeping a supplement-thick eating routine is basically significant for older adults due to the effect of food admission on well-being. Long periods of examination have shown that diet quality hugely affects the state of being, intellectual condition, bone well-being, eye well-being, vascular capacity, and the safe framework. However, this can be trying to accomplish for several reasons.
- As Pelchat talked about, maturing is frequently joined by a deficiency of hunger and changes in taste and smell, all of which can prompt more restricted food decisions and lower admission of refreshing food varieties (Wysocki & Pelchat, 1993).

- As Jensen talked about, maturing is additionally regularly joined by a broad oral well-being decrease and a diminished capacity to swallow, which can influence food decisions and admission (Jensen, 2019).
- Many older adults experience versatility limitations, which make it hard to search for food, lift substantial containers, open holders, and so forth.
- As both Wellman and Kinsella referenced, low income is common in maturing populaces, making it hard for some older adults to get to great food sources (i.e., on the grounds that those food sources will in general be more costly) (Pillsbury, 2010).

1.6 A MODIFIED FOOD GUIDE PYRAMID FOR OLDER ADULTS

Due to the changing dietary needs of older adults, Tucker's colleagues on the Jean Meyer USDA HRNCA evolved what they termed the modified food Pyramid for older adults (Schipper et al., 2008). Key changes to the original USDA meals manual pyramid consist of the placement of water at the bottom of the pyramid due to the fact many older adults no longer drink sufficient water to stay hydrated and the site of a flag at the top of the pyramid indicating the want for calcium, nutrition D, and vitamin B12 dietary supplements because many older adults do not get sufficient of these vitamins in a standard diet. After an update to the food manual pyramid befell for the general population, Tucker's colleagues also created a brand-new modified MyPyramid for older adults with illustrated examples of healthy meals in every food organization (Newby et al., 2003). Key changes to the original MyPyramid consist of the addition of examples of physical activity at the lowest of the pyramid. More physical activity lets in for the intake of larger portions of food, which in flip increases the likelihood that every one of the vital vitamins will be fed. Also, physical pastime enables keep muscle mass with aging.

1.7 DIETARY PATTERNS OF OLDER ADULTS

Of path, no longer all older adults follow the tips of the changed MyPyramid. Tucker mentioned the sort of methods that older adults consume. She and her colleagues had been inspecting dietary patterns in older adults as a part of the Baltimore Longitudinal look at growing older. They recognized five eating styles: "white bread" (human beings that reap significantly more power intake from white bread [16%, on average] relative to other patterns), "wholesome" (higher strength consumption from fruit, high fiber cereal, and entire grain bread), "meat" (better power intake from meat and potatoes), "alcohol" (better electricity consumption from alcohol), and "sweets" (better electricity intake from baked sweets) (Gilbert, 2000). As simply one example of the way eating regimen influences health, she confirmed information on waist circumference. Usually, as human beings age, their weight increases with the price of boom slowing down through the years; most of the won weight is deposited within the imperative location of the frame. Tucker and her colleagues discovered that older adults in the "white bread" group skilled an appreciably greater boom in weight circumference than older adults in the different eating groups. The "healthful" group confirmed the least advantage in weight circumference.

1.8 THE BEST ANTI-AGING SUPPLEMENTS

Maturing, which can be characterized as the "time-related crumbling of the physiological capacities vital for endurance and fruitfulness," is a cycle that the vast majority might want to slow (Liochev, 2015). A portion of its primary driver incorporate gathered cell harm brought about by responsive particles known as free radicals and the shortening of telomere, which are the constructions situated at the closures of chromosomes that assume a significant part in cell division (Liochev, 2015).

While maturing is inescapable, expanding the human life expectancy and easing back the maturing system has been a focal point of logical exploration for quite a long time.

Through that examination, researchers have recognized an enormous number of substances that have hostile to maturing properties, a considerable lot of which can be taken as enhancements by those searching for regular approaches to decelerate the maturing system and forestall age-related illness.

Note that this rundown isn't comprehensive, and numerous different enhancements may likewise offer an enemy of maturing impacts. The 12 enhancements with against maturing properties are discussed in subsections.

1.8.1 CURCUMIN

Researchers have shown that curcumin, the main active compound in turmeric, has anti-aging properties due to its potent antioxidant properties (He & Sharpless, 2017; Lee et al., 2019). Cellular senescence occurs when cells.

Cease to divide. Senescent cells accumulate as you age, which speeds up the aging process. Curcumin has been shown to activate certain proteins, including sirtuins and AMP-activated protein kinase (AMPK), which helps delay cellular senescence and improves longevity (Lee et al., 2019; Sundar et al., 2018; Bielak-Zmijewska et al., 2019). The compound curcumin has also been shown to treat some diseases associated with aging and to retard the onset of aging-related diseases, as well as to curb age-related symptoms (Sarker & Franks, 2018). There is some evidence that turmeric is linked to a reduced risk of mental decline as we age (Sarker & Franks, 2018). Use turmeric in recipes or take supplements to get more curcumin. A compound called curcumin is the main active ingredient in turmeric. As it activates certain proteins and protects cells against damage, it may slow down aging.

1.8.2 EGCG

Green tea polyphenol "epigallocatechin-gallate" (EGCG) is a well-known polyphenol component. It has amazing health benefits, with evidence indicating that it can lower the incidence of some malignancies as well as other diseases like heart disease (Legeay et al., 2015; Negri et al., 2018; Xiong et al., 2018). The ability of EGCG to increase longevity and guard against age-related illness development is one of its many potential health-promoting characteristics. EGCG has been shown to reduce the aging process by restoring mitochondrial activity in cells and working on age-related pathways, such as the AMPK signaling pathway. Autophagy, the process by which your body eliminates damaged cellular material, is also induced (Prasanth et al., 2019).

Green tea consumption has been linked to a lower risk of death from any cause, diabetes, stroke, and heart disease-related death. Furthermore, animal studies have demonstrated that it can protect against UV-induced skin aging and wrinkles (Bolke et al., 2019). Green tea or concentrated pills are also good sources of EGCG. Green tea contains EGCG, a polyphenol molecule that may boost mitochondrial function and promote autophagy. Green tea consumption has been associated with a lower risk of death from any cause.

1.8.3 COLLAGEN

Because of its ability to minimize the look of skin aging, collagen is marketed as a fountain of youth. It's a vital component of your skin that aids in the preservation of its structure. Collagen production slows with age, resulting in collagen loss in the skin, which increases indications of aging, such as wrinkles (Proksch et al., 2014). Collagen supplementation has been shown to minimize the indications of aging, such as wrinkles and dry skin, according to certain studies. For example, a 2019 research of 72 women found that taking a supplement containing 2.5 grams of collagen per day — together with numerous additional substances, including biotin — for 12 weeks improved skin moisture, roughness, and elasticity significantly (Hernández-Camacho et al., 2018).

Another trial of 114 women found that 8 weeks of treatment with 2.5 grams of collagen peptides decreased eye wrinkles and boosted skin collagen levels (Hernández-Camacho et al., 2018). Though these findings are encouraging, keep in mind that much collagen research is supported by collagen product manufacturers, which could skew the results. Collagen supplements come in a variety of forms, including powders and pills. Collagen is a common dietary supplement that can aid in the prevention of skin aging by increasing collagen levels in the skin.

1.8.4 CoQ10

Coenzyme Q10 (CoQ10) is an antioxidant produced by your body. It is required for energy production and protects cells against damage (Johansson et al., 2015). CoQ10 levels are thought to fall as people age, and supplementing with it has been demonstrated to benefit several aspects of health in the elderly. For example, supplementing with CoQ10 and selenium for four years increased overall quality of life, reduced hospital visits, and prevented the decline of physical and mental function in 443 older adults (González-Guardia et al., 2015). By reducing oxidative stress, CoQ10 supplements promote a healthier life by accumulating fewer free radicals and reducing age-related diseases, such as deterioration (Rissiek et al., 2020). There is more evidence needed before CoQ10 can be recommended as an anti-aging supplement. Consult your healthcare professional before using this supplement. A natural antioxidant, CoQ10 has been found to slow age-related physical decline in older adults and improve quality of life. Some studies suggest that supplementing with it may be beneficial.

1.8.5 NICOTINAMIDE RIBOSIDE AND NICOTINAMIDE MONONUCLEOTIDE (NMN)

The precursors to nicotinamide adenine dinucleotide (NAD+) are nicotinamide-riboside (NR) and nicotinamide mononucleotide (NMN). NAD+ is a molecule found in every cell of your body that plays a role in a variety of important functions such as energy metabolism, DNA repair, and gene expression (Yaku et al., 2019; Mills et al., 2016). NAD+ levels drop as people get older, and this drop is linked to rapid physical deterioration and the emergence of age-related illnesses, including Alzheimer's (Mills et al., 2016). Supplementing with the NAD+ precursors NMN and NR has been proven in animal trials to restore NAD+ levels and prevent age-related physical decline. For example, an oral NMN supplementation reduced age-related genetic alterations in mice and enhanced energy metabolism, physical activity, and insulin sensitivity (Elhassan et al., 2019). In addition, a 2019 study of 12 men aged 75 found that supplementing with 1 gram of NR daily for 21 days enhanced NAD+ levels in skeletal muscle and decreased inflammatory protein levels in the body (Han et al., 2019). However, one of the study's authors holds stock in and works as an advisor to the business that made the NR supplement in question, which could have skewed the results (Han et al., 2019). A number of other animal experiments have found that supplementing with both NR and NMN had good results. More human study is needed, however, before firm conclusions about the anti-aging effects of NR and NMN can be drawn (Han et al., 2019; Alavizadeh & Hosseinzadeh, 2014). NMR and NR supplementation may help enhance NAD+ levels in the body and prevent age-related genetic alterations.

1.8.6 CROCIN

Crocin is a yellow carotenoid pigment found in saffron, a popular and expensive spice used in Indian and Spanish cooking. Crocin has anticancer, anti-inflammatory, anti-anxiety, and antidiabetic properties, according to human and animal studies (Pitsikas, 2015). Apart from the qualities stated above, crocin has been studied for its ability to serve as an anti-aging compound and guard against age-related mental decline (Heidari et al., 2018). Crocin has been shown to help prevent age-related nerve damage in test tubes and rodents by decreasing the development of advanced glycation end products (AGEs) and reactive oxygen species (ROS), both of which contribute to the aging process (Fagot et al., 2018; Deng et al., 2018). Crocin has also been demonstrated to reduce inflammation and protect against UV-light-induced cellular damage in human skin cells (Deng et al., 2018). Given that saffron is the costliest spice on the planet, taking a concentrated saffron supplement is a more cost-effective method to increase your crocin intake.

Crocin, a pigment found in the spice saffron, appears to be a promising anti-aging supplement. It may reduce inflammation and prevent cellular damage, hence promoting longevity and preventing mental deterioration.

1.8.7 12 OTHER ANTI-AGING SUPPLEMENTS

Aside from the supplements mentioned above, the following supplements have anti-aging potential:

1. **Theanine:** Green tea contains the amino acid theanine, which is abundant in particular teas. It has been demonstrated to lengthen the lifetime of roundworms by roughly 5% (Zarse et al., 2012) and may help protect against mental decline.

2. **Rhodiola:** Anti-inflammatory and anti-aging effects are found in this medicinal plant. Treatment with Rhodiolarosea powder increased the longevity of fruit flies by 17% on average (Gospodaryov et al., 2013), according to one study.

3. **Garlic:** It is an excellent anti-inflammatory and antioxidant food. Supplementing with this bulb has been proven to reduce UV-light-induced skin aging and wrinkles in test tubes and rodents (Kim, 2016).

4. **Astragalus:** Traditional Chinese medicine (TMC) uses *Astragalus membranaceus as a stress-relieving* herb. By lowering oxidative stress, improving immunological function, and limiting cellular damage, it may aid in the prevention of aging (Liu et al., 2017).

5. **Fisetin:** It is a flavonoid molecule that can kill senescent cells, making it a senotherapeutic. It has been shown in rodent experiments to lower the number of senescent cells in tissues and to lengthen lifespan (Yousefzadeh et al., 2018).

6. **Resveratrol:** It is a polyphenol found in grapes, berries, peanuts, and red wine that activates genes called sirtuins, which may help people live longer. Fruit flies, yeasts, and nematodes have all been proven to live longer (Li et al., 2017).

Though these findings are encouraging, further human study is needed to completely understand how these supplements can help people live longer. L-theanine, Rhodiolarosea, garlic, *Astragalus membranaceus, fisetin, and resveratr*ol have all been proven to have anti-aging benefits in studies.

1.9 CONCLUSION

Certain Nutraceuticals can also assist sluggish the getting old process and sell durability and enhance health span via a couple of mechanisms, which include reducing oxidative strain, altering signaling pathways, influencing metabolism, and maintaining protein homeostasis. Curcumin, collagen, CoQ10, crocin, NMN, and fisetin are some substances proven to provide anti-aging consequences in research studies. Nonetheless, while a few research propose that taking certain dietary supplements can also help slow getting old, the best way to promote durability and standard fitness is to engage in wholesome practices like eating a nutritious diet, carrying out ordinary workouts, and lowering strain.

KEYWORDS

- **anti-aging**
- **anti-aging behaviors**
- **anti-aging nutrients**
- **anti-aging supplements**
- **diet quality**
- **error theories**
- **modern aging theories**
- **nutraceuticals**

REFERENCES

Alavizadeh, S. H., & Hosseinzadeh, H., (2014). Bioactivity assessment and toxicity of crocin: A comprehensive review. *Food Chem. Toxicol., 64*, 65–80. doi: 10.1016/j.fct.2013.11.016. Epub 2013 Nov 22. PMID: 24275090.

Bhowmik, D., Gopinath, H., Kumar, B. P., Duraivel, S., & Kumar, K. S., (2013). Nutraceutical-a bright scope and opportunity of the Indian healthcare market. *The Pharma Innovation, 1*(11), 29–41.

Bielak-Zmijewska, A., Grabowska, W., Ciolko, A., Bojko, A., Mosieniak, G., Bijoch, Ł., & Sikora, E., (2019). The role of curcumin in the modulation of aging. *Int. J. Mol. Sci., 20*(5), 1239. doi: 10.3390/ijms20051239. PMID: 30871021; PMCID: PMC6429134.

Bolke, L., Schlippe, G., Gerß, J., & Voss, W., (2019). A collagen supplement improves skin hydration, elasticity, roughness, and density: Results of a randomized, placebo-controlled, blind study. *Nutrients, 11*(10), 2494. doi: 10.3390/nu11102494. PMID: 31627309; PMCID: PMC6835901.

Bolognia, J. L., Jorizzo, J. L., & Schaffer, J. V., (2012). Evaluation of beauty and the aging face. *Dermatology* (3rd edn., pp. 2473–2478). Elsevier/Saunders: London.

Carocho, M., & Ferreira, I. C., (2013). A review on antioxidants, prooxidants and related controversy: Natural and synthetic compounds, screening and analysis methodologies and future perspectives. *Food and Chemical Toxicology, 51*, 15–25.

Chen, Y., Liu, X., Jiang, C., Liu, L., Ordovas, J. M., Lai, C. Q., & Shen, L. (2018). Curcumin supplementation increases survival and lifespan in *Drosophila* under heat-stress conditions. *Biofactors., 44*(6), 577–587. doi: 10.1002/biof.1454. Epub 2018 Nov 29. PMID: 30488487.

Chipalkatti, S., De, A. K., & Aiyar, A. S., (1983). Effect of diet restriction on some biochemical parameters related to aging in mice. *The Journal of Nutrition, 113*(5), 944–950.

Curtis, H. J., (1971). Genetic factors in aging. *Advances in Genetics, 16*, 305–324.

Deng, M., Li, D., Zhang, Y., Zhou, G., Liu, W., Cao, Y., & Zhang, W. (2018). Protective effect of crocin on ultraviolet B-induced dermal fibroblast photo aging. *Mol. Med. Rep., 18*(2), 1439–1446. doi: 10.3892/mmr.2018.9150. Epub 2018 Jun 11. PMID: 29901204; PMCID: PMC6072165.

Dhanjal, D. S., Bhardwaj, S., Sharma, R., Bhardwaj, K., Kumar, D., Chopra, C., Nepovimova, E., et al., (2020). Plant fortification of the diet for anti-aging effects: A review. *Nutrients, 12*(10), 3008.

Drewnowski, A., & Warren-Mears, V. A., (2001). Does aging change nutrition requirements? *The Journal of Nutrition, Health & Aging, 5*(2), 70–74.

Dumoulin, M., Gaudout, D., & Lemaire, B., (2016). Clinical effects of an oral supplement rich in antioxidants on skin radiance in women. *Clin. Cosmet. Investig. Dermatol., 9*, 315–324. doi: 10.2147/CCID.S118920.

Elhassan, Y. S., Kluckova, K., Fletcher, R. S., Schmidt, M. S., Garten, A., Doig, C. L., Cartwright, D. M., et al., (2019). Nicotinamide riboside augments the aged human skeletal muscle NAD+ metabolome and induces transcriptomic and anti-inflammatory signatures. *Cell Rep., 28*(7), 1717–1728.e6. doi: 10.1016/j.celrep.2019.07.043. PMID: 31412242; PMCID: PMC6702140.

Fagot, D., Pham, D. M., Laboureau, J., Planel, E., Guerin, L., Nègre, C., Donovan, M., & Bernard, B. A., (2018). Crocin is a natural molecule with potentially beneficial effects against skin aging. *Int. J. Cosmet. Sci., 40*(4), 388–400. doi: 10.1111/ics.12472. Epub 2018 Aug 9. PMID: 29893408.

Finch, C. E., (2010). *The Biology of Human Longevity: Inflammation, Nutrition, and Aging in the Evolution of Lifespans* (pp. 10–112). Elsevier/Saunders: London.

Ganceviciene, R., Liakou, A. I., Theodoridis, A., Makrantonaki, E., & Zouboulis, C. C., (2012). Skin anti-aging strategies. *Dermato-Endocrinol., 4*, 308–319. Link: https://goo.gl/BiYMTi. 2012.

Gilbert, S. F., (2000). *Developmental Biology* (6th edn.). Sunderland (MA): Sinauer Associates. Aging: The Biology of Senescence. Available from: https://www.ncbi.nlm.nih.gov/books/NBK10041/ (accessed on 24 June 2023).

Goldsmith, T. C., (2016). Evolution of aging theories: Why modern programmed aging concepts are transforming medical research. *Biochemistry (Moscow), 81*(12), 1406–1412.

González-Guardia, L., Yubero-Serrano, E. M., Delgado-Lista, J., Perez-Martinez, P., Garcia-Rios, A., Marin, C., Camargo, A., et al., (2015). Effects of the Mediterranean diet supplemented with coenzyme q10 on metabolomic profiles in elderly men and women. *J.*

Gerontol. A Biol. Sci. Med. Sci., *70*(1), 78–84. doi: 10.1093/gerona/glu098. Epub 2014 Jul 1. PMID: 24986061.

Gospodaryov, D. V., Yurkevych, I. S., Jafari, M., Lushchak, V. I., & Lushchak, O. V., (2013). Lifespan extension and delay of age-related functional decline caused by *Rhodiola rosea* depends on dietary macronutrient balance. *Longev. Healthspan*, *2*(1), 5, 1–14. doi: 10.1186/2046-2395-2-5. PMID: 24472572; PMCID: PMC3922952.

Hajhashemi, V., Vaseghi, G., Pourfarzam, M., & Abdollahi, A., (2010). Are antioxidants helpful for disease prevention? *Research in Pharmaceutical Sciences*, *5*(1), 1–8.

Han, X., Bao, X., Lou, Q., Xie, X., Zhang, M., Zhou, S., Guo, H., Jiang, G., & Shi, Q. (2019). Nicotinamide riboside exerts a protective effect against aging-induced NAFLD-like hepatic dysfunction in mice. *PeerJ.*, 7, e7568. doi: 10.7717/peerj.7568. PMID: 31523515; PMCID: PMC6717504.

He, S., & Sharpless, N. E., (2017). Senescence in health and disease. *Cell*, *169*(6), 1000–1011. doi: 10.1016/j.cell.2017.05.015. PMID: 28575665; PMCID: PMC5643029.

Heidari, S., Mehri, S., Shariaty, V., & Hosseinzadeh, H., (2018). Preventive effects of crocin on neuronal damages induced by D-galactose through AGEs and oxidative stress in human neuroblastoma cells (SH-SY5Y). *J. Pharmacopuncture.*, *21*(1), 18–25. doi: 10.3831/KPI.2018.21.003. Epub 2018 Mar 31. PMID: 30151301; PMCID: PMC6054079.

Hernández-Camacho, J. D., Bernier, M., López-Lluch, G., & Navas, P., (2018). Coenzyme Q10 supplementation in aging and disease. *Front Physiol.*, *9*, 44. doi: 10.3389/fphys.2018.00044. PMID: 29459830; PMCID: PMC5807419.

Himalian, R., Singh, S. K., & Singh, M. P. (2021). Ameliorative role of nutraceuticals on neurodegenerative diseases using the *Drosophila* melanogaster as a discovery model to define bio-efficacy. *Journal of the American College of Nutrition*, *3*, 1–29.

Jensen, M. E., (2019). *Dish it up! The Influence of Kind Dining™ & Green House Model Food and Food Service Innovations on the Resident Experience in Residential Care Facilities*. Doctoral dissertation, University of New Hampshire, Durham, Summer.

Johansson, P., Dahlström, Ö., Dahlström, U., & Alehagen, U., (2015). Improved health-related quality of life and more days out of hospital with supplementation with selenium and coenzyme Q10 combined. Results from a double-blind, placebo-controlled prospective study. *J. Nutr. Health Aging*, *19*(9), 870–877. doi: 10.1007/s12603-015-0509-9. PMID: 26482687.

Keservani, R. K., & Sharma, A. K., (2014). Flavonoids: Emerging trends and potential health benefits. *Journal of Chinese Pharmaceutical Sciences*, *23*(12), 815.

Keservani, R. K., Kesharwani, R. K., Sharma, A. K., Vyas, N., & Chadoker, A., (2010b). Nutritional supplements: An overview. *International Journal of Current Pharmaceutical Review and Research*, *1*(1), 59–75.

Keservani, R. K., Kesharwani, R. K., Vyas, N., Jain, S., Raghuvanshi, R., & Sharma, A. K., (2010a). Nutraceutical and functional food as future food: A review. *Der Pharmacia Lettre*, *2*(1), 106–116.

Keservani, R. K., Sharma, A. K., & Kesharwani, R. K. (2020). *Nutraceuticals and Dietary Supplements: Applications in Health Improvement and Disease Management*. CRC Press. ISBN: 9781771888738.

Kim, H. K., (2016). Garlic supplementation ameliorates UV-induced photoaging in hairless mice by regulating antioxidative activity and MMP expression. *Molecules*, *21*(1), 70. doi: 10.3390/molecules21010070. PMID: 26760989; PMCID: PMC6273408.

Kirkwood, T. B., & Austad, S. N., (2000). Why do we age? *Nature*, *408*(6809), 233–238.

Laughrea, M., (1982). On the error theories of aging, a review of the experimental data. *Experimental Gerontology*, *17*(4), 305–317.

Lee, S. H., Lee, J. H., Lee, H. Y., & Min, K. J., (2019). Sirtuin signaling in cellular senescence and aging. *BMB Rep.*, *52*(1), 24–34. doi: 10.5483/BMBRep.2019.52.1.290. PMID: 30526767; PMCID: PMC6386230.

Legeay, S., Rodier, M., Fillon, L., Faure, S., & Clere, N., (2015). Epigallocatechin gallate: A review of its beneficial properties to prevent metabolic syndrome. *Nutrients*, *7*(7), 5443–5468. doi: 10.3390/nu7075230. PMID: 26198245; PMCID: PMC4517007.

Li, H. J., Huang, X. M., & Zeng, F. Q., (2005). Analysis and evaluation of aging behavior of road asphalt [J]. *Journal of Highway and Transportation Research and Development*, *4*, 1–18.

Li, J., Zhang, C. X., Liu, Y. M., Chen, K. L., & Chen, G., (2017). A comparative study of anti-aging properties and mechanism: Resveratrol and caloric restriction. *Oncotarget.*, *8*(39), 65717–65729. doi: 10.18632/oncotarget.20084. PMID: 29029466; PMCID: PMC5630366.

Liochev, S. I., (2015). Which is the most significant cause of aging? *Antioxidants (Basel)*, *4*(4), 793–810. doi: 10.3390/antiox4040793. PMID: 26783959; PMCID: PMC4712935.

Liu, P., Zhao, H., & Luo, Y. (2017). Anti-aging implications of *Astragalus membranaceus* (Huang qi): A well-known Chinese tonic. *Aging and Disease, 8*(6), 868–886. doi: 10.14336/AD.2017.0816. PMID: 29344421; PMCID: PMC5758356.

Lyubomirsky, S., Tucker, K. L., Caldwell, N. D., & Berg, K., (1999). Why ruminators are poor problem solvers: Clues from the phenomenology of dysphoric rumination. *Journal of Personality and Social Psychology*, *77*(5), 1041–1060.

Mills, K. F., Yoshida, S., Stein, L. R., Grozio, A., Kubota, S., Sasaki, Y., Redpath, P., et al., (2016). Long-term administration of nicotinamide mononucleotide mitigates age-associated physiological decline in mice. *Cell Metab.*, *24*(6), 795–806. doi: 10.1016/j.cmet.2016.09.013. Epub 2016 Oct 27. PMID: 28068222; PMCID: PMC5668137.

Negri, A., Naponelli, V., Rizzi, F., & Bettuzzi, S., (2018). Molecular targets of epigallocatechin-gallate (EGCG): A special focus on signal transduction and cancer. *Nutrients.*, *10*(12), 1936. doi: 10.3390/nu10121936. PMID: 30563268; PMCID: PMC6315581.

Newby, P. K., Muller, D., Hallfrisch, J., Qiao, N., Andres, R., & Tucker, K. L., (2003). Dietary patterns and changes in body mass index and waist circumference in adults. *The American Journal of Clinical Nutrition*, *77*(6), 1417–1425.

Parrado, C., Mercado-Saenz, S., Perez-Davo, A., Gilaberte, Y., Gonzalez, S., & Juarranz, A., (2019). Environmental stressors on skin aging. Mechanistic insights. *Frontiers in Pharmacology*, *10*, 759.

Pillsbury, L., Miller, E. A., Boon, C., & Pray, L., (2010). *Providing Healthy and Safe Foods as We Age: Workshop Summary*. National Academies Press.

Pitsikas, N., (2015). The effect of *Crocus sativus* L. and its constituents on memory: Basic studies and clinical applications. *Evid. Based Complement Alternat Med.*, *926284*, 1–7. doi: 10.1155/2015/926284. Epub 2015 Feb 1. PMID: 25713594; PMCID: PMC4331467.

Prasanth, M. I., Sivamaruthi, B. S., Chaiyasut, C., & Tencomnao, T., (2019). A review of the role of green tea (*Camellia sinensis*) in antiphoto aging, stress resistance, neuroprotection, and autophagy. *Nutrients*, *11*(2), 474. doi: 10.3390/nu11020474. PMID: 30813433; PMCID: PMC6412948.

Proksch, E., Schunck, M., Zague, V., Segger, D., Degwert, J., & Oesser, S., (2014). Oral intake of specific bioactive collagen peptides reduces skin wrinkles and increases dermal matrix synthesis. *Skin Pharmacol. Physiol.*, *27*(3), 113–119. doi: 10.1159/000355523. Epub 2013 Dec 24. PMID: 24401291.

Rinninella, E., Cintoni, M., Raoul, P., Lopetuso, L. R., Scaldaferri, F., Pulcini, G., Miggiano, G. A., et al., (2019). Food components and dietary habits: Keys for a healthy gut microbiota composition. *Nutrients*, *11*(10), 2393.

Rissiek, B., Stabernack, J., Cordes, M., Duan, Y., Behr, S., Menzel, S., Magnus, T., & Koch-Nolte, F., (2020). Astrocytes and microglia are resistant to NAD+-mediated cell death along the ARTC2/P2X7 axis. *Frontiers in Molecular Neuroscience*, *12*, 330.

Russell, R. M., Rasmussen, H., & Lichtenstein, A. H., (1999). Modified food guide pyramid for people over seventy years of age. *The Journal of Nutrition*, *129*(3), 751–753.

Sarker, M. R., & Franks, S. F., (2018). Efficacy of curcumin for age-associated cognitive decline: A narrative review of preclinical and clinical studies. *Geroscience*, *40*(2), 73–95. doi: 10.1007/s11357-018-0017-z. Epub 2018 Apr 21. PMID: 29679204; PMCID: PMC5964053.

Schipper, J., Chanson, J. S., Chiozza, F., Cox, N. A., Hoffmann, M., Katariya, V., Lamoreux, J., et al., (2008). The status of the world's land and marine mammals: Diversity, threat, and knowledge. *Science*, *322*(5899), 225–230.

Shamloul, N., Hashim, P. W., Nia, J. J., Farberg, A. S., & Goldenberg, G., (2019). The role of vitamins and supplements on skin appearance. *Cutis*, *104*(4), 220–224.

Sikora, E., (2014). The evolution of aging theories: Why modern programmed aging concepts are transforming medical research. *Aging and Longevity*, *60*(2), 125–137.

Sundar, D. K. S., Houreld, N. N., & Abrahamse, H. (2018). Therapeutic potential and recent advances of curcumin in the treatment of aging-associated diseases. *Molecules*, *23*(4), 835. doi: 10.3390/molecules23040835. PMID: 29621160; PMCID: PMC6017430.

Wengreen, H. J., Neilson, C., Munger, R., & Corcoran, C., (2009). Diet quality is associated with better cognitive test performance among aging men and women. *The Journal of Nutrition*, *139*(10), 1944–1949.

Wysocki, C. J., & Pelchat, M. L., (1993). The effects of aging on the human sense of smell and its relationship to food choice. *Critical Reviews in Food Science and Nutrition*, *33*(1), 63–82.

Xiong, L. G., Chen, Y. J., Tong, J. W., Gong, Y. S., Huang, J. A., & Liu, Z. H., (2018). Epigallocatechin-3-gallate promotes healthy lifespan through mitohormesis during early-to-mid adulthood in *Caenorhabditis elegans*. *Redox Biol.*, 305–315. doi: 10.1016/j.redox.2017.09.019. Epub 2017 Sep 29. PMID: 28992589; PMCID: PMC5635249.

Yaku, K., Okabe, K., Gulshan, M., Takatsu, K., Okamoto, H., & Nakagawa, T., (2019). Metabolism and biochemical properties of nicotinamide adenine dinucleotide (NAD) analogs, nicotinamide guanine dinucleotide (NGD) and nicotinamide hypoxanthine dinucleotide (NHD). *Scientific Reports*, *9*(1), 1, 2.

Yousefzadeh, M. J., Zhu, Y., McGowan, S. J., Angelini, L., Fuhrmann-Stroissnigg, H., Xu, M., Ling, Y. Y., et al., (2018). Fisetin is a senotherapeutic that extends health and lifespan. *EBioMedicine*, *36*, 18–28. doi: 10.1016/j.ebiom.2018.09.015. Epub 2018 Sep 29. PMID: 30279143; PMCID: PMC6197652.

Zarse, K., Jabin, S., & Ristow, M., (2012). L-theanine extends the lifespan of adult *Caenorhabditis elegans*. *Eur. J. Nutr.*, *51*(6), 765–768. doi: 10.1007/s00394-012-0341-5. Epub 2012 Mar 16. PMID: 22422488; PMCID: PMC3419347.

CHAPTER 2

INTRODUCTION TO FLAVONOIDS

DEEPANSHU RANA,[1] NANCY GAUTAM,[2] KESHARI NANDAN,[3] and ANKIT SINGH[4]

[1]*Assistant Professor, Department of Microbiology, Sardar Bhagwan Singh University, Balawala, Dehradun, Uttarakhand, India*

[2]*Research Scholar, Department of Chemistry, Meerut College, Meerut, Uttar Pradesh, India*

[3]*Department of Chemistry, Gurukula Kangri (Deemed to be University), Haridwar, Uttarakhand, India*

[4]*Research Scholar, Department of Chemical Engineering, Rajiv Gandhi Institute of Petroleum Technology, Jais, Amethi, Uttar Pradesh, India*

ABSTRACT

Humans have been consuming plants and herbs for ages which are rich in phytonutrient compounds synthesized within these plants and herbs. Flavonoids are ubiquitous in the plant kingdom, having a characteristic flavan nuclei, giving rise to various functions, including UV protection, defense, auxin transport inhibition, allelopathy, and flower coloring. They are also responsible for various biological activities in plant, animal, and bacterial systems, and many groups have isolated and identified the structures of flavonoids possessing antifungal, antiviral, and antibacterial activity. Apart from this, the synergistic effects of flavonoids with existing chemotherapeutics are also evaluated. Hence, these compounds are becoming vital for nutraceutical, pharmaceutical, medicinal, cosmetic, and other applications.

Flavonoids as Nutraceuticals. Rajesh K. Kesharwani, Deepika Saini, Raj K. Keservani, and Anil Kumar Sharma (Eds.)

2.1 INTRODUCTION

Polyphenols are a group of chemical compounds produced during the secondary metabolism in specific plant organs like roots, stems, leaves, and fruits. These compounds serve as a large pool of bioactive chemicals displaying a variety of biological functions. Out of this large group, flavonoids are the major compounds having wide distribution in higher as well as lower plants (Karak, 2019).

Flavonoids generally have a low molecular weight (Fernandez et al., 2006; Heim et al., 2002; Karak, 2019) with an omnipresence in the photosynthesizing cells as glycosides or methylated derivatives (Karak, 2019). Although they are non-essential for the survival of plants still, they occur widely in the plant kingdoms (Buer et al., 2010; Cushnie & Lamb, 2005; Havsteen, 1983). These compounds are also present in fruit, vegetables, nuts, tea, wine, propolis, and honey making it an inevitable of human diet (Cushnie & Lamb, 2005; Grange & Davey, 1990; Harborne & Baxter, 1999; Middleton & Chithan, 1993).

The flavonoid research was pioneered by Albert Szent-Gyorgi in the year 1936 while he was working on lemon peels to establish the synergy between pure vitamin C and unidentified cofactors, which he named citrin and later vitamin P (Karak, 2019; Murray, 1998).

2.2 CHEMISTRY OF FLAVONOIDS

Flavonoids are the phytochemicals belonging to the class polyphenols which have been used in Chinese and ayurvedic medicines since ages. As per the Global health center, flavonoids not only exhibit antioxidant and anti-inflammatory activity but also demonstrate skin protection, brain function, blood sugar, and blood pressure regulation. In the year 1930, a substance originating from oranges was designated as vitamin P but later on recognized as flavonoid (rutin), and till now, more than 4,000 varieties have been identified (Karak, 2019; Middleton, 1998).

In plants, flavonoids can be present as glycosides, aglycones, or sometimes as methylated derivatives having structural diversity. The basic framework in the chemical structures of these compounds is of diphenyl propane comprising 15 carbon atoms in the primary nucleus having two six-membered rings linked with a three-carbon unit which may or may not be a part of a third ring (Middleton, 1984). This heterocyclic ring contains a pyrene ring having oxygen, which is linked with two benzene rings (ring A and ring B), referring to this as C6-C3-C6 labeled A, B, and C (Figure 2.1) (Karak, 2019; Pietta, 2000; Rice-Evans et al., 1976).

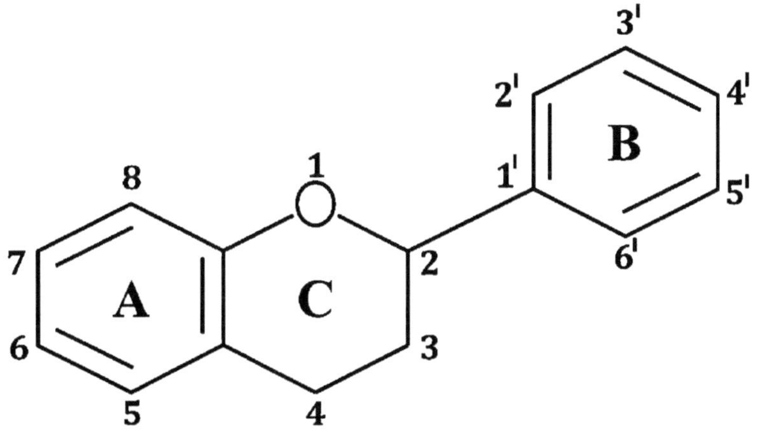

FIGURE 2.1 Basic framework of flavonoids.

As far as the naming of these magical compounds is concerned, it can be assigned in three ways (Cushnie & Lamb, 2005; Harborne & Baxter, 1999):

1. **Trivial Names:** They are broadly in use and may indicate class or plant source. For example, compounds belonging to the class anthocyanidin names ending with 'inidin,' and 'etin' belong to the class flavonol, whereas genera Triticum and Hypolaena contain compounds tricin and hypolaetin, respectively.
2. **Semi-Synthetic Names:** This naming scheming is based on trivial names depending on the parent structure. For example, flavone or chalcone, e.g., 3,5,7,3,4-pentahydroxy flavone or 3,3,4,5,7-pentahydroxyflavone.
3. **Synthetic Names:** This method is very rare and burdensome. For example, 3,4-dihydro-2-phenyl-2H-1-benzopyran for flavan.

Spectroscopic studies of flavonoids revealed that in most flavones and flavonols, absorption maxima of B ring lie in between the range of 320–385 nm while a maximum range of 250–285 nm is associated with ring A. Different functional groups might get attached to flavonoid, causing a shift in absorption peaks viz. kaempferol (367 nm), quercetin (371 nm), and myricetin (374 nm) (Kumar & Pandey, 2013; Yao et al., 2004). The one thing which distinguishes flavones from flavonols is the lack of 3-hydroxyl group. UV spectra of flavonones revealed the presence of a saturated heterocyclic C ring having no association between the A and B rings (Rice-Evans et al., 1996).

They exhibit maximum absorption at 270 nm because it has monosubstituted B ring but when di-, tri-, or o-substituted B ring are there, two peaks or one peak (258 nm) with a shoulder (272 nm) is present however anthocyanins has two distinctive bands 450–560 nm (hydroxyl cinnamoyl system-B ring) and 240–280 nm (benzoyl system-A ring) although, coloration of anthocyanins changes with the change in position of hydroxyl groups (Wollenweber & Dietz, 1981).

2.3 CLASSIFICATION OF FLAVONOIDS

Characterization of flavonoids is based on how the aromatic B ring is attached to the carbon of the benzopyran C ring, its degree of unsaturation as well as oxidation of the C ring (Panche et al., 2016). Flavonoids (2-phenylbenzopyrans), isoflavonoids (3-benzopyrans), and neoflavonoids (4-benzopyrans) are the three major classes of flavonoids depending upon the linkage of the aromatic ring to the benzopyrano (chromano) moiety. Flavonoids may be further sub-divided into the following groups based on the oxidation and saturation in the heterocyclic ring: flavan, flavanone, dihydro flavonol, flavonol, flavone, flavone-3-ol, and flavone-3,4-diol. Isoflavonoids and neoflavonoids are further classified into different groups and there are some minor flavonoids are also present in plants (Table 2.1) (Samanta et al., 2011; Stobiecki & Kachlicki, 2008). Table 2.2 comprises the data of different classes and subclasses of flavonoids with respect to their origin in natural sources and chemical structure (Panche et al., 2016).

TABLE 2.1 Sub-Division of Flavonoids

Isoflavonoids	Neoflavonoids	Minor Flavonoids
Isoflavan	4-arylcoumaril	2′-OH-chalcone
Isoflavone	3,4-dihydro-4-arylcoumarin	2′-OH-dihydrochalcone
Isoflavanone	Neoflavene	2′-OH-retro-chalcone
Isoflavan-3-ene	–	Aurone
Isoflavanol	–	Auronols
Rotenoid	–	–
Coumestane	–	–
3-arylcoumarin	–	–
Coumaronochromene	–	–
Coumaronochromone	–	–
Pterocarpan	–	–

TABLE 2.2 Flavonoid Classes, Structures, Subclasses, and Natural Sources

Classes	Structure	Subclasses	Sources	Parts	References
Flavones	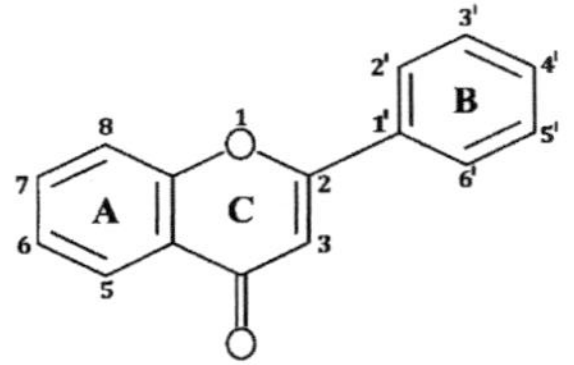	Luteolin, apigenin, tangeritin, tageretin, nobiletin, and sinensetin	Celery, parsley, red peppers, chamomile, mint, ginkgo biloba, citrus fruits	Leaves, flowers, fruits, and fruit peel	Manach et al. (2004)
Flavonols	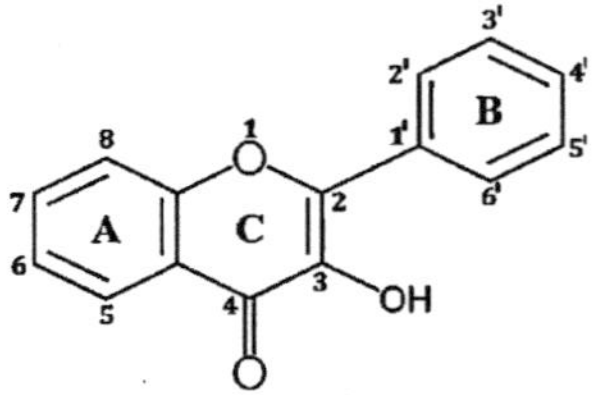	Kaempferol, quercetin, myricetin, rutin, morin	Onions, kale, lettuce, tomatoes, apples, grapes, and berries, tea, and red wine	Fruits, vegetables	Iwashina (2013)
Flavanones	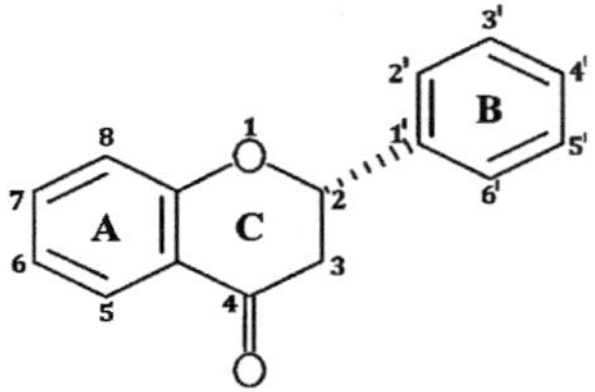	Hesperitin, naringenin, and eriodictyol, naringin, eriodictyol, hesperidin	Oranges, lemons, and grapes	Peel	Iwashina (2013)
Isoflavonoids	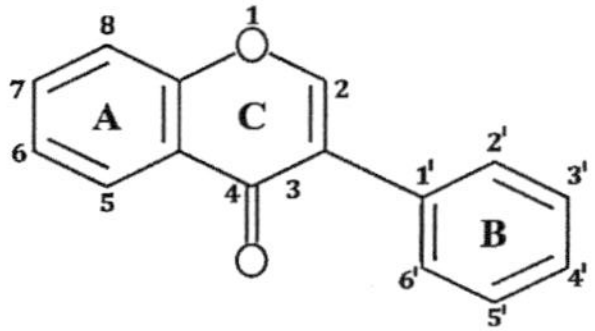	Genistin, genistein, daidzein, glycitein, daidzin	Leguminous plant like soyabean, microbes	Whole plant	Matthies et al. (2008); Panche et al. (2016)

TABLE 2.2 *(Continued)*

Classes	Structure	Subclasses	Sources	Parts	References
Neoflavonoids	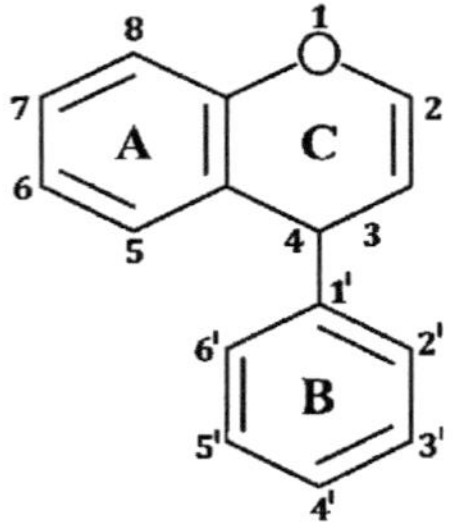	4-arylcoumarins (neoflavones), 4-arylchromanes, dalbergiones, and Dalbergia quinols	*Calophyllum inophyllum*, *Mesua thwaitesii*	Seeds, bark, and timber	Linuma et al. (1987); Nishimuta et al. (2000); Garazd et al. (2003)
Anthocyanins	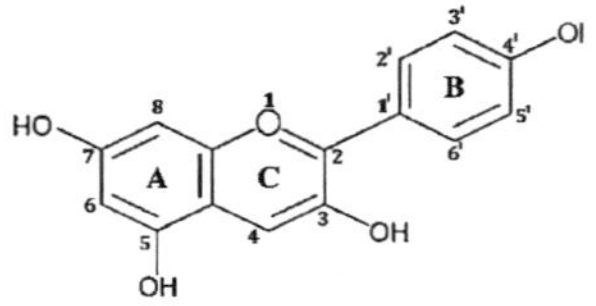	Cyanidin, malvidin, delphinidin, malvidin, pelargonidin, and peonidin	Cranberries, black currants, red grapes, merlot grapes, raspberries, strawberries, blueberries, bilberries, and blackberries	Plants, flowers, and fruits	Iwashina (2013)
Chalcones	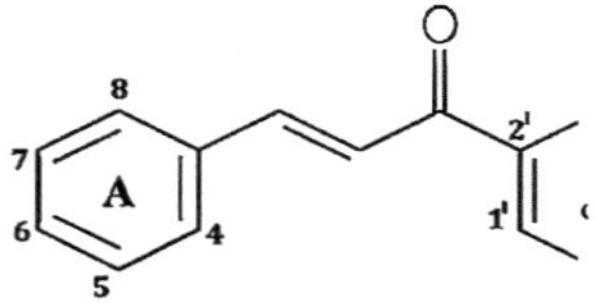	Phloretin, phloridzin, arbutin, and chalconaringenin	Tomatoes, pears, strawberries, bearberries, and certain wheat products	–	Panache et al. (2016)

2.4 SIGNIFICANCE OF FLAVONOIDS IN PLANTS

2.4.1 *SIGNALING MOLECULES*

These chemical compounds are synthesized by the root and shoot tissues of the plants, acting as an important signal molecule that plays a significant role in plant-microbe interaction (Dixon & Steele, 1999; Peer & Murphy, 2006) and the formation of root nodule also, which ultimately leads to nitrogen fixation (Fox et al., 2001; Sundaravarathan & Kannaiyan, 2002). Roots of legumes secret certain chemicals which not only attract the symbionts but also catch the attention of the pathogens. These chemicals will begin the nodulation process begin initiated by host-specific signal molecules, while flavonoid induction of specific pathogenicity genes and stimulation of development required for pathogenesis will lead to the development of infections in plants (Straney et al., 2002; Subramanian et al., 2007). Nodulation is governed by the nod gene, and flavonoids act as inducers of this particular gene accumulation of this gene will trigger flavonoid production in the root, which again induces the nod gene (Steinkellner et al., 2007; Tsai & Phillips, 1991). Flavonoids also act as eco-sensing molecules, which accomplishes symbiotic mutualisms (Ndakidemi & Dakora, 2003).

2.4.2 *PHYTOALEXINS*

Plants always encounter several pathogens, and as a defense mechanism, they produce some chemicals known as phytoalexins mainly contain phenolics, stilbenoids, alkaloids, terpenoids, coumarins, and polyacetylenes (Fawe et al., 1998; Iwashina, 2003; McNally et al., 2003). However, in legumes, isoflavonoids are involved in defense response in which migration of phenyl ring is obtained. One such compound is vesitol belonging to the class of isoflavones synthesized in some species of lotus (Lanot & Morris, 2005). Although aglycone rhamnetin is a flavonoid which was first reported, phytoalexin isolated from cucumber was previously believed to be nonexistent in this family (Fawe et al., 1998). Stilbene is a phytoalexin produced in the family VItaceae and was found to be active against several phytopathogens (Jeandet et al., 2002) because these compounds have low molecular weight and exhibit antimicrobial activities, thus becoming a vital factor against plant pathogens (Bajaj, 1996). Although still, there is still a scarcity availability of data related to the role of these compounds in the interaction between rhizobia and their legume host (Parkniske et al., 1991).

2.4.3 DETOXIFYING AGENT

Reactive oxygen produced from ultraviolet rays is toxic to different cells, like chloroplast. However, superoxides cannot readily diffuse into vacuoles, but peroxides can diffuse across membranes. Although, scavenging of ROS by phenolic or flavonoids may act as detoxifying agents (Jansen et al., 2001; Michalak, 2006; Yamasaki et al., 1997). Different flavonoids like myricetin, quercetin, and kaempferol showed inhibitory activity against AOX (Shimoji & Yamasaki, 2005). Leaves of *Ligustrum vulgare*, when exposed to high-intensity of sunlight, it was observed that ortho-dihydroxy B-ring substituted flavonoids like quercetin and luteolin accumulated in the mesophyll and epidermal tissue of exposed leaves (Agate et al., 2009). The rates of oxidation were in the order quercetin > kaempferol > quercetin glycoside >> kaempferol glycoside (Yamasaki et al., 1997). Kolaviron, a flavonoid extracted from Garcinia kola seeds, showed antioxidant and scavenging properties by inhibiting the peroxidation of linoleic acid. It also exhibited a scavenging effect on superoxide and hydrogen peroxide by inhibiting deoxyribose oxidation induced by a Fenton-type reaction system (Farombi et al., 2002). Redox reaction governed by flavonoids is considered as an alternative system to detoxify H_2O_2 in grapevine leaves (Perez et al., 2002).

2.4.4 SEED GERMINATION

Flavonoids are also playing an important role in seed germination (Gould & Lister, 2006; Shirley, 1998). Flavonoids are stored in tapetosomes which is a specialized organ present in tapetum cells situated in inner most anther coat layer, and during the development of pollen, these chemicals released and allow the growth of pollen tubes for germination by inducing pollen-specific genes (Poureel & Grotewold, 2009). Flavonoids not only play an important role in seed germination but also play several roles like protection from pathogens and predators, seed maturation, dormancy, protection from UV light, and decreased oxygen supply catalyzed by phenol oxydoreductases (Kubasek et al., 1992; Poureel & Grotewold, 2009; Shirley, 1998). Flavan-3-ol polymers may form a barrier for important processes to continue the dormancy of seed; the main effects produced by the seed coat are interference with water uptake, mechanical resistance to radicle protrusion; interference with gas exchange, particularly oxygen and carbon dioxide; prevention of inhibitor leakage from the embryo and light filtration. Many studies have reported that flavonoids possess an inhibitory effect on seed germination (Debeaujon et al., 2007; Poureel & Grotewold, 2009).

2.4.5 ATTRACTING THE POLLINATOR

Flavonoids like anthocyanins act as an attractant for pollinators via flower color and for seeds dispersal agents via brightly colored food (Deroles, 2009; Iwashina, 2003). Pollination was identified as the trigger for rapid anthocyanin synthesis (Ram & Mathur, 1984). Flavones and flavonols also act as pollinator attractants in addition to visible anthocyanins (Kaplan et al., 2004). In angiosperms, flavonoids are concerned with sexual reproduction via pollination, seed thaliana, anthocyanins are found in sporophytic tissue cell vacuum (Marinova et al., 2007). In angiosperms, color play an important role in attracting pollinators such as bees, butterfly, birds, and insects; and this elegant plant-animal co-evolution is found to occur in the orchid genus; Ophrys apply scent, shape, and color to mimic female bees, causing the male bee to attempt copulation and thereby pollination takes place (Davies, 2004). The production of these co-pigment-aluminum anthocyanin complexes depends not only on pH but also on the type of organic acids which constitute the buffering system (Jurd & Asen, 1966).

2.5 BIOPOTENTIALS

2.5.1 ANTI-CHOLINESTERASE ACTIVITY

Alzheimer's disease is a disruptive neurological disorder which generally affects the central nervous system, and the enzyme acetylcholinesterase (AChE) is responsible for that, while inhibition of this enzyme is one of the therapies for symptomatic relief of mild to moderate AD (Perry et al. 1978). The *in vitro* inhibitory studies done on various flavonoids showed that quercetin and macluraxanthone possess a concentration-dependent inhibition ability against AChE and butyrylcholinesterase (BChE) (Khan et al., 2009). Sheng et al. (2009) performed the study on different flavonoids as a potential inhibitors of AChE and found that isoflavone derivative 10d inhibits AChE with an IC_{50} of 4 nM, showing a high BChE:AChE inhibition ratio (4575-fold), superior to donepezil (IC50 = 12 nM, 389-fold). Molecular studies revealed flavonols and flavones farobin-A, gericudranin-B, glaziovianin-A, rutin, and xanthotoxin containing a 2,3-double bond may act as preferential inhibitors of COX-2 (D'Mello et al., 2011; Madeswaran et al., 2011).

2.5.2 STEROID-GENESIS MODULATORS

Three enzymes like 3β-hydroxysteroid dehydrogenase (HSD), 17β-HSD, and aromatase are responsible for steroid synthesis, and Abyssinones and related flavonoids can be used as potential steroid-genesis modulators. The flavonones possess consistent binding affinity to all the three enzymes used and are better steroidogenesis modulators in hormone-dependent cancer (Hatti et al., 2009).

2.5.3 XANTHINE OXIDASE (XO) MODULATORS

Hyperuricemia is a condition that arises due to the increased level of uric acid in blood serum, causing gout and stones in the kidney. This elevation in uric acid is caused by an enzyme called XO which converts the hypoxanthine to xanthine, followed by the synthesis of uric acid (Borges et al., 2002; Hille, 1996). Glycyrrhiza glabra is an excellent source of licoisoflavone-A which comes out to be a potent inhibitor of XO (Alnajjar, 2008). Umamaheswari et al. (2011) *in silico* study of different flavonoids and found that all are potent inhibitors of XO. Ethanolic extract of Gnaphalium contains four flavonoids having potential as XO inhibitors, whereas Flavonoids isolated Apigenin, luteolin, and 5-hydroxy-6,7,3′,4′-tetramethoxyflavone also contributed to the inhibitory effect of Gnaphalium affine extract on XO activity (Lin et al., 2014).

2.5.4 DISEASE-COMBATING ACTIVITY

Diabetes mellitus is a chronic disorder in which a deficiency of insulin production occurs, which ultimately leads to the unavailability of glucose to the body cells. Among some flavonoids, epicatechin, which is usually found in green tea leaves, acts as insulin mimetics acting as insulin receptor activator, reducing the harmful effect of diabetes, while extract rich in tannin showed inhibitory activity against α-amylase (da Silva et al., 2014; Ganugapati et al., 2011). Another major disease among human beings is cardiovascular disease (CVD), and a diet rich in flavonoids reduces the risk of these diseases (Kim et al., 2016). Some flavonoids from citrus fruits are reportedly observed to modulate the lipid metabolism reducing the effect of CVD and also decrease the risk of hypertension (Hügel et al., 2016; Mulvihill et al., 2016). These compounds also exhibit immune-regulatory activity as well as reduce the risk of Alzheimer's disease (AD) by decreasing the production of Alzheimer's amyloid protein (Aβ) (Cardenas et al., 2016; Paris et al., 2011). Recently it has been reported that an apple of the type pelingo is rich in food components that can markedly inhibit *in vitro* tumorigenesis and the growth of human breast cancer cells (Schiavano et al., 2015).

2.5.5 RADICAL SCAVENGING

Free radicals usually cause injury to the cells, and flavonoids can prevent this by direct scavenging of free radicals. Most of the flavonoids contain the hydroxyl group, which neutralizes the harmful effect of reactive oxygen by stabilizing or making them inactive (Korkina & Afanasev, 1996). Superoxides can be scavenged directly by flavonoids. Some of them scavenge highly reactive oxygen-derived radicals called peroxynitrite. Such as epicatechin and rutin are remarkable scavengers; rutin scavenge the radicals by inhibiting enzyme XO (Hanasaki et al., 1994). Some workers suggest that the scavenging properties of flavonoids are due to the inhibition of LDL oxidation which, theoretically, prevents action against atherosclerosis (Kerry & Abbey, 1997).

2.5.6 ANTI-INFLAMMATION

In a natural system, inflammation generally starts with the release of arachidonic acid, and from this, neutrophils create chemotactic compounds using lipoxygenase. Some flavonoids, when administered orally in micronized form, leads towards the suppression of leucocyte, which may relate decline in total serum complement, creating a protective mechanism against inflammation that arises during reperfusion injury (Friesenecker et al., 1994; Friesenecker & Tsai, 1995). Some researchers have also worked on flavonoids and observed that some of them inhibit the degranulation of neutrophils, and some inhibit the metabolism of arachidonic acid, establishing the anti-inflammatory and anti-thrombogenic activities of flavonoids (Alcaraz & Ferrandiz, 1987; Ferrándiz et al., 1996).

2.6 RECENT TRENDS

Flavonoids have a variety of potential benefits, making them very useful to improve human health by fortifying the diet with them. Having diverse molecular structure and scarcity of data on its uses makes its study cumbersome. Furthermore, concrete research is required to discover Nobel flavonoids from untapped sources which can replace the synthetic medicines. In this context, there is a need for research and development programs involving *in vivo* studies which will give a hopeful and safe picture for the future. Currently, the intake of fruit, vegetables, and refreshments containing flavonoids is recommended, although it is very premature to make recommendations on daily flavonoid intakes (Panche et al., 2016).

KEYWORDS

- **anti-cholinesterase activity**
- **biopotentials**
- **flavonoids**
- **phytoalexins**
- **pollinator**
- **polyphenols**
- **signaling molecules**

REFERENCES

Agate, G., Stefano, G., Biricolti, S., & Tattini, M., (2009). Mesophyll distribution of 'antioxidant' flavonoid glycosides in *Ligustrum vulgare* leaves under contrasting sunlight irradiance. *Ann. Bot.*, *104*(5), 853–861.

Alcaraz, M. J., & Ferrandiz, M. L., (1987). Modification of arachidonic metabolism by flavonoids. *J. Ethnopharmacol.*, *21*(3), 209–229.

Alnajjar, B., (2008). *Computational Studies of Natural Flavonoids Towards the Discovery of a Potential Xanthine Oxidase Inhibitor*. MSc Thesis, Universiti Sains, Malaysia.

Bajaj, Y. P. S., (1996). *Biotechnology in Agriculture and Forestry 37: Medicinal and Natural Plants IX* (p. 104). Germany, Springer-Verlag Berlin Heidelberg.

Borges, F., Fernandes, E., & Roleira, F., (2002). Progress towards the discovery of xanthine oxidase inhibitors. *Curr. Med. Chem.*, *9*(2), 195–217.

Buer, C. S., Imin, N., & Djordjevic, M. A., (2010). Flavonoids: New roles for old molecules. *J. Integr. Plant Biol.*, *52*(1), 98–111.

Cardenas, H., Arango, D., Nicholas, C., Duarte, S., Nuovo, G. J., He, W., Voss, O. H., et al., (2016). Dietary apigenin exerts immune-regulatory activity *in vivo* by reducing NF-κB activity, halting leukocyte infiltration, and restoring normal metabolic function. *Int. J. Mol. Sci.*, *17*(3), 323.

Cushnie, T. T., & Lamb, A. J., (2005). Antimicrobial activity of flavonoids. *Int. J. Antimicrob. Agents.*, *26*(5), 343–356.

D'Mello, P. R., Gadhwal, M. K., Joshi, U. R., & Shetgiri, P. R., (2011). Modeling of COX-2 inhibitory activity of flavonoids. *Int. J. Pharm. Pharmac. Sci.*, *3*(4), 33–40.

Da Silva, S. M., Koehnlein, E. A., Bracht, A., Castoldi, R., De Morais, G. R., Baesso, M. L., Peralta, R. A., et al., (2014). Inhibition of salivary and pancreatic α-amylases by a pinhão coat (*Araucaria angustifolia*) extract rich in condensed tannin. *Food Res. Int.*, *56*, 1–8.

Davies, K. M., (2004). An introduction to plant pigments in biology and commerce. In: Davies, K., (ed.), *Plant Pigments and Their Manipulation: Annual Plant Reviews* (pp. 1–22). CRC Press Blackwell Publishing Ltd, Oxford, UK.

Debeaujon, I., Lepiniec, L., Poureel, L., & Routaboul, J. M., (2007). Seed coat development and dormancy. In: Bradford, K., & Nonogaki, H., (eds.). *Seed Development, Dormancy and Germination: Annual Plant Reviews* (pp. 25–49). Blackwell Publishing Ltd.

Deroles, S., (2009). Anthocyanin biosynthesis in the plant cell cultures: A potential source of natural colorants. In: Gould, K., Davies, K., & Winefield, C., (eds.), *Anthocyanins: Biosynthesis, Functions, and Applications* (pp. 107–168). Springer Science and Business Media, New York, USA.

Dixon, R. A., & Steele, C. L., (1999). Flavonoids and isoflavonoids are a gold mine for metabolic engineering. *Trends Plant Sci.*, *4*(10), 394–400.

Farombi, E. O., Akanni, O. O., & Emerole, G. O., (2002). Antioxidant and scavenging activities of flavonoid extract (kolaviron) of *Garcinia kola* seeds. *Pharm. Biol.*, *40*(2), 107–116.

Fawe, A., Abou-Zaid, M., Menzies, J. G., & Belanger, R. R., (1998). Silicon-mediated accumulation of flavonoid phytoalexins in cucumber. *Phytopathol.*, *88*(5), 396–401.

Fernandez, S. P., Wasowski, C., Loscalzo, L. M., Granger, R. E., Johnston, G. A. R., Paladini, A. C., & Marder, M., (2006). Central nervous system depressant action of flavonoid glycosides. *Eur. J. Pharmacol.*, *539*, 168–176.

Ferrándiz, M. L., Gil, B., Sanz, M. J., Ubeda, A., Erazo, S., González, E., Negrete, R., et al., (1996). Effect of bakuchiol on leukocyte functions and some inflammatory responses in mice. *J. Pharm. Pharmacol.*, *48*(9), 975–980.

Fox, J. E., Starcevic, M., Kow, K. Y., Burow, M. E., & McLachlan, J. A., (2001). Nitrogen fixation: Endocrine disrupters and flavonoid signaling. *Nature*, *413*(6852), 128–129.

Friesenecker, B., & Tsai, A. G., (1995). Cellular basis of inflammation, edema, and the activity of daflon 500 mg. *Int. J. Microcirc.*, *15*(1), 17–21.

Friesenecker, B., Tsai, A. G., Allegra, C., & Intaglietta, M., (1994). Oral administration of purified micronized flavonoid fraction suppresses leukocyte adhesion in ischemia-reperfusion injury: *In vivo* observations in the hamster skin fold. *J. Vasc. Res.*, *14*(1, 2), 50–55.

Ganugapati, J., Mukkavalli, S., & Sahithi, A., (2011). Docking studies of green tea flavonoids as insulin mimetics. *Int. J. Comput. Appl.*, *30*(4), 48–52.

Garazd, M. M., Garazd, Y. L., & Khilya, V. P., (2003). Neoflavones. 1. Natural distribution and spectral and biological properties. *Chem. Nat. Compd.*, *39*(1), 54–121.

Gould, K. S., & Lister, C., (2006). Flavonoid functions in plants. In: Andersen, Ø. M., & Markham, K. R., (eds.), *Flavonoids* (pp. 397–442). CRC Press, Boca Raton.

Grange, J. M., & Davey, R. W., (1990). Antibacterial properties of propolis (bee glue). *J. R. Soc. Med.*, *83*, 159, 160.

Hanasaki, Y., Ogawa, S., & Fukui, S., (1994). The correlation between active oxygens scavenging and antioxidative effects of flavonoids. *Free Radic. Biol. Med.*, *16*(6), 845–850.

Harborne, J. B., & Baxter, H., (1999). *The Handbook of Natural Flavonoids.* John Wiley and Sons: Chichester, UK.

Hatti, K. S., Diwakar, L., Rao, G. V., Kush, A., & Reddy, G. C., (2009). Abyssinones and related flavonoids as potential steroidogenesis modulators. *Bioinformation*, *3*(9), 399.

Havsteen, B., (1983). Flavonoids are a class of natural products of high pharmacological potency. *Biochem. Pharmacol.*, *32*, 1141–1148.

Heim, K. E., Tagliaferro, A. R., & Bobliya, D. J., (2002). Flavonoids antioxidants: Chemistry, metabolism and structure-activity relationships. *J. Nutr. Biochem.*, *13*, 572–584.

Hille, R., (1996). The mononuclear molybdenum enzymes. *Chem. Rev.*, *96*(7), 2757–2816.

Hügel, H. M., Jackson, N., May, B., Zhang, A. L., & Xue, C. C., (2016). Polyphenol protection and treatment of hypertension. *Phytomedicine.*, *23*(2), 220–231.

Iwashina, T., (2003). Flavonoid function and activity to plants and other organisms. *Biol. Sci. Space.*, *17*(1), 24–44.

Iwashina, T., (2013). Flavonoid properties of five families newly incorporated into the order *Caryophyllales*. *Bull. Natl. Mus. Nat. Sci.*, *39*, 25–51.

Jansen, M. A., Van, D. N. R. E., Tan, M. A., Prinsen, E., Lagrimini, L. M., & Thorneley, R. N., (2001). Phenol-oxidizing peroxidases contribute to the protection of plants from ultraviolet radiation stress. *Plant Physiol.*, *126*, 1012–1023.

Jeandet, P., Douillet-Breuil, A. C., Bessis, R., Debord, S., Sbaghi, M., & Adrian, M., (2002). Phytoalexins from the Vitaceae: Biosynthesis, phytoalexin gene expression in transgenic plants, antifungal activity, and metabolism. *J. Agric. Food Chem.*, *50*(10), 2731–2741.

Jurd, L., & Asen, S., (1966). The formation of metal and Øco-pigmentØ complexes of cyanidin 3- glucoside. *Phytochem.*, *5*(6), 1263–1271.

Kaplan, F., Kopka, J., Haskell, D. W., Zhao, W., Schiller, K. C., Gatzke, N., Sung, D. Y., & Guy, C. L., (2004). Exploring the temperature-stress metabolome of Arabidopsis. *Plant Physiol.*, *136*, 4159–4168.

Karak, P. (2019). Biological activities of flavonoids: An overview. *Int. J. Pharm. Sci. Res*, *10*(4), 1567–1574.

Kerry, N. L., & Abbey, M., (1997). Red wine and fractionated phenolic compounds prepared from red wine inhibit low-density lipoprotein oxidation *in vitro*. *Atherosclerosis*, *135*(1), 93–102.

Khan, M. T., Orhan, I., Şenol, F. S., Kartal, M. U., Şener, B., Dvorská, M., Šmejkal, K., & Šlapetová, T., (2009). Cholinesterase inhibitory activities of some flavonoid derivatives and chosen xanthone and their molecular docking studies. *Chem. Biol. Interact.*, *181*(3), 383–389.

Kim, K., Vance, T. M., & Chun, O. K., (2016). Greater flavonoid intake is associated with improved CVD risk factors in US adults. *Br. J. Nutr.*, *115*(8), 1481–1488.

Korkina, L. G., & Afanas'Ev, I. B., (1996). Antioxidant and chelating properties of flavonoids. *Adv. Pharmacol.*, *38*, 151–163.

Kubasek, W. L., Shirley, B. W., McKillop, A., Goodman, H. M., Briggs, W., & Ausubel, F. M., (1992). Regulation of flavonoid biosynthetic genes in germinating Arabidopsis seedlings. *The Plant Cell*, *4*, 1229–1236.

Kumar, S., & Pandey, A. K., (2013). Chemistry and biological activities of flavonoids: An overview. *Sci. World J.*, *2013*, 1–16.

Lanot, A., & Morris, P., (2005). Elicitation of isoflavan phytoalexins. In: Marquez, A. J., (ed.), *Lotus Japonicus Handbook* (pp. 355–361). Springer, The Netherlands.

Lin, W. Q., Xie, J. X., Wu, X. M., Yang, L., & Wang, H. D., (2014). Inhibition of xanthine oxidase activity by *Gnaphalium affine* extract. *Chin. Med. Sci. J.*, *29*(4), 225–230.

Linuma, M., Tanaka, T., Hamada, K., Mizuno, M., Asai, F., Reher, G., & Kraus, L., (1987). Revised structure of neoflavone in *Coutarea hexandra*. *Phytochemistry.*, *26*(11), 3096–3097.

Madeswaran, A., Umamaheswari, M., Asokkumar, K., Sivashanmugam, T., Subhadradevi, V., & Jagannath, P., (2011). Docking studies: In silico lipoxygenase inhibitory activity of some commercially available flavonoids. *Ban. J. Pharmacol.*, *6*(2), 133–138.

Manach, C., Scalbert, A., Morand, C., Rémésy, C., & Jiménez, L., (2004). Polyphenols: Food sources and bioavailability. *Am. J. Clin. Nutr.*, *79*(5), 727–747.

Marinova, K., Pourcel, L., Weder, B., Schwarz, M., Barron, D., Routaboul, J. M., Debeaujon, I., & Klein, M., (2007). The *Arabidopsis* MATE transporter TT12 acts as a vacuolar flavonoid/H1-antiporter active in proanthocyanidin-accumulating cells of the seed coat. *The Plant Cell, 19*, 2023–2038.

Matthies, A., Clavel, T., Gütschow, M., Engst, W., Haller, D., Blaut, M., & Braune, A., (2008). Conversion of daidzein and genistein by an anaerobic bacterium newly isolated from the mouse intestine. *Appl. Environ. Microbiol.*, *74*(15), 4847–4852.

McNally, D. J., Wurms, K. V., Labbe, C., & Belanger, R. R., (2003). Synthesis of C-glycosyl flavonoid phytoalexins as a site-specific response to fungal penetration in cucumber. *Physiol. Mol. Plant Path.*, *63*(6), 293–303.

Michalak, A., (2006). Phenolic compounds and their antioxidant activity in plants growing under heavy metal stress. *Polish J. of Environ. Stud.*, *15*(4), 523–530.

Middleton, E. J., (1998). Effect of plant flavonoids on immune and inflammatory cell function. *Adv. Exp. Med. Biol.*, *439*, 175–182.

Middleton, E., (1984). The flavonoids. *Trends. Pharmacol. Sci.*, *5*, 335–338.

Middleton, Jr. E., & Chithan, K., (1993). The impact of plant flavonoids on mammalian biology: Implications for immunity, inflammation and cancer. In: Harborne, J. B., (ed.), *The Flavonoids: Advances in Research Since 1986* (pp. 619–652). Chapman and Hall; London, UK.

Mulvihill, E. E., Burke, A. C., & Huff, M. W. (2016). Citrus flavonoids as regulators of lipoprotein metabolism and atherosclerosis. *Annu. Rev. Nutr.*, *36*, 275–299.

Murray, M. T., (1991). *Quercetin: Nature's Antihistamine.* Better nutrition 1998. NTP technical report (No. 409) on the toxicology and carcinogenesis studies of quercetin in F344/N rats (NIH Publication No. 91–3140 1991). U.S. Department of Health and Human Services, Public Health Service, National Toxicology Program, Research Triangle Park, NC.

Ndakidemi, P. A., & Dakora, F. D., (2003). Legume seed flavonoids and nitrogenous metabolites as signals and protectants in early seedling development. *Funct. Plant Biol.*, *30*(7), 729–745.

Nishimuta, S., Taki, M., Takaishi, S., Iijima, Y., & Akiyama, T., (2000). Structures of 4-aryl-coumarin (neoflavone) dimers isolated from *Pistacia chinensis* BUNGE and their estrogen-like activity. *Chem. Pharm. Bull.*, *48*(4), 505–508.

Panche, A. N., Diwan, A. D., & Chandra, S. R. (2016). Flavonoids: An overview. *J. Nutr. Sci.*, 5.

Paris, D., Mathura, V., Ait-Ghezala, G., Beaulieu-Abdelahad, D., Patel, N., Bachmeier, C., & Mullan, M., (2011). Flavonoids lower Alzheimer's Aβ production via an NFκB dependent mechanism. *Bioinformation, 6*(6), 229.

Parkniske, M., Ahlborn, B., & Werner, D., (1991). Isoflavonoid-Inducible resistance to the phytoalexin glyceolin in soybean rhizobia. *J. Bacteriol.*, *173*(11), 3432–3439.

Peer, W. A., & Murphy, A. S., (2006). Flavonoids as signal molecules: Targets of flavonoid action. In: Peer, W. A., & Murphy, A. S., (eds.), *The Science of Flavonoids* (pp. 239–268). Springer, New York.

Perez, F. J., Villegas, D., & Mejia, N., (2002). Ascorbic acid and flavonoid-peroxidase reaction as a detoxifying system of H_2O_2 in grapevine leaves. *Phytochemistry*, *60*(6), 573–580.

Perry, E. K., Tomlinson, B. E., Blessed, G., Bergmann, K., Gibson, P. H., & Perry, R. H., (1978). Correlation of cholinergic abnormalities with senile plaques and mental test scores in senile dementia. *Br. Med. J.*, *2*(6150), 1457–1459.

Pietta, P., (2000). Flavonoids as antioxidants. *J. Nat. Prod.*, *63*, 1035–1042.

Poureel, L., & Grotewold, E., (2009). Participation of phytochemicals in plant development and growth. In: Osbourn, A. E., & Lanzotti, V., (eds.), *Plant-Derived Natural Product* (pp. 269–282). Springer, Dordrecht, Heidelberg, London, New York.

Ram, H. Y. M., & Mathur, G., (1984). Flower color changes in *Lantana camara*. *J. Exp. Bot.*, *35*(11), 1656–1662.

Rice-Evans, C. A., Miller, N. J., & Paganga, G., (1996). Structure antioxidant activity relationships of flavonoids and phenolic acids. *Free Radic. Biol. Med.*, *20*(7), 933–956.

Samanta, A., Das, G., & Das, S. K., (2011). Roles of flavonoids in plants. *Int. J. Pharm. Sci. Tech.*, *6*(1), 12–35.

Schiavano, G. F., De Santi, M., Brandi, G., Fanelli, M., Bucchini, A., Giamperi, L., & Giomaro, G., (2015). Inhibition of breast cancer cell proliferation and *in vitro* tumorigenesis by a new red apple cultivar. *PLoS One*, *10*(8), e0135840.

Sheng, R., Lin, X., Zhang, J., Chol, K. S., Huang, W., Yang, B., He, Q., & Hu, Y., (2009). Design, synthesis, and evaluation of flavonoid derivatives as potent AChE inhibitors. *Bioorg. Med. Chem.*, *17*(18), 6692–6698.

Shimoji, H., & Yamasaki, H., (2005). Inhibitory effects of flavonoids on alternative respiration of plant mitochondria. *Biologia. Plantarum.*, *49*(1), 117–119.

Shirley, B. W., (1998). Flavonoids in seeds and grains:physiological function, agronomic importance and the genetics of biosynthesis. *Seed Sci. Res.*, *8*, 415–422.

Steinkellner, S., Lendzemo, V., Langer, I., Schweiger, P., Khaosaad, T., Toussaint, J. P., & Vierheilig, H., (2007). Flavonoids and strigolactones in root exudates as signals in symbiotic and pathogenic plant-fungus interactions. *Molecules*, *12*, 1290–1306.

Stobiecki, M., & Kachlicki, P., (2008). Isolation and identification of flavonoids. In: Grotewold, E., (ed.), *The Science of Flavonoids* (pp. 47–70). Springer Science + Business Media I.I.C.; New York, USA.

Straney, D., Khan, R., Tan, R., & Bagga, S., (2002). Host recognition by pathogenic fungi through plant flavonoids. In: Buslig, B. S., & Manthey, J. A., (eds.), *Flavonoids in Cell Function: Advances in Experimental Medicine and Biology* (pp. 9–22). Kluwer Academic/Plenum Publishers, New York, USA.

Subramanian, S., Stacey, G., & Yu, O., (2007). Distinct, crucial roles of flavonoids during legume nodulation. *Trends Plant Sci.*, *12*(7), 282–285.

Sundaravarathan, S., & Kannaiyan, S., (2002). Role of plant flavonoids as signal molecules to rhizobium. In: Kannaiyan, S., (ed.), *Biotechnology of Biofertilizers* (pp. 144–164). Co-published by Kluwer Academic Publisher, The Netherlands.

Tsai, S. M., & Phillips, D. A., (1991). Flavonoids released naturally from alfalfa promote the development of symbiotic glomus spores *in vitro*. *Appl. Environ. Microbiol.*, *57*(5), 1485–1488.

Umamaheswari, M., Madeswaran, A., Asokkumar, K., Sivashanmugam, T., Subhadradevi, V., & Jagannath, P., (2011). Discovery of potential xanthine oxidase inhibitors using in silico docking studies. *Der. Pharma. Chemica.*, *3*(5), 240–247.

Wollenweber, E., & Dietz, V. H., (1981). Occurrence and distribution of free flavonoid aglycones in plants. *Phytochemistry*, *20*(5), 869–932.

Yamasaki, H., Sakihama, Y., & Ikehara, N., (1997). Flavonoid-peroxidase reaction as a detoxification mechanism of plant cells against H_2O_2. *Plant Physiol.*, *115*, 1405–1412.
Yao, L. H., Jiang, Y. M., Shi, J., Tomas-Barberan, F. A., Datta, N., Singanusong, R., & Chen, S. S., (2004). Flavonoids in food and their health benefits. *Plant Foods Hum. Nutr.*, *59*(3), 113–122.

CHAPTER 3

IMPORTANCE OF FLAVONOIDS IN AGRICULTURE

NEHA SAINI, RITU KATARIA, ITTISHREE BHARDWAJ, and PREETI PANCHAL

G.V.M. College of Pharmacy, Sonipat, Haryana, India

ABSTRACT

Flavonoids are important not only for their roles in floras but similarly for their medicinal and nutraceutical applications. Flavonoids are a family of plants that can be found in various concentrations in an inclusive range of plant species. Flavonoids are secondary plant metabolites that play a key part in the biological activity processes of plants. They are dependable for the color characteristic of flowers and fruits. They also take part in symbiosis between plants and microbes. Flavonoids are a class of bioactive chemicals present in a wide range of plant-based foods. Flavonoids are classified into subgroups that establish their chemical arrangement, containing flavones, flavanones, flavonols, flavanonols, anthocyanins, and isoflavones. The determinable and subjective examination of flavonoids has been supported by the metabolomics technique. Flavonoids are a broad set of compounds found in agricultural crops, and researchers have used metabolomics technologies to examine flavonoid composition using various analytical approaches. Allelopathy is currently studied as an organic management of weeds and insect pests, as well as a way to alleviate stress and contamination in consideration of boosting yield output in order to tackle many problems in agriculture. The main objective of this study is to give an outline of their uses and demonstrate that metabolomics is a useful means for studying flavonoid metabolism in various agricultural crops.

Flavonoids as Nutraceuticals. Rajesh K. Kesharwani, Deepika Saini, Raj K. Keservani, and Anil Kumar Sharma (Eds.)

3.1 INTRODUCTION

The name "flavonoid" was coined from "flavous," a Latin word that means "yellow," which resembles the color of flavonoid in ecology. Although their name, many other flavonoids are white, and the most important flavonoid-related anthocyanins are purple, red, or blue in color. Flavonoids, also known as bioflavonoids, are a class of secondary metabolites found in plants and fungi (Rana & Gulliya, 2019).

Secondary plant metabolites called flavonoids are essential to the biological activity of plants. They can be trusted to accurately predict the colour of fruits and flowers. Additionally, they participate in microbial and plant symbiosis. These connections will be utilized to restrain weeds and insects organically, as well as minimize stress and illness, to boost crop yields (Tenango et al., 2017).

Flavonoids are low molar mass compounds with a polyphenolic structure that are elaborate in photosynthesis and other biological functions in plants. They frequently show defensive properties against biotic and abiotic stresses such as Ultraviolet-B radiation, soil salinity, and water stress, detoxifying reactive oxygen species (ROS), at least in part developed in Crops in stressful circumstances (Shojaie et al., 2016).

Flavonoids are a class of unaffected essence that belongs to a group of plant subordinate metabolites accompanying a polyphenolic form that is possibly placed in the crop, edible part of the plant, and few liquors. They bear a sort of advantageous biochemical and antioxidant possessions connected to afflictions like malignant growth, Mental disorder, and so forth (Panche et al., 2016).

The research for novel molecules with important physiological qualities led to the study of flavonoid chemistry, which occurred for most natural products (Maraise et al., 2006). Flavonoids are a wide set of polyphenolic chemicals with a benzo-pyrone structure that remain throughout plants. The phenylpropanoid pathway is responsible for their production. Secondary phenolic metabolites, such as flavonoids, are thought to be responsible for a wide range of pharmacological effects, according to research. Flavonoids are phenolic hydroxylated compounds that are identified as generated by plants in retort to contagious infection (Mondal & Rahman, 2020).

HPLC coupled to UV rays, mass, or nuclear magnetic resonance (NMR) detectors can distinguish, evaluate, and identify flavonoids in a single operation. The technique of capillary electrophoresis (CE) has recently gained popularity (https://prezi.com/p/xrxs28nhrbvf/flavonoids/?fallback=1).

Flavonoids are a class of bioactive chemicals present in a wide range of plant-based foods. Flavonoids are classified into subgroups that establish their chemical arrangement, containing flavones, flavanones, flavonols, flavanonols, anthocyanins, and isoflavones (Kazlowska & Wegierek, 2014).

3.2 CLASSIFICATION

Flavonoids various subclasses are explained below (Figure 3.1) with few examples:

1. **Flavonols:** Flavonoids with a ketone group are recognized as flavonols. Proanthocyanins are completed by these basic units. Flavonols can be created in abundance in an extensive range of fruits and vegetables. Kaempferol, quercetin, Rutin, myricetin, and fisetin are the most explored flavonols. Flavonols are abundant in onions, broccoli, lettuce, tomato, apples, grapes, and berries. Flavonols can also be found in tea and red wine, in addition to fruits and vegetables.
2. **Flavanones:** These are also a major class of complexes that can be present in some citrus fruits, including oranges, lemons, and grapes. This set of flavonoids includes hesperidin, naringenin, and eriodictyol. Because of their superoxide radicals' characteristics, flavanones are associated with a variety of advantages that are related to health. These compounds are present in citrus fruit juice and peel, giving them a bitter taste. Citrus flavonoids have medicinal properties that include antioxidants, anti-inflammatory, lipid levels, and cholesterol-lowering.
3. **Isoflavones:** Isoflavonoids are a subgroup of flavonoids that make up a significant and distinct subgroup. Isoflavonoids are mostly found in soybeans and other leguminous plants, and their availability in the kingdom Plantae is restricted. Certain isoflavonoids have also been discovered in microbes. They are also identified to have an important role as precursors for the synthesis of phytoalexins during plant-microbe interactions. Isoflavonoids offer a lot of potential for treating diseases. Isoflavones like genistein, Glycetin, and daidzein are commonly referred to as phytoestrogens because of their estrogenic action.
4. **Flavones:** These are among the greatest prominent flavonoid subclasses. Flavones are originated as glucosides in leaves, flowers, and fruits of plants. Flavones can be found in parsley, celery, red peppers, tea, peppermint, and ginkgo. This group of flavonoids includes luteolin, apigenin, and tangeritin.

5. **Anthocyanins:** These are pigments found in plants, flowers, and fruits that give them their color. Cyanidin, delphinidin, Petunidin, malvidin, proanthocyanidins (PAC), and peonidin are the most studied anthocyanins. They're typically located in the cranberry's outer cell layers. They are also present in red grapes, grapes, raspberries, strawberries, berries, and blackberries.
6. **Flavan-3-Ols:** Flavanonols are the 3-hydroxy precursors of flavanones, usually known as dihydroflavonols or catechins. They are a multi-substituted and decidedly diverse subclass and include Epicatechin Gallate, Catechin. Because the hydroxyl group is always attached to the third position of the C ring, flavonols are also known as flavan-3-ols. Bananas, apples, berries, apricots, and pears are high in flavan-3-ols (Panche et al., 2016).

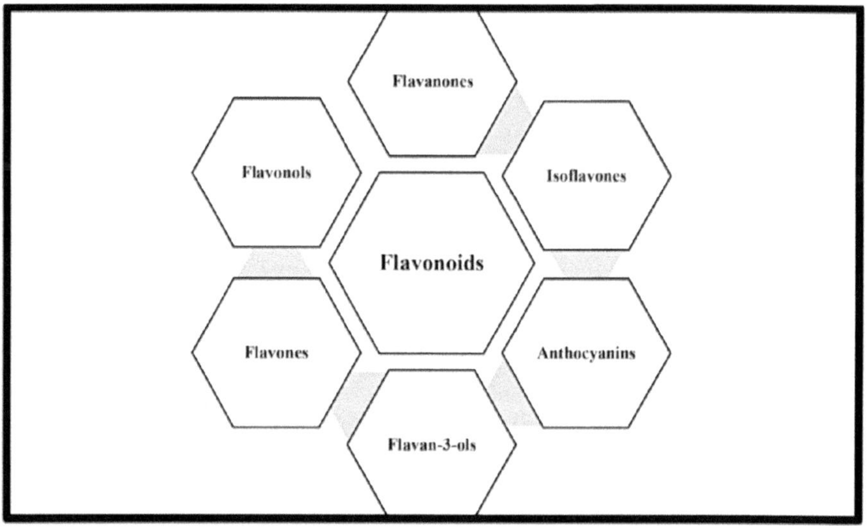

FIGURE 3.1 Classes of flavonoids.

Source: Adapted from: https://www.researchgate.net/figure/Classification-and-example-of-flavonoids-and-their-chemical-structures-Flavonoids-are_fig3_301332394.

3.3 FUNCTION OF FLAVONOIDS IN PLANTS

Flavonoids are involved in a variety of biological processes in plants. Embryo growth and development, fruit expansion and maturation, pollen tube germination, and hormonal transmission are all aided by them. Flavonoids have antioxidant qualities in response to biotic and abiotic stimuli; they protect against destruction caused by fungi, viruses, parasites, and Producers or herbivores. They also have a role in floral, fruit, and seed pigment and color variances. Flavonols, for example, are associated with the yellowish color, flavanols with the color ochre to brownish, and anthocyanins with the color reddish to purple. Anthocyanins and PACs are primarily being the reason for the pigments in corn kernels and petunia flowers. These develop in the nucleus or on the cell membrane. Anthocyanin build-up occurs in corn due to vacuole sequestration (Tenango et al., 2017).

Flavonoids trigger a cascade of activities containing stress-persuaded morphogenesis that defend plants from a variety of unanticipated lesions. Flavonoids are vital for catalyzing electron transference and eliminating reactive oxygen, notably in the manner of superoxide anions, hydroxyl radicals, lipid peroxides, or hydroperoxides, as well as protecting life whole from the detrimental belongings of oxidative processes in contact macromolecules. They prevent the harmful effects of toxic chemicals on cells in this way (Wagh et al., 2017).

3.4 ROLE OF FLAVONOIDS IN PLANTS (FIGURE 3.2)

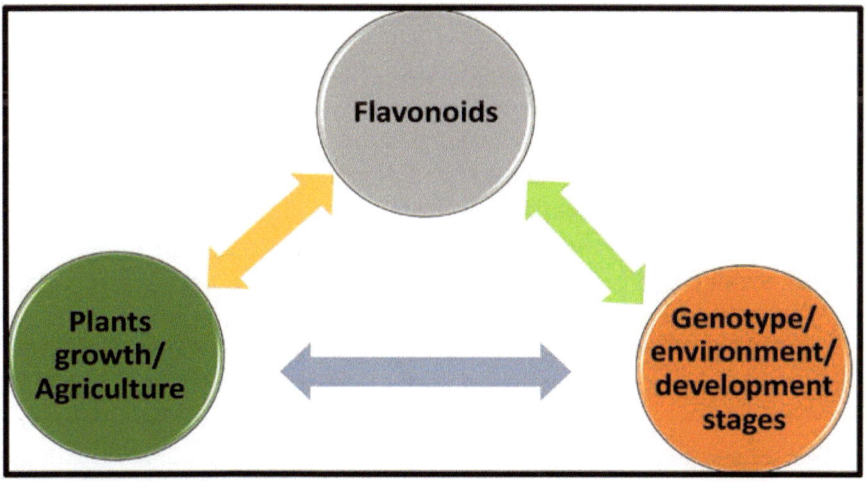

FIGURE 3.2 Role of flavonoids in different fields.

Source: Reprinted from: Weston & Mathesius (2013). Copyright © 2013, Springer Science Business Media

3.4.1 AS GROWTH REGULATOR

Flavonoids provide important functional processes in plant-environment relationships. Auxin transport and its catabolism may be regulated by flavonoids in the micromolar range. Flavonoids tendency to develop auxin gradients results in phenotypic variants with various morphoanatomical characteristics. Controlling auxin transport with flavonoids could be extremely useful in stress-induced proliferation and differentiation responses in plants. When comparing dihydroxy flavonoid-rich species to monohydroxy flavonoid-rich species, phenotypes with markedly distinct morphological features develop. In sunny areas, dwarfed hairy variants accompanying slight, tiny, and dense leaves to direct light part of daytime transmittance are common, protecting plants below in the canopies from light-induced intracellular homeostatic disruptions. Shaded plants, on the other hand, have extended Internodes and wide leaf lamina, as well as leaf senescence, since they are rich in flavonoid compounds, i.e., apigenin, and have insignificant quantities of quercetin derivatives (Kumar et al., 2013).

3.4.2 CROP YIELD AND PLANT GROWTH

Soil productivity has a direct relationship with plant development and production. In agronomic terms, it refers to the soil's ability to produce a certain yield of agricultural crops. Though, a variety of aspects, including soil physicochemical qualities and management-related factors, influence soil ability. It is the ratio of inputs to outputs, which is connected to water and fertilizer supply (inputs) vs crop yield in agronomic settings (output) (Mia et al., 2010). These are in addition to these components, the phytomicrobiome is increasingly being recognized as a vital component of crop productivity. The land is home to a diverse range of microbial species, including unrestricted, symbiotic organisms, host-specific and non-host-specific microbes, all of which can actually impact plant growth and productivity. About 1.0 g of soil is predicted to have 1 billion bacterial cells with 10,000 different genes. Microbes obtain decreased carbon-rich food items secreted from plant roots, while plants aid microorganisms in nutrient uptake, resistance to disease, and stress management by direct and indirect mechanisms in the phytomicrobiome. Numerous studies have established the significance of the phytomicrobiome to crop yield and productivity.

Plants have a biosynthetic mechanism that produces hundreds of biocompounds that are needed to carry out essential tasks. Sugars, amino acids, organic acids, phenolics, enzymes, and growth regulators are among the substances released by the root system. Many microbes linked with the plant root system use these secretions as a carbon and energy source (Hijri, 2016).

3.4.3 COMBATING OXIDATIVE STRESS

Flavonoids have long been considered to serve a diversity of function in plants. Various abiotic and natural determinants influence the result of sensitive oxygen variety (ROS) in plants, resulting in oxidative stress. Plants' flavonoid production is almost entirely boosted by oxidative stress. They bear the strength to take in ultimate forceful sunlight wavelengths (UV-B and UV-A), limit ROS production, and quench ROS after they have formed. Though early implant proceeds from the water to the land, flavonoids performed key UV-B screening roles. The character of replacement ahead of unconnected rings of flavonoids decide in consideration of antioxidant ability to perform and talent to physically take in liquid UV wavelengths. Flavonoids accompanying a dihydroxy B ring substituted bear a higher antioxidant volume, while those accompanying a monohydroxy B ring substituted bear a higher capability to consume UV wavelengths.

3.4.4 FLAVONOIDS IN RHIZOSPHERE

The rhizosphere is by far the most complicated and intense region for plants to connect with their surroundings. The secretion of metabolic byproducts comprising varied molecular weight organic and inorganic compounds such as ions, phenolics, enzymes, secondary metabolites, and celluloses may be required for maximum biological activity, nutrients uptake, and plants–microbes interaction (Pathan et al., 2010). Flavonoids are probable to be radiated from plant root systems and have implicit effects on plant development by facilitating rhizospheric relations, such as fascinating compatible rhizosphere-dwelling rhizobia, stimulating mycorrhizal growth and hyphal branching, increasing nutrient solubility, including phosphorus and iron, and repelling pests and root pathogens (Mandal et al., 2010; Buer et al., 2007).

Flavonoid secretions from roots are thought to be accepted by ATP-dependent active transport mediated by ABC transporters (Badri et al., 2008). Flavonoids not only help to defend against damaging abiotic factors, but they can also help to boost the concentration and bioavailability of soil nutritional components in a low-nutrient environment. When nutrients are scarce in the soil, ABC transporters can release flavonoids into the rhizosphere, where they can bind with metals required for plant development and survival (Shaw et al., 2006). Flavonoid production can be passive also as a result of root cap and epidermal cell degradation (Sugiyama et al., 2007). Flavonoid permanence and mobility in the rhizosphere may be controlled by its solubility, structure, microbial accessibility, and binding affinity, as these substances can be adsorbed to soil or cell wall cation binding sites. Flavonoid glycosides are water-insoluble and are believed to be less adsorbed to binding sites allowing for greater flexibility and accessibility (Shaw et al., 2006).

3.4.5 FLAVONOIDS AND LEGUME-RHIZOBIUM INTERACTION

Many studies have tried explaining and streamlining the connections that occur among particular or specified plants and their symbionts. They have attempted to clarify the contacts that occur between an individual or specific species and their symbiotic organisms. In fact, these relationships are significantly more complicated, including a variety of microorganisms linked with a single plant that exchange chemical signals. These interactions, on the other hand, benefit plants in a variety of ways. One of the most important services offered by soil bacteria is soil fertility and nutrient uptake (Hassan et al., 2012).

Nitrogen deficit is a major issue in crop yields because of rapid nitrogen loss from the soil due to leaching, denitrification, and immobilization. Atmospheric nitrogen is fixed and becomes part of the soil nitrogen replenishment by biological and synthetic mechanisms of fixation from the atmosphere. However, researchers and farmers have taken notice of the natural biological fixation of atmospheric nitrogen (N_2), which accounts for around 60% of total atmospheric nitrogen fixation (Davidson, 2009). Signal exchange is used as a form of interaction between the host and the symbiont for mutual advantage in below-ground interactions that contribute to the creation of legume nitrogen-fixing symbioses. Root tips are at the source of rhizobium attachment and infection, which frequently discharge the highest quantities of flavonoids. These secondary metabolites operate as signaling chemicals, attracting rhizobia to plant roots and activating nod genes in the rhizobia, which initiate the legumes modulation mechanism. Nitrogen-fixing microorganisms, primarily *Rhizobium* and *Bradyrhizobium*, fix over 50% of the nitrogen necessary in legume crops, and the balance is provided by fertilizer additives (Hartwig et al., 1991).

3.4.6 PLANT ABIOTIC STRESSES

Plants are vulnerable to a variety of unfavorable environmental circumstances since they are sessile. Plant hemostasis, physiology, and growth are all affected by environmental fluctuations and harsh growing circumstances, resulting in smaller and undersized plants. A variety of negative biotic and abiotic stresses pose a threat to sustainable agriculture and are frequently responsible for lower crop yields. Plant organs and tissues coordinate their mechanisms for dealing with abiotic stressors. Chemical signals are used. The majority of plant responses to stress situations are unknown. Biosynthetic flavonoids have gained significant attention for their capacity to promote resistance to biotic and abiotic stresses. Abiotic stressors, including UV radiation, salt, and drought tolerance, have significantly aided flavonoids as follows in the below points (Shah & Smith, 2020). Plants regulate gene expression and provide a comprehensive method of control that further regulates developmental stages when both biotic and abiotic stressors occur. Transcription factors (TF) and their actions are one of the variables that contribute to and aid gene transcription (Muhammad et al., 2019).

3.4.6.1 UV SCAVENGERS: FLAVONOIDS

UV light is invisible, has a short wavelength, and is very powerful. These wavelengths have quite enough energy to break chemical bonds in plants, causing damage and deformities through photochemical processes. UV-A, B, and C are separated on the basis of different wavelengths of the light. UV-C is the most energetic and can ionize specific compounds. UV-B can disturb plant metabolism by influencing photosynthesis, starch concentration, and transpiration, as well as causing cellular damage. For Example, Tomato flower/fruit coordination was shown to be increased under strong radiation with negligible influence on vegetative plant components in a study conducted in sterile environmental conditions. There was also an improvement in UV-B receptors and chlorophyll content, as well as triterpenoid molecules that are UV absorption by-products of antioxidant pathways. By stimulating oxidative pathways in plants, UV-A/B can be employed as an abiotic stimulus to improve fruit quality.

3.4.6.2 ROLE OF FLAVONOIDS IN THE MANAGEMENT OF SALT AND DROUGHT STRESS

Salinity is a key limitation to worldwide crop productivity and is one of the most serious abiotic stresses. Excess soluble salts have an impact on nearly 20% of irrigated land. Natural and Anthropogenic activities have caused deteriorated and unrestrained formerly productive agriculture fields, resulting in increased build-up of soluble salts in the root zone, primarily Sodium Chloride. Osmotic potential and ion toxicity are two methods through which surplus soluble salts in soil solution can impede plant growth. To cope with the stress circumstances, plants undergo various changes in response to salt stress. Several metabolic pathways sustained osmotic potential, ion modularity, and poison ion exclusion are all involved in these adaptations. Specific hazardous chemicals, such as superoxide, singlet oxygen, and hydrogen peroxide, induce oxidative damage in cells as a result of ionic toxicity, which is characterized by a secondary impact of saline stress.

Drought confrontation in wheat is linked to enhanced flavonoid accretion, according to research on flavonoid production and accumulation in wheat leaves during drought stress. Flavonoid deposition was seen throughout the year; however, during periods of intense dryness, a considerable increase was found. In *Arabidopsis thaliana*, drought extenuation by flavonoids and flavonoid byproducts has been established. Individual flavonoids' roles were unknown, but enhanced flavonoid synthesis in plants and associated drought tolerance were proven (Shah & Smith, 2020).

3.5 ROLE OF FLAVONOIDS IN THE TRANSPORT OF MECHANISM

Flavonoid deposition may be cell-specific and tissue-specific, but there is proof of flavonoid inter-cellular and intracellular mobility by active transport mechanisms. Membrane-bound vesicle transporters such as the multidrug and lethal compound expulsion and ATP-binding cassette groups are most likely to transfer intracellularly. Flavonoids have already been found in a few layers of *Arabidopsis* shoot and root cells, demonstrating that they are easily transferred in plant tissue cells. Because ABC transporter blockers reduce auxin transport, this long-distance movement may be mediated by ABC transporter groups. Passive flavonoid transfer to the rhizosphere can also arise through the breakdown of root boundary and root cap cells. ATP-binding cassette transporters have been linked to root secretion composition variations in mutated *Arabidopsis*, which include flavonoids such as phytoalexins, as well as carbohydrates and organic acids (Badri et al., 2012).

3.6 PLANT METABOLOMICS AS A TOOL IN AGRICULTURE

Metabolomics is the science of profiling and characterization of metabolites. Small organic compounds generated by protein catalysis are known as plant metabolites. Plant metabolite research may reveal a lot about the plant's biosynthetic and catabolic pathways; however, it is so vital to understand the biochemical and physiological functioning of plant cells. Furthermore, as plant metabolites are direct indicators of plant growth and yield, they may be used to accelerate the design and selection of biochemical markers, resulting in useful tools for improving plant agronomic features. Plant metabolite profiling begins with the isolation of metabolites from samples, followed by analytical procedures that characterize the metabolite components qualitatively or quantitatively as mentioned below (Figure 3.3). The determinable and subjective examination of flavonoids has been supported by the metabolomics technique utilized to resolve this challenge, as well as the liquid chromatography-mass spectrometry. Furthermore, by merging simple, high-performance liquid chromatography (HPLC) and NMR technologies, simple and efficient methods for quick flavonoid identification have been developed, enabling the investigation of flavones and flavanones (Seger & Sturm, 2007).

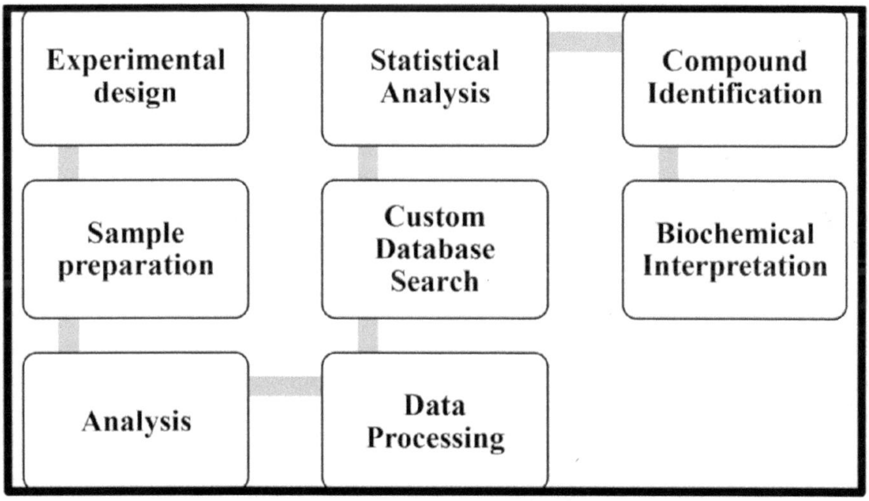

FIGURE 3.3 Workflow of plant metabolomics in agriculture.

Source: Reprinted with permission from: Ciasca et al. (2020).

Flavonols are components of the flavonoid biosynthetic way, in addition to producing anthocyanins and concentrated tannins in crops (Mattivi et al., 2006). Flavonoids are a broad set of compounds found in agricultural crops, and researchers have used metabolomics technologies to examine flavonoid composition using various analytical approaches (Table 3.1).

TABLE 3.1 Metabolomics of Flavonoids in Different Plant Organs of Agricultural Crops (Shah & Smith, 2020)

Sl. No.	Crop	Plant Organs	Analytical Techniques	Compounds
1.	Soybean	Leaves	(Reverse phase)-high-performance liquid chromatography and nuclear magnetic resonance.	Naringenin, rutin, quercetin, kaempferol, and its glucosides, and total flavonoids.
2.	Red Tomato	Fruit	HPLC and LC/NMR, LC/MS, and LC/MS/MS	Naringenin chalcone and rutin
3.	Grapes	Berries	LC-MS	Quercetin and derivatives, kaempferol, and isorhamnetin.
4.	Rice	Leaf and bran	LC-QTOF-MS	Tricon, Tricon 7-O rutinoside.
5.	Maize	Kernels	LC-MS/MS	Apigenin, luteolin, methyl, malvidin, pentose, rhamnose, tricin, chalcone, synthase, chalcones isomerase.

3.7 FLAVONOIDS: ALLELOPATHY AND ITS APPLICATIONS IN AGRICULTURE

The straight or secondary influence of subordinate chemicals caused by a contributor plant ahead of a receiver plant is known as allelopathy. This form of interaction has the potential to be both beneficial and detrimental. In order to address various issues in agriculture, allelopathy is being researched as a natural method of controlling weeds and insect pests, as well as a means of reducing stress and contamination [18]. Weeds are the class that conflict containing crops and unaffected herbicides from molasses, sunflower, eucalyptus, and edible grain are used to control them when used together and they have a higher efficacy than when used separately. Example: Exudates from the roots of rice plants have been shown to lessen the infestation of fungi of the genus Fusarium on melon. Furthermore, including *Brassica napus* L. plants in the loam reduces the inhabitants of some nematode worms in orchards. Allelochemicals have direct and indirect effects on plants, including soil alteration, physicochemical qualities, microbial population alterations, and nutrient availability differences are indirect effects. The physiologic and biochemical transitions that take place all along plant progress and incident are the direct action (Muzell et al., 2016).

Weed management approaches have always been an important part of agricultural systems, but they have evolved significantly over time due to the availability of tools and techniques, as well as environmental and sustainability concerns, ranging from ancient techniques such as hand pulling and soil tilling with simple tools to the current use of herbicides and mechanized conventional tillage (Li et al., 2010).

Allelopathy's importance has grown in recent years, particularly in agriculture. Beyond the reduction or restriction of weed growth, plant-plant interactions can influence or decide the variety, productivity, and reproduction of a plant community (Erica et al., 2017). Wheat, rice, rye, barley, maize, and sugarcane are all known to exhibit allelopathic properties due to phytochemical exudation. Sunflower cultivars have exhibited significant allelopathic effects on herbicides. However, there was heterogeneity among cultivars, implying that phytotoxic effects, as well as weed control, differ by cultivar or genotype. Furthermore, the sunflower cultivars reduced total weed concentration and weight by 10–87% and 34–81%, respectively. The secondary metabolite flavan-3-ol (–)-catechin was known to be involved in *Centaurea maculosa*'s invasive nature and phytotoxicity.

In order to ensure agricultural efficiency and sustainability in agroecosystems, studies on the use of allelopathy as an alternative to pesticides in the control of pests, diseases, and weeds are rising (Jabran et al., 2015). Planting a sort alongside allelopathic properties as a hide crop, in a capable of rotating order, or as a leftover part or protective covering, exceptionally in reduced-till settings, can potentially be utilized to suppress weeds (James et al., 2016). Allelopathic plants can be employed as crop rotation alternation plants, plant extracts, and biopesticides, as well as being included in the intercropping method and as a cover crop plant (Fulya, 2020).

3.8 SIGNIFICANCE OF FLAVONOIDS IN PEST MANAGEMENT

The need for natural goods derived from vegetation to be utilized as pest control agents rises daily basis. Flavonoids are being used to develop new insecticides as a substitute for artificial pesticides. They can stop the development of larvae of certain insect species by inhibiting enzymatic activity (Kim et al., 2000). Some flavonoids prevent the creation of hormone, which is involved in the ecdysis and reproduction of a variety of insects (Oberdorster,

2001). Flavonoids are crucial secondary metabolites that originate in floras that protect them against a variety of stressors, including predatory insects. In response to insect weed growth, diseases, and other stresses, plants create a variety of secondary metabolites which serves as plant defensive mechanism, such as phenols. By altering the steroid hormone systems, flavonoids and isoflavonoids have a direct impact on insect behavior, growth, and development. They are strong antibiotics that form complexes with numerous enzymes, limiting insect pests' access to food proteins. Flavonoids produced by plants in response to insect pests will act as key biological markers for inducing insect pest susceptibility and the identification of resistant lines in breeding strategies (Abdul et al., 2016; Shirley, 1998).

3.9 FUTURE CHALLENGES AND RESEARCH

Climate change is the greatest threat to contemporary human civilization. Humanity is in danger due to rising worldwide food demand and ever-increasing global warming. It is important to note the several obstacles to executing the proposed study model: first, flavonoids have extremely complicated conjugate profiles after consumption, making it difficult to isolate, identify, and measure any specific structure. Furthermore, because of the scarcity of analytical standards, this will be a major project that will necessitate a synthetic or biosynthetic focus. Second, demonstrating bioavailability in most situations necessitates expensive human experiments, including labeled chemicals, which are intrinsically difficult to synthesize and costly to make/purchase. Third, there are a plethora of prospective endpoints for demonstrating biological impacts, many of which have yet to be standardized (Kay, 2010).

Future polyphenol bioactivity research will necessitate a better knowledge of their intake, bioavailability, and metabolism. Flavonoid research in the future will require a multidisciplinary strategy that includes epidemiology, human intervention, and molecular research. It is in need of research and development programs that include *in vivo* studies and provide a positive and secure future vision. Flavonoid-rich fruits, vegetables, and drinks are currently advised for consumption (Panche et al., 2016).

The present state of research is about the primary factors that influence the contents of flavonoids in onions and several ways that are used to boost the gathering of these chemicals. Example: Red varieties have the highest flavonoid levels, and resistant onions have higher flavonoid levels than susceptible onions. In terms of soil management, nitrogen fertilizer levels should be kept to a minimum to favor flavonoid levels. Organically cultivated onions also have advanced levels of flavonoids and antioxidant activity than conventionally grown onions. Color, flavor, bitterness, and texture are all sensory qualities that phenolic chemicals can change, influencing consumer perception. Identification of particular chemicals in various onion cultivars and agronomic approaches might help researchers better understand the physiological reactions to onion eating. This would be helpful with the expansion of onion-making systems that offer more well-being benefits, as well as the development of intake guidelines for these compounds. Clarifying the connections between the genetic constitution and agro-environmental aspects of the flavonoid configuration in onions is an important and demanding aspect of future work (Rodrigues et al., 2017). Traditional Mediterranean diets include a high intake of flavonoid-rich plant items. For the first time, correlations between the flavonoids of *Rhamnus davurica* and their antiproliferative activities were discovered in this study, indicating that perhaps the fingerprint description of flavonoids and their anticancer actions can provide important information on quality control for this herbal medicine and its derived natural remedies. A premium marketing technique may benefit from the manufacture of fresh "useful food" with specific health rights. Plant foods' minimal quality could be defined in the future based on their gratified bioactive mechanisms (Chen et al., 2016; Ros et al., 2010).

One of the plans that have been funded over the centuries is presented under the fruit and vegetable research population's ability to submit effective proposals in the EU Commission's: "Flavonoids in fruits and vegetables: their impact on food quality, nutrition, and human health," according to the FLAVO project. The study focuses on fruits that are commonly available in Europe, such as apples, grapes, and strawberries, as well as their by-products. The goal of FLAVO was to track flavonoids in fruits and vegetables and optimize their health benefits. The European Fruit Research Institutes Network (EUFRIN) promoted this action, which would be beneficial to the study of consumer behavior about new products, the collection of enhanced plant foods through breeding, and the selection of agronomical techniques.

3.10 CONCLUSION

The significance of understanding flavonoids as a large set of natural compounds found in various species has been underlined throughout this chapter. Their biosynthesis, investigative techniques for their examination, and biological action have all been studied extensively. They are chemicals that have varied sensitivities to biotic and abiotic conditions and can be used in agriculture as pollinator attractants, floral and fruit pigments, allelochemical functions, beneficial organism symbiosis, and pest control.

Flavonoids are important not only for their roles in plants but also for their medicinal and nutraceutical applications. Flavonoids are a family of plants that can be found in various concentrations in an inclusive range of plant species. Furthermore, its health benefits have been established, with one of the most noteworthy instances being its antioxidant activity. Agriculture-related data, on the other hand, is dispersed. As a result, the goal of this chapter is to give an overview of their uses and demonstrate that metabolomics is a useful tool for studying flavonoid metabolism in various agricultural crops.

KEYWORDS

- **abiotic stresses**
- **allelopathy**
- **flavones**
- **flavonoids**
- **legume-rhizobium interaction**
- **metabolism**
- **metabolomics technique**
- **oxidative stress**
- **rhizosphere**

REFERENCES

Abdul, R. W., Suraj, P. S., & Hari, C. S., (2016). Differential induction of flavonoids in groundnut in response to *Helicoverpa armigera* and *Aphis craccivora* infestation. *Int. J. Insect. Sci., 8*, 55–64. Published online. doi: 10.4137/IJIS.S39619.

Afzal, A., & Bano, A., (2008). Rhizobium and phosphate solubilizing bacteria improve the yield and phosphorus uptake in wheat (*Triticum aestivum*). *Int. J. Agric. Biol., 10*, 85–88.

Available from: https://prezi.com/p/xrxs28nhrbvf/flavonoids/?fallback=1 (accessed on 24 June 2023).

Available from: https://www.researchgate.net/figure/Classification-and-example-of-flavonoids-and-their-chemical-structures-Flavonoids-are_fig3_301332394 (accessed on 24 June 2023).

Badri, D. V., Chaparro, J. M., Manter, D. K., Martinoia, E., & Vivanco, J. M., (2012). Influence of ATP-binding cassette transporters in root exudation of phytoalexins, signals, and in disease resistance. *Frontiers in Plant Science, 3*, 149.

Badri, D. V., Loyola-Vargas, V. M., Broeckling, C. D., De-La-Peña, C., Jasinski, M., Santelia, D., Martinoia, E., et al., (2008). Altered profile of secondary metabolites in the root exudates of *Arabidopsis* ATP-binding cassette transporter mutants. *Plant Physiology, 146*, 762–771.

Buer, C. S., Muday, G. K., & Djordjevic, M. A., (2007). Flavonoids are differentially taken up and transported long distances in Arabidopsis. *Plant Physiology, 145*, 478–490.

Chen, G., Li, X., Saleri, F., & Guo, M., (2016). Analysis of flavonoids in *Rhamnus davurica* and its antiproliferative activities. *Molecules, 21*, 1275. https://doi.org/10.3390/molecules21101275.

Ciasca, B., Lanubile, A., Marocco, A., Pascale, M., Logrieco, A. F., & Lattanzio, V. M. T., (2020). Application of an integrated and open source workflow for LC-HRMS plant metabolomics studies. case-control study: Metabolic changes of maize in response to *Fusarium verticillioides* infection. *Front. Plant Sci., 11*, 664.

Davidson, E. A., (2009). The contribution of manure and fertilizer nitrogen to atmospheric nitrous oxide since 1860. *Nat. Geosci., 2*, 659–662.

Erica, L. S., Beatriz, H. L. N. S. M., Aurea, P. F., & Sirlei, D. T., (2017). Flavonoids: Classification, biosynthesis, and chemical ecology. In: Goncalo, C. J., (ed.), *Flavonoids – from Biosynthesis to Human Health.* IntechOpen. doi: 10.5772/67861. Available from: https://www.intechopen.com/chapters/54574 (accessed on 24 June 2023).

Farooq, M., Bajwa, A. A., Cheema, S. A., & Cheema, Z. A., (2013). Application of allelopathy in crop production. *International Journal of Agriculture and Biology, 15*, 1367–1378.

Fulya, B. (2020). Plant allelochemical interference: Allelopathy. *Mod. Concept. Dev. Agrono., 7*(3), MCDA. 000663.

Hartwig, U. A., Joseph, C. M., & Phillips, D. A., (1991). Flavonoids released naturally from alfalfa seeds enhance the growth rate of *Rhizobium meliloti. Plant Physiol., 95*, 797–803.

Hassan, S., & Mathesius, U., (2012). The role of flavonoids in root-rhizosphere signaling: Opportunities and challenges for improving plant-microbe interactions. *J. Exp. Bot., 63*, 3429–3444.

Hijri, M. (2016). Analysis of a large dataset of mycorrhiza inoculation field trials on potatoes shows highly significant increases in yield. *Mycorrhiza, 26*, 209–214.

Jabran, K., Mahajan, G., Sardana, V., & Chauhan, B. S., (2015). Allelopathy for weed control in agricultural systems. *Crop Protection, 72*, 57–65.

James, J. F., Bala, R., & Carlene, A. C., (2016). *Allelopathy: How Plants Suppress Other Plants.* HS944, 2016-08-25. doi: doi.org/10.32473/edis-hs186-2013.

Kay, C., (2010). The future of flavonoid research. *British Journal of Nutrition, 104*(S3), S91–S95. doi: 10.1017/S000711451000396X.

Kim, J. S., Kwon, C. S., & Son, K. H., (2000). Inhibition of ά-glucosidase and ά-amylase by luteolin, a flavonoid. *Bioscience, Biotechnology, and Biochemistry, 64*, 2458–2461.

Kozlowska, A., & Wegierek, D. S., (2014). *Flavonoids-Food Sources and Health Benefits* (Vol. 65, No. 2, pp. 79–85). National Institute of Public Health and Hygiene.

Kumar, S., & Pandey, A. K., (2013). Chemistry and biological activities of flavonoids: An overview. *The Scientific World Journal, 162750,* 16.

Li, Z. H., Wang, Q., Ruan, X., Pan, C. D., & Jiang, D. A., (2010). Phenolics and plant allelopathy. *Molecules, 15*(12), 8933–8952.

Mandal, S. M., Chakraborty, D., & Dey, S., (2010). Phenolic acids act as signaling molecules in plant-microbe symbioses. *Plant Signal. Behav., 5*, 359–368.

Marais, J. P. J., Deavours, B., Dixon, R. A., & Ferreira, D., (2006). The stereochemistry of flavonoids. *The Science of Flavonoids* (pp. 1–72). The Ohio State University Columbus, Ohio, USA.

Mattivi, F., Guzzon, R., Vrhovsek, U., Stefanini, M., & Velasco, R., (2006). Metabolite profiling of grape: Flavonols and anthocyanins. *J. Agric. Food Chem., 54*(20), 7692–7702. doi: 10.1021/jf.061538c. PMID: 17002441.

Mia, M. B., & Shamsuddin, Z., (2010). Rhizobium as a crop enhancer and biofertilizer for increased cereal production. *Afr. J. Biotechnol., 9*, 6001–6009.

Mondal, S., & Rahaman, S. T. (2020). Flavonoids: A vital resource in health care and medicine. *Pharmacy and Pharmacology International Journal, 8*(2), 91–102.

Muhammad, K., et al., (2019). Role of flavonoids in plant interactions with the environment and against human pathogens: A review. *Journal of Integrative Agriculture, 18*(1), 211–230.

Muzell, T. M., Vidal, R. A., Balbinot, J. A., Von, H. B., & Da Silva, S. F., (2016). Allelopathy: Driving mechanisms governing its activity in agriculture. *Journal of Plant Interactions, 11*(1), 53–60.

Oberdorster, E., Clay, M. A., Cottam, D. M., Wilmot, F. A., McLachlan, J. A., & Milner, M. J., (2001). Common phytochemicals are ecdysteroid agonists and antagonists: A possible evolutionary link between vertebrate and invertebrate steroid hormones. *Journal of Steroid Biochemistry and Molecular Biology, 77*, 229–238.

Panche, A. N., Diwan, A. D., & Chandra, S. R. (2016). Flavonoids: An overview. *Journal of International Science, 5*(47), 1–15.

Panche, A., Diwan, A., & Chandra, S. (2016). Flavonoids: An overview. *Journal of Nutritional Science, 5*, E47. doi: 10.1017/jns.2016.41.

Pathan, S. I., Ceccherini, M. T., Sunseri, F., & Lupini, A., (2020). Rhizosphere as a hotspot for plant-soil-microbe interaction. In: *Carbon and Nitrogen Cycling in Soil* (pp. 17–43). Springer: Berlin/Heidelberg, Germany.

Rana, C. A., & Gulliya, B., (2019). Chemistry and pharmacology of flavonoids: A review. *Indian Journal of Pharmaceutical Education and Research, 53*(1), 367–389.

Rodrigues, A. S., Almeida, D. P. F., Gándara, J. S., & Pérez-Gregorio, M. R., (2017). *Onions: A Source of Flavonoids* (pp. 439–471). Intech.

Ros, R. Z., Lacueva, C. A., Lamuela-Raventós, R. M., Tormo, M. J., Ramón, Q. J., & González, C. A., (2010). *Estimation of Dietary Sources and Flavonoid Intake in a Spanish Adult Population (EPIC-Spain), 110*(3), 390–398. doi: https://doi.org/10.1016/j.jada.2009.11.024.

Seger, C., & Sturm, S., (2007). Analytical aspects of plant metabolite profiling platforms: Current standings and future aims. *J. Proteome. Res., 6*(2), 480–497.

Shah, A., & Smith, D. L. (2020). Flavonoids in agriculture: Chemistry and roles in biotic and abiotic stress responses, and microbial associations. *Agronomy, 10*, 1209.

Shaw, L. J., Morris, P., & Hooker, J. E., (2006). Perception and modification of plant flavonoid signals by rhizosphere microorganisms. *Environ. Microbiol., 8*, 1867–1880.

Shirley, B., (1998). Flavonoids in seeds and grains: Physiological function, agronomic importance and the genetics of biosynthesis. *Seed Science Research, 8*(4), 415–422. doi: 10.1017/S0960258500004372.

Shojaie, B., Ajeran, A. M., & Ghannadian, M., (2016). Flavonoids dynamic response to different drought conditions: Amount, type and localization of flavonoids in roots and shoots of *Arabidopsis thaliana* L. *Turkish Journal of Biology, 14*, 612–622.

Sugiyama, A., Shitan, N., & Yazaki, K., (2007). Involvement of a soybean ATP-binding cassette-type transporter in the secretion of genistein, a signal flavonoid in legume-rhizobium symbiosis. *Plant Physiol., 144*, 2000–2008.

Tenango, M. P., Hernandez, M. S., & Hernandez, E. A., (2017). *Flavonoids in Agriculture*. INTECH. Accessed from http://dx.doi.org/10.5772/intechopen.68626 (accessed on 24 June 2023).

Waghmare, P. R., Takdhat, P., & Moon, A. G., (2017). Flavonoids. *World Journal of Pharmaceutical Research, 6*(6), 335–347.

Weston, L. A., & Mathesius, U., (2013). Flavonoids: Their structure, biosynthesis, and role in the rhizosphere, including allelopathy. *Journal of Chemical Ecology, 39*(2), 283–297.

CHAPTER 4

THERAPEUTIC ANTIVIRAL POTENTIAL OF FLAVONOIDS

RUCHI RANI, MANDAR BHUTKAR, and SHAILLY TOMAR

Department of Bioscience and Bioengineering, Indian Institute of Technology, Roorkee, Uttarakhand, India

ABSTRACT

Flavonoids are well-known naturally occurring biomolecules founds to be effective antivirals. These biomolecules can act directly and indirectly at various steps of viral infection. Flavonoids from plants serve as an untapped reservoir of therapeutically active constituents that needs to be explored as potential antiviral candidates against RNA and DNA viruses. Further, structure-based studies can play a crucial role in identifying antiviral activity showing flavonoids. Despite many efforts, very few flavonoids are in clinical trials for antiviral treatment against various virus infections. Here, flavonoids as evidence-based natural sources of antivirals against a number of classes of viruses have been discussed.

4.1 INTRODUCTION

Infectious diseases have been known to humans since ancient times. The emergence, dissemination, and persistence of these diseases remain a pressing worldwide health concern. These diseases are caused by various microbes such as viruses, bacteria, protozoa, fungi, and others. Humans and viruses are always at odds, and viruses are constantly upgrading their attack and defense strategies against humans, which is associated with high mortality and morbidity. The current global context implies that viral disease increases rapidly due to unprecedented climate change and globalization.

Flavonoids as Nutraceuticals. Rajesh K. Kesharwani, Deepika Saini, Raj K. Keservani, and Anil Kumar Sharma (Eds.)

In recent years, there has been significant progress in understanding the genetic basis and molecular mechanism of many infectious diseases. The prevalence of viral infections, which impact several million individuals each year, has heightened interest in the antiviral potential of naturally occurring flavonoids. Synthetic antiviral molecules frequently have limited efficacy and substantial side effects, while flavonoids being natural molecules have no side effects. With growing worldwide awareness of nutrition and health, the inclusion of naturally occurring flavonoids in the diet provides health benefits against various viral diseases and helps in improving human health (Keservani & Sharma, 2014; Keservani et al., 2010a, b, 2020).

Flavonoids are naturally occurring secondary plant metabolites with a polyphenolic structure in barks, roots, stems, and plant by-products (Middleton, 1998). Long before flavonoids were identified as the active ingredients, these natural chemicals were known for their health benefits. Flavonoids have been placed in over 4,000 different types, many of which are responsible for the appealing colors of flowers, fruit, and leaves (De Groot & Rauen, 1998). Flavonoids are benzo-γ-pyrone derivatives structures that are ubiquitously present in plants. They have two aromatic rings (A and B) connected by a heterocyclic pyran-4 ring (C) (Figure 4.1). The C2–C3 double bond and the C ring's 4-oxo functional group are essential components in flavonoids' biological activity (Takano-Ishikawa et al., 2006). Flavonoids are classified into subgroups based on the degree of unsaturation and oxidation of the C-ring and the B-ring linkage to the carbon of the C-ring. Flavonoids in which the B ring is attached to the C3 and C4 position of the C ring are known as isoflavones and neoflavonoids, respectively. In contrast, the B ring attached to the C2 position of the C ring is further classified into many subgroups based on the structural features of the C ring. The following subgroups are flavonols, flavanonols, flavones, flavanones, catechins, chalcones, and anthocyanins.

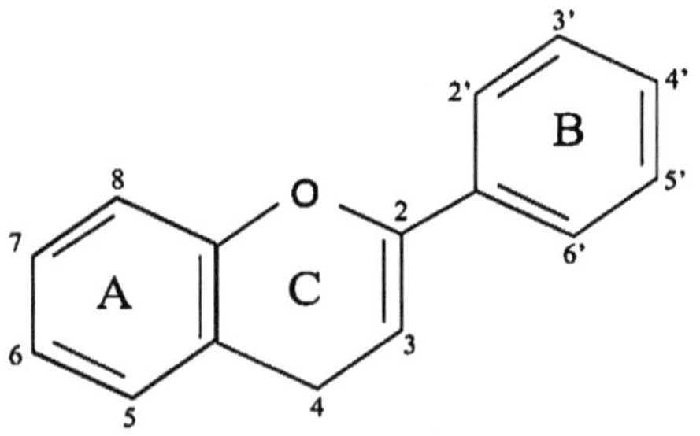

FIGURE 4.1 The basic structure of flavonoids.

Flavonoids' chemical properties are determined by their degree of hydroxylation, structural class, degree of polymerization, and various substitutions and conjugations (Heim et al., 2002). Even though flavonoids possess numerous health benefits and are widely available in human diets, researchers face obstacles to using these natural compounds as therapeutic choices in the clinical context. The absorption and bioavailability of flavonoids in humans are the main hindrances influenced by factors such as glycosylation, pKa, esterification, molecular weight, lipophilicity, and interactions with enteric bacteria, and other metabolic conjugations along the alimentary tract (Cook & Samman, 1996; Hollman & Katan, 1998; Jaganath et al., 2006; Makino et al., 2009; Hollman et al., 1999; Scalbert et al., 2002; Yao et al., 2004). As a result, efforts to improve the bioavailability of flavonoids when consumed by humans are critical for turning these natural molecules into viable antiviral therapeutics. To improve their bioavailability, researchers have tried several approaches to increase the solubility of substances or to move the absorption site in the gut. Structural changes in flavonoids that resulted in a shift in the site of the large-to-small intestine for absorption increased the intake of flavonoids in individuals. In addition to solubility, the approach that boosts the bioavailability of the flavonoids is by increasing the dissolving rate and permeability of flavonoids, limiting their degradation and metabolism in the gastrointestinal tract, and directly delivering the flavonoids to their physiological targets (Puranik et al., 2019; Scalbert et al., 2002).

Recent curiosity in these substances has been sparked by the potential health benefits arising from the antiviral (Dejani et al., 2021; Ninfali et al., 2020; Kaul et al., 1985; Zandi et al., 2011), antimicrobial (Gutiérrez-Venegas et al., 2019; Cushnie & Lamb, 2005), anti-parasitic (Liu et al., 1992; Mead & McNair, 2006; Stermitz et al., 2002), antioxidant (Forester & Lambert, 2011; Heim et al., 2002; Pietta, 2000), anti-inflammatory (Maleki et al., 2019), anti-carcinogenic (Adhami & Mukhtar, 2006; Forester & Lambert, 2011; Seelinger et al., 2008; Imran et al., 2019; Qadir, 2017; Yang & Wang, 2011), anti-mutagenic (Miyazawa et al., 1999, 2000, 2001; Miyazawa & Hisama, 2003) as well as other beneficial activities of these polyphenolic compounds.

4.2 FLAVONOIDS AS ANTIVIRAL

Flavonoids (Keservani & Sharma, 2014) are effective against DNA as well as RNA viruses. The virus is an obligatory parasite that is incapable of propagating on its own. However, once a virus infects a susceptible cell, it can direct its machinery to produce more viruses. Viruses can infect various cells, including animals, bacteria, plants, humans, yeast, protozoa, and archaea. The diversity of nucleic acids found in viruses is astonishing. The size and complexity of viruses' nucleic acid vary greatly from few as 2,000 bases to more than 2×10^6 bases, leading to coding capacities extending from 2 to over 2,000 proteins. If we compare it with the human genome, it is 3.2×10^9 bases long, or almost 2×10^5 times longer. Among these different classes of viruses, (+) RNA viruses are the largest group of eukaryotic viruses that can infect insect, mammalian, and plant hosts.

In this chapter, the antiviral potential of flavonoids has been discussed. Albert Szent-Gyorgyii presented the first evidence of flavonoids' biological activity in 1938, demonstrating that orange peel flavonoids decrease capillary bleeding and fragility associated with scurvy (Samuelsson, 1999). After that, a broad spectrum of biological activities has been defined for flavonoids. The research for antiviral compounds extracted from plants began in the 1950s when extracts of 288 plants were shown to have antiviral action against the influenza A virus in embryonated eggs (Chantrill et al., 1952). Flavones' antiviral activity has been examined and documented since the 1990s when synergistic antiviral potential of apigenin and 5-ethyl-2′-deoxyuridine on the multiplication of herpes simplex virus type 1 (HSV-1) and pseudorabies virus (Mucsi, 1984) were studied *in vitro*. While in 1992, apigenin exhibited a synergistic effect with acyclovir in cell culture on HSV-1 and HSV-2 (Beladi, 1992). Although, apigenin structure was initially recognized in 1900 (Li et al., 1997) and synthesized in 1939 (Hutchins & Wheeler, 1939). Aside from these viruses, apigenin has been shown to have antiviral activity against the African swine fever virus (ASFV), DNA virus (Hakobyan et al., 2016), and picornaviruses, RNA virus (Lv et al., 2014; Qian et al., 2015). In addition, the antivirals have been intensively explored for other flavones like baicalein (Johari et al., 2012; Zandi et al., 2012), hydroxy flavone (Wang et al., 2013, 2014; Kawser et al., 2014), luteolin (Fan et al., 2016; Peng et al., 2017, 2018; Shadrack et al., 2021; Shawan et al., 2021), and wogonin (Choi et al., 2015; Guo et al., 2007; Seong et al., 2018; Chu et al., 2020).

Recently, flavonols are also reported to be effective against poliovirus 1, HSV-1, and 2 (Amoros et al., 1992; Lyu et al., 2005) and respiratory syncytial virus (RSV) (Barnard et al., 1993). While one of the potential flavonols, quercetin showing activity against different viruses such as influenza A virus (Ha et al., 2014; Wu et al., 2016), HSV-1, HSV-2 (Hung et al., 2015), Japanese Encephalitis Virus (JEV) (Johari et al., 2012), ZIKA virus (ZIKV) (Wong et al., 2017), Ebola virus (Qiu et al., 2016), hepatitis B virus (HBV) (Cheng et al., 2015), hepatitis C virus (HCV) (Gonzalez et al., 2009), murine coronavirus (CoV) and dengue virus (DENV) infection (Chiow et al., 2016), porcine epidemic diarrhea virus (Choi et al., 2009), anti-Mayaro virus (MAYV) (dos Santos et al., 2014), etc. Other flavonols and their derivatives, such as glycoside rutin, sulfated rutin, kaempferol, fisetin, etc., also act as antivirals against different viruses.

4.2.1 MOLECULAR INHIBITION TARGETS

Flavonoids with antiviral activity should have the following properties: no toxicity in healthy cells, high efficacy for numerous viral diseases, oral consumption capability, and inexpensive cost. Different flavonoids have been discovered to inhibit the virus by various mechanisms. Flavonoids can function as therapeutic inhibitors or indirect immune system inhibitors based on antiviral modes of action. The following categories can be found in flavonoids that can suppress viral activity (i) They can suppress viruses inhibition from attaching to or entering the host cells (Table 4.1); (ii) inhibitors of viral replication in the early stages (Table 4.2); (iii) blockers for transcription and translation (Table 4.3); (iv) blocking late phases of maturation, such as assembly, packaging, and release (Table 4.4); and (v) flavonoids that can prevent viral infections by interfering with host components necessary for infection or by regulating the immune system to decrease viral titer. To multiply and thrive, viruses rely on the metabolism of their hosts and their surroundings. So, they misuse and take over the host's cellular machinery and spread throughout the body (Table 4.5).

TABLE 4.1 Viruses Targeted by Different Flavonoids Attachment or Entry Stage

Flavonoids	Virus	*In Vitro/In Vivo*	References
Quercetin	JEV	Vero cells	Johari et al. (2012)
	Rhinovirus	BEAS-2B and C57BL/6 mice	Ganesan et al. (2012)
	Influenza A virus	293T cells, Madin Darby Canine Kidney (MDCK) cells, and human lung epithelial A549 cells.	Wu et al. (2016)
	HCV	Human embryonic kidney (HEK) 293T.	Rojas et al. (2016)
	HSV-1 and HSV-2	Vero cells	Lyu et al. (2005)
	HSV-1	Raw 264.7 cells and Vero cells	Lee et al. (2017)
Luteolin	Epstein-Barr virus (EBV)	rAkata cells with TW01 cells	Wu et al. (2017)
	DENV	AG129 mice	Peng et al. (2017)

TABLE 4.2 Viruses Targeted by Different Flavonoids at the Early Viral Replication Stage

Flavonoids	Virus	*In Vitro/In Vivo*	References
Quercetin	DENV-2	Vero cells	Zandi et al. (2011b)
	HCV	Human embryonic kidney (HEK) 293T.	Rojas et al. (2016)
	HSV-1	Raw 264.7 cells and Vero cells.	Lee et al. (2017)
	HBV	HepG2.2.15 cells and BALB/c mice.	Bai et al. (2016)
	JEV	Vero cells	Johari et al. (2012)
	Mayaro viruses	Vero cells	dos Santos et al. (2014)
Luteolin	Enterovirus 71 and Coxsackievirus a16	293T cells, RD cells (human embryonal rhabdomyosarcoma), and Vero cells.	Xu et al. (2014)
	JEV	A549 cells	Fan et al. (2016)
Baicalein	JEV	Vero cells	Johari et al. (2012)
	Influenza A/FM1/1/47 (H1N1) virus	BALB/c mice	Xu et al. (2010)
	Influenza A/FM1/1/47 (H1N1) virus.	MDCK cells	Chen et al. (2011)
Apigenin	EV71	Vero cells	Lv et al. (2014)
	Vaccinia virus	HeLa cells, MDCK cells, HuH7 cells, and Con1 cells.	Chang et al. (2009)

TABLE 4.3 Viruses Targeted by Different Flavonoids at Transcription and Translation Stage

Flavonoids	Virus	*In Vitro/In Vivo*	References
Quercetin	HCV	*In silico* studies	Fatima et al. (2014)
	SARS-CoV	*In silico* study	Ryu et al. (2010)
Apigenin	Enterovirus (EV71)	Vero cells	Lv et al. (2014)
	Foot-and-mouth disease (FMD)	BHK-21 cells	Qian et al. (2015)
	EBV	rAkata cells	Wu et al. (2017)
	SARS-CoV	*In silico* study	Ryu et al. (2010)
Luteolin	SARS-CoV	*In silico* study	Ryu et al. (2010)

TABLE 4.4 Viruses Targeted by Different Flavonoids at the Assembly, Packaging, and Release Stage

Flavonoids	Virus	*In Vitro/In Vivo*	References
Quercetin	HCV	Humanembryonic kidney (HEK) 293T	Rojas et al. (2016)
	Canine distemper virus	Vero cells	González-Búrquez et al. (2018)
	HCV	Human embryonic kidney (HEK) 293T.	Rojas et al. (2016)
Luteolin	DENV 1–4	AG129 mice	Peng et al. (2017)
Baicalin	Strain A/Thailand/K (H3N2)	*Insilco* studies	Veerasamy & Rajak. (2021)
Biochanin A and baicalein	H5N1 influenza A virus	A549 cells and Vero cells	Sithisarn et al. (2013)

TABLE 4.5 Viruses Targeted by Different Flavonoids by Interfering with the Host Components

Flavonoids	Virus	*In Vitro/In Vivo*	References
Quercetin	HSV-1	Raw 264.7 cells and Vero cells	Lee et al. (2017)
Luteolin	JEV	BHK-21 and Raw264.7 cells	Li et al. (2014)
Biochanin A and baicalein	H5N1 influenza A virus	A549 cells and Vero cells	Sithisarn et al. (2013)
Apigenin	EV71	Vero cells	Lv et al. (2014)
	Hepatitis C virus	Huh7 and 293T cells	Shibata et al. (2014)

4.2.2 FLAVONOIDS AGAINST VIRUS CLASSES

Emerging and re-emerging viral pathogens have become increasingly important throughout the world in recent decades, as they have substantial impacts on human health and economic wealth. Infectious diseases can jump from animal to human or human to human either directly through contact or indirectly through contaminated inanimate objects, intermediate hosts, bites of insect vectors, etc. (Parrish et al., 2008). According to the Baltimore classification, viruses can be placed in one of the seven following groups mentioned in Table 4.6. Viruses can have circular and linear double-stranded deoxyribonucleic acid (dsDNA), circular or linear single-stranded (ss) DNAs, ds ribonucleic acid (RNAs), positive and negative-strand RNAs ('+' strand and '–' strand RNAs, respectively). The primary distinction between

'+' strand RNAs and '-' strand RNA viruses is that '+' strand RNAs viruses can be directly translated into proteins, whereas '-' strand RNA viruses refer to the encoding of the viral proteins by the complementary strand. Some RNA viruses are ambisense, which means they contain both the RNAs, '+' and '-' strands. While some individual virus particles contain multiple RNA fragments, which form the viral genome and are all required for successful virus replication. Although some viral particles have RNA, when the virus infects a new host cell, this RNA is transcribed into DNA.

TABLE 4.6 Baltimore Classification of Viruses with Their Examples

Classes	Viruses	Mode of mRNA Synthesis	Examples
Class I	dsDNA	dsDNA is used to replicate the viral genome that is further used for mRNA synthesis.	Adenoviruses, Papillomaviruses, polyomaviruses, Herpesviruses, Pox viruses
Class II	ssDNA viruses	ssDNA genome replicates to dsDNA, then mRNA is synthesized.	Parvoviruses
Class III	dsRNA viruses	dsRNA genome transcribed to mRNA.	Reoviruses
Class IV	(+) ssRNA viruses	+ssRNA virus genome functions as mRNA	Togaviruses, Coronaviruses, Picornaviruses
Class V	(–) ssRNA viruses	–ssRNA is complementary to the viral mRNA.	Orthomyxoviruses, Rhabdoviruses
Class VI	ssRNA-reverse transcriptase (RT) viruses	Genome first converted to dsDNA viruses through reverse transcription. Further, dsDNA is integrated into the host cell.	Retroviruses
Class VII	dsDNA-RT viruses	Replicate through an RNA intermediate.	Hepadnaviruses, Caulimovirus

4.2.2.1 FLAVONOIDS AGAINST DSDNA VIRUSES

Herpesviridae family consists of human dsDNA viruses such as HSV, varicella-zoster virus (VZV), HCMV, EBV. HSV infections include gingivostomatitis, herpes genitalis, herpetic keratitis, and dermal whitlows, while VZV infection causes chickenpox in children and further reactivation of the latent virus in adults causes herpes zoster called shingles. EBV and HCMV cause infectious mononucleosis; however, EBV is associated with Burkitt's lymphoma and other malignancies (Whitley, 1996) (virus pathogen database and analysis resource (ViPR) – *Herpesviridae*). Quercetin and Isoquercitrin strongly suppressed the expression of VZV and HCMV immediate-early (IE) genes (Kim et al., 2020). Dihydromyricetin from *Ampelopsis grossedentata* has shown the Anti-HSV-1 effect via the toll-like receptor 9 (TLR9) dependent anti-inflammatory pathway (Zhou et al., 2020). Various flavonoids such as epicatechin, epigallocatechin (flavonols), genistein (isoflavone), naringenin (flavanone), and quercetin (flavonol) showed a high level of inhibitory activity for HSV-1 and HSV-2 (Lyu et al., 2005). Moreover, (–)-Epigallocatechin-3-gallate (EGCG) inhibits EBV spontaneous lytic infection via ERK1/2 and PI3-K/Akt signaling (Liu et al., 2013). Furthermore, quercetin-induced apoptosis prevents EBV infection (Lee et al., 2015) (Table 4.7).

TABLE 4.7 dsDNA Viruses and Flavonoids

Family	Viruses	Flavonoids	Mechanism of Action	References
Herpesviridae	VZV and HCMV	Quercetin and Isoquercitrin	Expression of immediate-early (IE) genes	Kim et al. (2020)
	HSV-1	Dihydromyricetin	TLR9-dependent anti-inflammatory pathway.	Zhou et al. (2020)
	HSV-1 and HSV-2	Epicatechin, epigallocatechin, genistein, naringenin, and quercetin.	Not mentioned	Lyu et al. (2005)
	EBV	(–)-Epigallocatechin-3-gallate	ERK1/2 and PI3-K/Akt signaling.	Liu et al. (2013)
	EBV	Quercetin	Apoptosis pathway	Lee et al. (2015)

4.2.2.2 FLAVONOIDS AGAINST SSDNA VIRUSES

Parvovirus B19 (B19V) is an ssDNA virus from the *Parvoviridae* family. It is a human pathogenic virus responsible for many clinical manifestations. Usually, infections are mild, self-limiting, and controlled by developing a specific immune response. Still, in many cases, clinical situations can be more complex and require therapy (Manaresi & Gallinella, 2019). In the *in vitro* study, various flavonoid structures inhibited endonuclease activity of the non-structural protein 1 (nsP1) of Parvovirus B19 (Xu et al., 2019) (Table 4.8).

TABLE 4.8 ssDNA Viruses and Flavonoids

Family	Viruses	Flavonoids	Mechanism of Action	References
Parvoviridae	Parvovirus B19	Flavonoid-like structure	Not mentioned	Xu et al. (2019)

4.2.2.3 FLAVONOIDS AGAINST DSRNA VIRUSES

Rotaviruses are members of the family *Reoviridae*, which are non-enveloped, triple-layered icosahedral viruses containing a genome of dsRNA. Rotavirus spreads quickly among infants and young children. The virus can cause severe watery diarrhea, vomiting, fever, and abdominal pain (*Rotavirus Vaccination |CDC*). Epigallocatechin gallate (EGCG) of green tea also has an antiviral activity for Rotaviruses (Lipson et al., 2017). Another flavonoid, genistein, inhibits rotavirus replication and upregulates aquaporin 4 (AQP4) expression in rotavirus-infected $CaCO_2$ cells (Huang et al., 2015).

Similarly, baicalin inhibits rotavirus via the gluconeogenesis-related p-JNK-PDK1-AKT-SIK2 signaling pathway (Song et al., 2021). Furthermore, diosmin and hesperidin had the most effective inhibitory activity on rotavirus infection (Bae et al., 2000). On the other hand, theaflavins neutralize bovine rotavirus and bovine CoV infections (Clark et al., 1998) (Table 4.9).

TABLE 4.9 dsRNA Viruses and Flavonoids

Family	Viruses	Flavonoids	Mechanism of Action	References
Reoviridae	Rotavirus	EGCG	Not mentioned	(*Rotavirus Vaccination* | *CDC*)
		Genistein	Replication	Lipson et al. (2017)
		Baicalin	Gluconeogenesis-related p-JNK-PDK1-AKT-SIK2 signaling pathway	Song et al. (2021)
		Diosmin and hesperidin	Not mentioned	Song et al. (2021)
	Bovine rotavirus	Theaflavins	Not mentioned	Clark et al. (1998)

4.2.2.4 FLAVONOIDS AGAINST POSITIVE-SENSE SSRNA VIRUSES

Picornaviruses belong to the family *Picornaviridae*, which are small, nonenveloped, icosahedral viruses that possess a positive-strand genomic RNA of ~7.5 kb. It consists of various viruses such as poliovirus, human rhinovirus (HRV), enterovirus, coxsackievirus B3 (CV-B3), etc. In the *in vitro* study of Poliovirus, 3-Methylquercetin inhibits at late replication stages, blocks genomic RNA synthesis, and reduces viral protein and RNA synthesis (González et al., 2016; Vrijsen et al., 1987). Similarly, 3-Methylkaempferol impedes Poliovirus-1 replication by inhibition of positive-strand of viral RNA (Robin et al., 2016). 5,30-Dihydroxy-3,6,7,8,40-pentamethoxyflavone and 5-hydroxy-3,6,7,30,40-pentamethoxyflavone deter Poliovirus-1 replication via inhibition of cellular processes (apoptosis and downstream signaling pathways) (Thomas Ortega et al., 2019). Furthermore, Chrysosplenol C, Luteolin, Pachypodol (RO 09-0179) restrict Poliovirus replication (Desideri et al., 2016; Semple et al., 1999; Xu et al., 2014). In addition to this, Pachypodol (RO 09-0179) inhibits Poliovirus in late replication via blocking the synthesis of (+) strand RNA; conversely, it inhibits CV by Interference with viral replications between the uncoating and RNA synthesis stage (Robin et al., 2016). Similarly, 6-Chloro-40-oxazolinylflavanone deters poliovirus and HRV replication (Desideri et al., 2016). Moreover, EGCG has shown its virucidal effect on Poliovirus (Polansky & Itzkovitz, 2013). Likewise, Kaempferol-3-O-[2″,6″-di-O-Z-p-coumaroyl]-d-glucopyranoside and derivatives inhibit replication of CV-B3, HRV (Kim et al., 2019).

Togaviridae is an enveloped virus with (+) ssRNA genomes of 10–12 kb. Within the family, the genus Alphavirus includes many diverse species, e.g., chikungunya virus (CHIKV), Semliki-forest virus (SFV), MAYV. Proanthocyanidin (PAC), a dimer containing epicatechin, was observed to reduce virus yields when adding PAC at different moments after infection. The set of results indicates that PAC binds to viral and non-cellular elements and may inactivate the MAYV (Ferraz et al., 2019). On the other hand, Baicalein, fisetin, and quercetagetin displayed potent inhibition of CHIKV infection *in vitro* (Oo et al., 2018; Lani et al., 2016). In addition to this, the *in-vitro* study fisetin inhibited CHIKV replication via inhibition of NS protein 1 and 3 and downregulation of E2 protein and its precursor pE (Lani et al., 2016). Moreover, nobiletin inhibited CHIKV infection during the translation/replication stages and viral entry, making nobiletin a potential clinical antiviral agent in the prevention and post-exposure treatment (Lin et al., 2017). Furthermore, silymarin exhibited significant antiviral activity against CHIKV, reducing CHIKV replication efficiency and downregulating the production of viral proteins involved in replication (Lin et al., 2015). Similarly, EGCG and Suramin have shown synergistic antiviral activity against the CHIKV (Lu et al., 2017). *In vitro* study of hesperetin and hesperidin as inhibitors of ZIKV and CHIKV proteases (Eberle et al., 2021). The green tea catechin EGCG inhibits CHIKV infection (Weber et al., 2015).

Flaviviruses consist of positive-stranded RNA viruses from the family *Flaviviridae*. The genus incorporates HCV, DENV, ZIKA, JEV, West Nile virus (WNV), tick-borne encephalitis virus (TBEV), yellow fever virus (YFV), etc. HCV causes an inflammation of the liver, both acute and chronic, ranging in severity from a mild illness to a severe and lifelong illness, including liver cirrhosis and cancer. It is a bloodborne virus, and most infections occur through exposure to unsafe blood-related practices, injection drug use, sexual practices, etc. Currently, an estimated 58 million people have chronic HCV infection worldwide, with about 1.5 million new infections occurring per year (*Hepatitis C*, 2021). miR-122 is specifically and abundantly expressed in hepatocytes; apigenin inhibits micro-RNA122, positively regulating HCV replication (Shibata et al., 2014).

Similarly, naringenin reduces HCV secretion in infected cells by 80% via inhibition of HCV replication. Moreover, Quercetin inhibits HCV in various ways, such as interfering with transcription by inactivating the NS3 helicase and NS5 protease, inhibiting viral genome replication, hampering the production of infectious HCV particles, and decreasing the specific infectivity of the newly produced viral particles (Wagoner et al., 2010). On the other hand, silymarin had antiviral effects against HCV cell culture infection, including inhibition of virus entry, RNA and protein expression, and infectious virus production (Wagoner et al., 2010). Moreover, EGCG shows an anti-HCV activity by reducing cellular infectivity via enhancing miR-548m expression and repressing the CD81 receptor, as well as it also acts as an entry inhibitor (Calland et al., 2012; Mekky et al., 2019). Likewise, a cinnamon-derived procyanidin type A compound inhibits HCV cell entry (Fauvelle et al., 2017). More on this, deguelin impedes HCV replication in human hepatoma cells by suppressing cellular autophagy via downregulation of Beclin1 expression (Liao et al., 2020). Additionally, theaflavins, Rutin, acts as an entry inhibitor of the HCV in cell culture (Bose et al., 2017; Chowdhury et al., 2018).

DENV and ZIKV are mosquito-borne viral infections found in tropical and sub-tropical climates worldwide, primarily urban and semi-urban areas. There are four serotypes of the DENV. The infections cause only mild illness; can cause an acute flu-like illness. Sometimes this develops into a potentially lethal complication called severe dengue. There is no specific treatment for dengue/severe dengue, nor vaccine is available. There are an estimated approximately 100–400 million infections each year. Dengue prevention and control depends on effective vector control measures. Continuous public involvement can improve vector control efforts substantially (*Dengue and Severe Dengue*, 2021). In an *in-vitro* experiment, baicalin and its metabolite act as entry inhibitors for the DENV (Moghaddam et al., 2014). Diisopropyl chrysin-7-yl phosphate also inhibits replication of viral RNA and the expression of viral protein (Du et al., 2016).

Similarly, quercetin and fisetin exhibited significant inhibitory activity against DENV-2 by inhibition of replication (Zandi et al., 2011). Furthermore, naringenin inhibits DENV-2 and DENV-4 via a reduction in RNA levels (Frabasile et al., 2017; Zandi et al., 2011). Moreover, the highly biologically active green tea component EGCG inhibited DENV infection of all serotypes (Raekiansyah et al., 2018). On the other hand, glabranine and 7-O-methyl-glabranine exert a dose-dependent inhibitory effect *in vitro* on the DENV (Sánchez et al., 2000).

ZIKV disease contains symptoms such as mild and include fever, rash, conjunctivitis, muscle, and joint pain, malaise, or headache. Symptoms typically last for 2–7 days. Most people with ZIKV infection do not develop symptoms. ZIKV infection during pregnancy can cause infants to be born with microcephaly and other congenital malformations, known as congenital ZIKA syndrome. Infection with the ZIKV is also associated with other pregnancy complications, including preterm birth and miscarriage. There is no vaccine or approved drug available to treat and prevent infections by ZIKV.

The green tea molecule EGCG inhibits ZIKV entry, indicating that this drug might be possibly used to prevent infections (Carneiro et al., 2016). Moreover, Pinocembrin, a flavanone found in honey, tea, and red wine, acts on post-entry processes of the ZIKV replication cycle via inhibition of viral RNA production and envelope protein synthesis (Lee et al., 2019).

Similarly, Silymarin has shown an antiviral effect against ZIKV *in vitro* assays (da Silva et al., 2020). Sophoraflavenone G restricts both DENV and ZIKV infection via RNA polymerase interference (Sze et al., 2017). Furthermore, quercetin-3-β-O-D-glucoside has demonstrated antiviral activity in ZIKV infection *in vivo* and *in vitro* experiments (Wong et al., 2017).

JEV replication was deterred by luteolin after the entry stage (Fan et al., 2016). Similarly, in *In vitro* experiments, baicalein and quercetin have shown antiviral activity against JEV (Johari et al., 2012).

Coronavirus disease (COVID-19) is an infectious disease caused by the SARS-CoV-2 virus. COVID-19 affects different people in different ways. Most infected people will develop mild to moderate illness and recover without hospitalization. The most common symptoms include fever, cough, tiredness, loss of taste or smell. Less common symptoms include sore throat, headache, aches, and pains, diarrhea, a rash on the skin, discoloration of fingers or toes, and red or irritated eyes. Severe symptoms include difficulty breathing or shortness of breath, speech or mobility loss, confusion, and chest pain (Coronavirus, 2021). *Coronaviridae* family consists of viruses such as SARS-CoV, SARS-CoV-2 virus, and MERS-CoV. The genome comprises positive sense ssRNA (Naqvi et al., 2020). In FRET-based enzymatic assay, dihydromyricetin has shown inhibitory activity on SARS-CoV-2 Mpro. Molecular docking helped to identify the binding pose of dihydromyricetin with SARS-CoV-2 Mpro protease. Moreover, the effects of dihydromyricetin were protective against BLM-induced pulmonary inflammation and fibrosis in C57BL6 mice (Xiao et al., 2021) (Table 4.10).

4.2.2.5 FLAVONOIDS AGAINST NEGATIVE SSRNA VIRUSES

Arenaviridae family consists of enveloped viruses with negative-sense ssRNA, such as Lassa virus, Lymphocytic choriomeningitis virus (LCMV), etc. These viruses cause hemorrhagic fever, and it is endemic in western Africa. Genistein and tyrphostin are kinase inhibitors that have shown anti-viral activity against the Lassa virus (Kolokoltsov et al., 2012). Tangeretin blocks cellular entry of LCMV that causes viral hemorrhagic fever (Tang et al., 2018). Influenza viruses belong to the family *Orthomyxoviridae*. These are enveloped viruses with negative-sense RNA genomes that cause highly contagious respiratory disease with potentially fatal outcomes. Influenza viruses are assumed to be transmitted predominantly by aerosol infection symptoms, including fever, headache, cough, sore throat, nasal congestion, sneezing, and body aches. Influenza viruses also cause local epidemics or pandemics with a significant infection rate (Blut & Krankheitserreger, 2009). Baicalin attaches to non-structural protein 1-p85β (RNA binding domain) in turn down-regulates IFN-ɣ. It activates the JAK/STAT1 pathway and indirectly reduces influenza A virus load in the *in-vitro* assay (Chu et al., 2015; Nayak et al., 2014). It has been reported that green tea extract epigallocatechin inhibits the growth of the influenza virus A, B (Imanishi et al., 2002). Ginkgetin inhibits the influenza virus sialidase. Ginkgetin-sialic acid conjugates showed a significant survival effect in the influenza-virus-infected mice (Miki et al., 2007). Similarly, quercetin has shown inhibitory activity in early influenza virus infection in *in-vitro* assays (Wu et al., 2016). Furthermore, Silymarin has shown an inhibitory effect in late viral RNA synthesis of the Influenza A virus (Song & Choi, 2011) (Table 4.11).

TABLE 4.10 Sense ssRNA Viruses and Flavonoids

Family	Viruses	Flavonoids	Mechanism of Action	References
Picornaviridae	Poliovirus	3-Methylquercetin	Late replication stages	González et al. (2016); Vrijsen et al. (1987)
	Poliovirus	3-Methylkaempferol; 5,30-Dihydroxy-3,6,7,8,40-pentamethoxyflavone; 5-hydroxy-3,6,7,30,40-pentamethoxy flavone; Chrysosplenol C; Luteolin; Pachypodol; 6-Chloro-40-oxazolinylflavanone; EGCG	Replication	Desideri et al. (2016); Thomas Ortega et al. (2019)
	HRV	6-Chloro-40-oxazolinylflavanone.	Replication	Desideri et al. (2016)
	HRV and CV-B3	Kaempferol-3-O-[2″,6″-di-O-Z-p-coumaroyl]-d-glucopyranoside and derivatives.	Replication	Kim et al. (2019)
Togaviridae	MAYV	Proanthocyanidin	Various moments after infection.	Ferraz et al. (2019)
	CHIKV	Silymarin	Down-regulation of viral proteins production.	Liu et al. (2015)
		EGCG and Suramin	Not mentioned	Lu et al. (2017)
		Hesperetin and Hesperidin	Virus proteases	Eberle et al. (2021)
		EGCG	Not mentioned	Weber et al. (2015)
		Fisetin	Inhibition of NS protein 1 and 3 and downregulation of E2 protein and its precursor pE2.	Lani et al. (2016)
Flaviviridae	HCV	Apigenin	Micro-RNA122 inhibition.	Shibata et al. (2014)
		Naringenin	Replication	Wagoner et al. (2010)
		Quercetin	Inactivating the NS3 helicase and NS5 protease.	Wagoner et al. (2010)

TABLE 4.10 *(Continued)*

Family	Viruses	Flavonoids	Mechanism of Action	References
		Silymarin	Inhibition of virus entry, RNA, and protein expression, and infectious virus production.	Wagoner et al. (2010)
		EGCG	Inhibition of cell entry, reducing cellular infectivity via enhancing miR-548m expression and repressing the CD81 receptor.	Calland et al. (2012); Mekky et al. (2019)
		Procyanidin type A	Inhibition of cell entry.	Fauvelle et al. (2017)
		Deguelin	Suppression of cellular autophagy via downregulation of Beclin1 expression.	Liao et al. (2020)
		Theaflavins, Rutin	Entry inhibitor	Bose et al. (2017); Chowdhury et al. (2018)
	DENV	Baicalin and its metabolite	Virus entry	Moghaddam et al. (2014)
		Diisopropyl chrysin-7-yl phosphate.	Replication and the expression of viral protein.	Du et al. (2016)
		Quercetin, fisetin, and naringenin.	Replication	Frabasile et al. (2017); Zandi et al. (2011, b)
		EGCG, Glabranine, and 7-O-methyl-glabranine.	Not mentioned	Raekiansyah et al. (2018); Sánchez et al. (2000)
	ZIKV	EGCG	Virus entry	Carneiro et al. (2016)
		Pinocembrin	Replication	Lee et al. (2019)
		Silymarin	Not mentioned	ds Silva et al. (2020)

TABLE 4.10 *(Continued)*

Family	Viruses	Flavonoids	Mechanism of Action	References
	ZIKV and DENV	Sophoraflavenone G	RNA polymerase interference	Sze et al. (2017)
	ZIKV	Quercetin-3-β-O-D-glucoside	Not mentioned	Wong et al. (2017)
	JEV	Luteolin	Post entry	Fan et al. (2016)
	JEV	Baicalein and quercetin	Not mentioned	Johari et al. (2012)
Coronaviridae	SARS-CoV-2	Dihydromyricetin	SARS-CoV-2 M protease.	Xiao et al. (2021)
			And BLM-induced pulmonary inflammation and fibrosis.	

TABLE 4.11 Negative Sense ssRNA Viruses and Flavonoids

Family	Viruses	Flavonoids	Mechanism of Action	References
Arenaviridae	Lassa virus	Genistein and tyrphostin	Not mentioned	Kolokoltsov et al. (2012)
		Tangeretin	Virus entry	Tang et al. (2018)
Orthomyxoviridae	Influenza virus	Baicalin	Non-structural protein 1-p85β.	Chu et al. (2015); Nayak et al. (2014)
		Epigallocatechin	Not mentioned	Imanishi et al. (2002)
		Ginkgetin	Sialidase	Miki et al. (2007)
		Quercetin	The early stage of infection.	Wu et al. (2016)
		Silymarin	Late viral RNA synthesis.	Song & Choi (2011)

4.2.2.6 FLAVONOIDS AGAINST RETROVIRUSES

Retroviruses are enveloped, positive senses RNA viruses which replicate by converting the RNA genome into the DNA intermediate. These viruses belong to the *Retroviridae* family, and it includes viruses such as the human immunodeficiency virus (HIV) and human T-cell lymphotropic virus type (HTLV). HIV causes asymptomatic infection, acute infection with symptoms that may include fever, sweats, myalgia or arthralgia, sore throat, lymphadenopathy, nausea, vomiting, diarrhea, headaches, and rash, and acquired immune deficiency syndrome (AIDS). HTLV causes adult T-cell leukemia (ATL) (pre-adult T-cell leukemia, chronic, acute, and lymphoma forms); and tropical spastic paraparesis, a neurologic disease (Cloyd, 1996; Ryu, 2017). Baicalin inhibits the fusion of virus envelope protein with T cells and monocytes expressing CD4/CXCR4 or CD4/CCR5 of HIV cells (Li et al., 2000).

Similarly, EGCG inhibits HIV entry into cells via direct binding to $CD4^+$ T-cells and blocks the binding of envelope protein gp120 to cells, directly binding to $CD4^+$ T-cells (Kawai et al., 2003). Furthermore, genistein acts at the assembly and release level against HIV via inhibition of Vpu protein involved in forming ion channels in infected cells (Sauter et al., 2014). Moreover, sulfated Rutin acts as a fusion inhibitor by inhibiting glycoprotein-mediated cell-cell fusion (Tao et al., 2007) (Table 4.12).

TABLE 4.12 Retroviruses and Flavonoids

Family	Viruses	Flavonoids	Mechanism of Action	References
Retroviridae	HIV	Baicalin	Fusion of virus	Li et al. (2000)
		EGCG	Entry inhibitor	Kawai et al. (2003)
		Genistein	Assembly and release	Sauter et al. (2014)
		Sulfated Rutin	Fusion inhibitor	Tao et al. (2007)

4.2.2.7 FLAVONOIDS AGAINST HEPADNAVIRIDAE

Hepadnaviridae is a family of small, enveloped viruses with partially double-stranded DNA with RNA intermediate in the life cycle (*Hepadnaviridae – Reverse Transcribing DNA and RNA Viruses – ICTV*). HBV belongs to the *Hepadnaviridae* family and causes diseases such as hepatitis B, hepatocellular carcinomas (chronic infections), and cirrhosis. The virus spreads through blood, semen, or other body fluids from an infected person. This is a vaccine-preventable disease (*Hepatitis B – FAQs, Statistics, Data, and Guidelines | CDC*, 2021). Nobiletin acts as a novel inhibitor that inhibits HBsAg production and HBV replication (Hu et al., 2020). Likewise, EGCG inhibits HBV gene expression and replication via various ways such as ERK1/2-mediated downregulation of HNF4α, inhibition of entry of HBV into hepatocytes, inhibition of HBV DNA synthesis, opposing HBV-induced incomplete autophagy by enhancing lysosomal acidification (Huang et al., 2014; Pang et al., 2014; Zhong et al., 2015; He et al., 2011; Wang et al., 2020). EGCG inhibits HBV infection in human liver chimeric mice (Lai et al., 2018). Similarly, quercetin inhibits HBV antigen secretion and genome replication in human hepatoma cell lines (Cheng et al., 2015). Furthermore, flavocoxid is a proprietary blend of two flavonoids, baicalin and catechin, inhibits HBV replication by targeting multiple steps of the viral life cycle (Huang et al., 2017; Pollicino et al., 2018) (Table 4.13).

TABLE 4.13 *Hepadnaviridae* and Flavonoids

Family	Viruses	Flavonoids	Mechanism of Action	References
Hepadnaviridae	HBV	Nobiletin	Replication	Hu et al. (2020)
		EGCG	Entry, gene expression, and replication	Huang et al. (2014); Pang et al. (2014); Zhong et al. (2015); He et al. (2011); Wang et al. (2020)
		Quercetin	Replication	Cheng et al. (2015)
		Flavocoxid	Replication	Huang et al. (2017); Pollicino et al. (2018)

4.2.3 *TARGETING HOST MACHINERY EXPLOITED BY VIRUSES*

The interplay between virus and host determines viral infection resistance and recovery. The host's defenses may act directly and indirectly on virus reproduction by modifying or killing the infected cell. Non-specific host defenses work early in the virus encounter to avoid or restrict infection, whereas specific host defenses work after infection to restore immunity or develop memory cells for future threats.

Cellular metabolism is one factor that is increasingly recognized for virus-host interactions. Nowadays, viruses' co-existence with their host can be viewed as a molecular race between the virus and the host elimination mechanism. Continued interaction between the host and pathogen during their co-evolution has shaped the immune system. In turn, the viruses have manipulated the host control mechanism to facilitate their own replication, transcription, and translation. This chapter investigates how host cell metabolic pathways are imitated, exploited, or disrupted by different classes of viruses to bypass/escape immune responses. This includes a brief synopsis of the virus's introduction to the signaling pathways and examples of the virus's strategies for controlling the host cell metabolic activities.

The inter-related pathways studied for their significant role in viral infection are glycolysis, Krebs cycle, pentose phosphate pathway, β-oxidation, amino-acid and fatty acid synthesis pathway, and the host immune systems. T. Murayama studied in 1998 that the production of interleukin 8 (IL-8) during HCMV infection increased infectious viral replication. They conclude that the progression of HCMV infection is linked to the virus replication and the host immune system (Murayama, 1998). While Munger et al. first characterized the effect of HCMV-infected cells in the host metabolic environment, concluding that the levels of several metabolic pathways, such as the Krebs cycle, as the infection progresses glycolysis and amino acids biosynthesis, upsurge. They evaluated 63 different intracellular metabolites during infection compared to normal fibroblast development, indicating that the virally infected cells considerably impacted metabolic homeostasis (Munger et al., 2006). Viruses such as HCV, DENV, HSV 1, HIV-1, rubella virus, rhinovirus, influenza virus, adenovirus, and CoV have been shown to initiate a host cell response characterized by an elevated level of glucose, resulting in a hospitable intracellular environment for viral replication and inflammatory cytokine expression (Codo et al., 2020; Deng et al., 2011; Fischl & Bartenschlager, 2011; Lee et al., 2020; Logette et al., 2021; Loisel-Meyer et al., 2012; Prusinkiewicz et al., 2020; Bilz et al., 2018; Ren et al., 2021; Vafeiadou et al., 2009; Weng et al., 2021; Wu et al., 2020). Nowadays, several viruses have been demonstrated to disrupt numerous major metabolic pathways, and the number of signaling pathways has been examined (Du et al., 2021; Birungi et al., 2010; Gou et al., 2021; Lin et al., 2010; Manchester & Anand, 2017; Romagnolo & Carvalho, 2021; Xu et al., 2021).

Mitochondria are vital organelles that generate adenosine triphosphate (ATP) via oxidative phosphorylation and govern cell cycle and differentiation, manage calcium signaling, and generate reactive oxygen species (ROS) (Wallace, 2012). The mitochondria are directly targeted by viral proteins or impacted by physiological modifications to the cellular environment, such as disrupted calcium homeostasis, endoplasmic reticulum stress, oxidative stress, and hypoxia during viral pathogenesis. An example of this virus-mitochondria interaction is the blocking of mitochondria-associated antiviral signaling. Khan et al. revealed that HBV and HCV promote mitophagy and mitochondrial fission, downregulating apoptosis and boosting viral persistence (Kim et al., 2014). Kaposi's sarcoma-associated herpesvirus (KSHV) encodes microRNAs that have a detrimental influence on mitochondrial biogenesis and function, perhaps through an interaction with the heat shock proteins. However, it is unclear how and why KSHV affects host cell metabolism (Yogev et al., 2014). The HCMV protein also affects the reticular mitochondrial network and inhibits apoptotic signals in mitochondria. It directly suppresses apoptotic signaling pathways by interfering with the mitochondria-localized inhibitor of apoptosis and viral inhibitor of caspase-8-induced apoptosis proteins (Goldmacher, 2005). However, how viral proteins that target mitochondrial dynamics affect core mitochondrial metabolic activity is poorly unknown. Further research into the involvement of mitochondrial dynamics in viral infection will help us better understand virus-host interaction and their significance in pathogenesis.

Other than these metabolic processes, viruses also aim for polyamine pathways. Polyamines are positively charged small molecules found in all cells. Cellular functions such as nucleic acid binding, cell cycle, and membrane fluidity are all affected by polyamine levels in the cells (Frugier et al., 1994). Gibson et al.'s studies have revealed that viral capsids of herpesvirus include substantial quantities of polyamines, which are thought to neutralize charges on viral DNA to aid compaction and encapsidation (Gibson & Roizman, 1971). Bacteriophage R17 (Fukuma & Cohen, 1975) and Vaccinia virus (Lanzer & Holowczak, 1975) also incorporate polyamine into virions. According to Mounce et al., polyamines are necessary for both the transcription and translation of RNA viruses, i.e., CHIKV and ZIKV. They demonstrated that CHIKV and ZIKV replication is restricted when polyamines are depleted via type I interferon signaling-mediated activation of spermidine/spermine N1-acetyltransferase (SAT1) (Mounce et al., 2016). *In vivo* studies in mice, *Drosophila melanogaster*, and zebrafish showed that polyamines are necessary for efficient viral replication in multiple hosts (Mounce et al., 2016). So, the potential role of polyamines in virus replication for a broader range of RNA viruses highlights as a promising drug target.

Similarly, some studies show the crucial role of flavonoids against these viruses by shielding the host cell factors. Reid et al. in 2014 show ER chaperon (HSPA5) inhibitor EGCG play a protective role against the Ebola virus (Reid et al., 2014). Moreover, naringenin inhibits HCV production mediated by the activation of peroxisome proliferator-activated receptor (PPARα), resulting in a decrease in very-low-density lipoprotein production (Goldwasser et al., 2011). Furthermore, Wang et al.'s studies showed the role of flavopiridol in inhibiting viral replication. They conclude that the host RNA polymerase activity is inhibited by flavopiridol, ,which causes a decrease in viral mRNA production (Wang et al., 2012). A few examples of viruses that target the different host target by various viruses are given in Table 4.14.

TABLE 4.14 Different Host Pathways Targeted by Viruses

Pathways	Virus	*In Vitro*/ *In Vivo*	References
Glycolytic pathway	KSHV	*In vitro*	Yogev et al. (2014)
	HCMV	*In vitro*	Munger et al. (2008); Vastag et al. (2011)
	Adenovirus	*In vitro*	Thai et al. (2014)
	EBV	*In vitro*	Xu et al. (2014)
	DENV	*In vitro*	Fontaine et al. (2015); Lee et al. (2020); Romagnolo & Carvalho (2021)
	HCV	*In vitro*	Ramière et al. (2014); Deng et al. (2011); Shoji et al. (2015)
	Influenza A	*In vitro*	Munger et al. (2008); Ren et al. (2021)
	HSV-1	*In vitro*	Vastag et al. (2011)
Mitochondrial pathway	KSHV	*In vitro*	Yogev et al. (2014)
	HCV	*In vitro*	Kim et al. (2014)
	Hepatitis B	*In vitro*	Kim et al. (2013)
	HCMV	*In vitro*	Arnoult et al. (2004); Goldmacher (2005); Pauleau et al. (2007)
	Classical swine fever virus (CSFV)	*In vitro*	Gou et al. (2017)

TABLE 4.14 *(Continued)*

Pathways	Virus	*In Vitro/ In Vivo*	References
TCA cycle	HCMV	*In vitro*	Munger et al. (2008); Spencer et al. (2011); Vastag et al. (2011)
	HSV-1	*In vitro*	Vastag et al. (2011)
Lipid metabolism	Hepatitis C	*In vitro*	Filipe & McLauchlan (2015); Herker et al. (2010); Hofmann et al. (2018); Lee et al. (2019); Romagnolo & Carvalho (2021); Vieyres & Pietschmann (2019)
	DENV	*In vitro*	Cloherty et al. (2020); Romagnolo & Carvalho (2021); Samsa et al. (2009); Zhang et al. (2018)
	ZIKV	*In vitro*	Cloherty et al. (2020)
Fatty acid synthesis pathway.	HCMV	*In vitro*	Munger et al. (2008); Purdy et al. (2015)
	CSFV	*In vitro*	Liu et al. (2021)
	DENV	*In vitro*	Tang et al. (2014)
Polyamine pathway	HSV	*In vitro*	Gibson & Roizman (1971)
	CHIKV	*In vitro* and *in vivo*	Mounce et al. (2016, 2017)
	CV B3	*In vitro*	Dial et al. (2019); Hulsebosch & Mounce (2021); Kicmal et al. (2019)
	ZIKV	*In vitro* and *in vivo*	Mounce et al. (2016); Routhu et al. (2018)
	Bunyavirus	*In vitro*	Mastrodomenico et al. (2019)
	SARS-CoV-2	*In vitro*	Firpo et al. (2021)

4.3 STRUCTURE-BASED STUDY OF FLAVANOIDS AGAINST VARIOUS VIRUSES

Structure-based drug design is becoming a crucial tool for faster and more cost-efficient lead discovery than the conventional method (Aggarwal et al., 2014, 2015, 2017; Fatma et al., 2020; Kumar et al., 2021; Sharma et al., 2016, 2018). Structural studies have provided a plethora of new targets and opportunities for future drug discovery against viruses (Saha et al., 2018; Kumar et al., 2021; Narwal et al., 2018; Rani et al., 2020; Choudhary et al., 2020; Singh et al., 2018). Experimental high-throughput screening is time-consuming and expensive. Hence, it is essential to overcome the boundaries of the conventional drug discovery methods with efficient, low-cost, and broad-spectrum computational alternatives. Here, some examples of structure-based flavonoid drug discovery have been given, which can be used as potential therapeutics.

Human papillomavirus (HPV) is a dsDNA virus from the *Papovaviridae* family. Early protein 6 (E6) has a vital role in replication and oncogenesis in cervical cancers. The new class of luteolin disrupts the E6/E6AP interaction. Hence, these compounds may play a role in developing antiviral therapy in treating HPV infection and cervical cancer (Cherry et al., 2013). *In silico* study of 5,7,40-Trihydroxy-30-methoxy flavone inhibits Rhinovirus (HRV) entry inhibition by binding to the HRV protein grid (Kant et al.).

Similarly, Silymarin interacts with the binding pocket of DENV nsP4B, including potential antiviral activity against DENV (Qaddir et al., 2017). Puranik et al. studied the halogenated dihydro-rugosa flavonoids against non-structural protein 3 (nsP3) protein of CHIKV, which shows antiviral activity based on the in *in-silico* as well as in *in vitro* studies (Puranik et al., 2019). Coumarin, Coumestan, and Neoflavone derivatives have shown interaction with HCV NS5B Polymerase, indicating potential antiviral activity (Nichols et al., 2013; Kaushik et al., 2008). Tripathi et al. investigated the role of naringin flavanoid against CHIKV nsP2. They found that the structural changes in nsP2 caused by naringin binding are expected to disrupt the enzyme's normal activity during the CHIKV viral life cycle (Tripathi et al., 2020). Despite the fact that the ZIKV is a well-known and widespread pandemic. Furthermore, Catoline et al. presented the pedalitin flavanoid as a possibility for hit-to-lead (H2L) optimization studies for antiviral candidates against ZIKV infections (Lima et al., 2021). In addition, citrus flavanone naringenin showed antiviral action against ZIKV in *in-silico* and *in-vitro* studies, suggesting that it inhibits viral reproduction or its assembly (Cataneo et al., 2019).

In silico docking study revealed that the herbacetin and pectolinarin interact with the domain of severe acute respiratory syndrome coronavirus 2 (SARS-CoV-2) main protease (3CLpro), indicating that it has antiviral potential. The research provides crucial scaffolds for developing 3CLpro inhibitors as antiviral to combat SARS-CoV-2 infection (Jo et al., 2020).

Susmit et al. studied several SARS-CoV-2 proteins involved in viral replication, transcription, and translation as potential therapeutic targets. In a docking investigation against 3Clpro, spike receptor-binding domain (RBD), nsp3 (PLpro), nsp12 (RNA dependent RNA polymerase; RdRp), and Angiotensin-converting enzyme 2 (ACE2) receptor, EGCG and theaflavin digallate showed excellent binding against all the proteins of virus (Jang et al., 2021; Mhatre et al., 2021). In a similar study, researchers showed that EGCG forms favorable interactions with the spike protein and can potentially impair the function of the SARS-CoV-2 UK variant (Mhatre et al., 2021). Mohmoud et al. tested a database of 2017 flavone compounds against SARS-CoV-2 Mpro virtually. Based on the docking scores, they discovered rutin's dynamics and energetics against SARS-CoV-2, which may be investigated *in vitro* and *in vivo* to combat the epidemic (Ibrahim et al., 2021). *In vitro* investigations against SARS-CoV-2 showed that baicalein and baicalin have considerable antiviral efficacy. Furthermore, cell-based and biochemical investigations revealed that both drugs directly block the function of SARS-CoV-2 RdRp (Zandi et al., 2021). Quercetin and its seleno functionalized derivative (8-(p-tolylselenyl) quercetin) blocks SARS-CoV-2 replication in virus-infected cells by inhibiting Mpro. Moreover, Seleno derivation increases Mpro activity by forming a hydrogen bond among the selenium atom and Gln189 residue in the catalytic pocket. Similarly, Rutin, a glycosylated conjugate of quercetin, increases bioavailability at low micromolar concentrations (Rizzuti et al., 2021).

The principles and methods discussed here highlight the strategies *in silico* approaches that have been applied to identify flavonoid compounds as antivirals. Undoubtedly, challenges remain to show antiviral activity. Nevertheless, in the current scenario, structure-based drug identification has a crucial role in drug discovery. As shown in the above studies, structure-based drug identification has been able to identify promising flavonoid compounds that might represent future solutions in critical areas of antivirals.

4.4 CLINICAL STUDIES OF FLAVANOIDS AGAINST VARIOUS VIRUSES

Clinical trials research studies new tests and treatments and assesses their effects on human health outcomes. There are four phases in which people volunteer to participate in clinical trials to test medical interventions. Flavonoids have been used to treat virus infections in clinical trials few of the trials have been mentioned below.

Clinical trials have shown that Gene-Eden-VIR/Novirin is a safe and effective treatment against many viruses, including HCV, HPV, HCMV, HSV, and EBV (Polansky et al., 2016, 2017, 2018; Polansky & Itzkovitz, 2013). In the Phase II Clinical trial, intravenously injection of Silibinin in HCV-infected pegylated interferon/ribavirin therapy non-responders have shown potent antiviral activity. It has been postulated that silibinin's antiviral activity is facilitated by immune-mediated IFN-JAK/STAT independent antiviral mechanisms by regulating Toll-like receptor 7, interferon regulatory factor 3, and p38 protein kinase pathways (Ferenci et al., 2008). Silibinin monotherapy has effective antiviral activity in established HCV recurrence patients where the graft is not responding to standard therapy and confirms the therapy has high safety and tolerability without interaction with immunosuppressive drugs (Rendina et al., 2014).

4.5 CONCLUSION

Since the beginning, viral infections have affected the whole ecosystem. The present scenario reiterates that viral disease spreads worldwide rapidly. Pandemics and epidemics caused by these virus infections influence billions of lives directly and indirectly. Drug discovery itself is a challenging and time-consuming job. Structure-based studies have an essential role in the process of identifying novel molecules and postulating interactions. Various flavonoids have shown antiviral activity via manipulating host as well as viral factors.

Furthermore, clinical trials are helping researchers to understand the flavonoids impact on human health. As synthetic antiviral molecules usually have limited efficacy due to their substantial side effects. Overall, naturally occurring flavonoids in the diet can be seen as a beacon of hope.

KEYWORDS

- **antiviral retroviruses**
- **flavonoids**
- **Hepadnaviridae**
- **molecular inhibition**
- **retroviruses**
- **RNA and DNA viruses**
- **ssRNA viruses**

REFERENCES

Adhami, V. M., & Mukhtar, H., (2006). Polyphenols from green tea and pomegranate for prevention of prostate cancer. *Free Radic. Res.*, *40*(10), 1095–1104. https://doi.org/10.1080/10715760600796498.

Aggarwal, M., Dhindwal, S., Kumar, P., Kuhn, R. J., & Tomar, S., (2014). Trans-protease activity and structural insights into the active form of the alphavirus capsid protease. *J. Virol.*, *88*(21), 12242. https://doi.org/10.1128/JVI.01692-14.

Aggarwal, M., Kaur, R., Saha, A., Mudgal, R., Yadav, R., Dash, P. K., Parida, M., Kumar, P., & Tomar, S., (2017). Evaluation of antiviral activity of piperazine against chikungunya virus targeting hydrophobic pocket of alphavirus capsid protein. *Antiviral Res.*, *146*, 102–111. https://doi.org/10.1016/J.ANTIVIRAL.2017.08.015.

Aggarwal, M., Sharma, R., Kumar, P., Parida, M., & Tomar, S., (2015). Kinetic characterization of trans-proteolytic activity of chikungunya virus capsid protease and development of a FRET-based HTS assay. *Sci. Rep.*, *5*. https://doi.org/10.1038/SREP14753.

Amoros, M., Simões, C. M., Girre, L., Sauvager, F., & Cormier, M., (1992). Synergistic effect of flavones and flavonols against herpes simplex virus type 1 in cell culture. comparison with the antiviral activity of propolis. *J. Nat. Prod.*, *55*(12), 1732–1740. https://doi.org/10.1021/NP50090A003.

Arnoult, D., Bartle, L. M., Skaletskaya, A., Poncet, D., Zamzami, N., Park, P. U., Sharpe, J., et al., (2004). Cytomegalovirus cell death suppressor VMIA blocks bax- but Not Bak-mediated apoptosis by binding and sequestering Bax at mitochondria. *Proc. Natl. Acad. Sci. U. S. A.*, *101*(21), 7988–7993. https://doi.org/10.1073/PNAS.0401897101.

Bae, E. A., Han, M. J., Lee, M., & Kim, D. H., (2000). *In vitro* inhibitory effect of some flavonoids on rotavirus infectivity. *Biol. Pharm. Bull.*, *23*(9), 1122–1124. https://doi.org/10.1248/BPB.23.1122.

Bai, L., Nong, Y., Shi, Y., Liu, M., Yan, L., Shang, J., Huang, F., et al., (2016). Luteolin inhibits hepatitis B virus replication through extracellular signal-regulated kinase-mediated down-regulation of hepatocyte nuclear factor 4α expression. *Mol. Pharm.*, *13*(2), 568–577. https://doi.org/10.1021/acs.molpharmaceut.5b00789.

Barnard, D. L., Huffman, J. H., Meyerson, L. R., & Sidwell, R. W., (1993). Mode of inhibition of respiratory syncytial virus by a plant flavonoid, SP-303. *Chemotherapy*, *39*(3), 212–217. https://doi.org/10.1159/000239128.

Mucsi, I., Gyulai, Z., & Béládi, I. (1992). Combined effects of flavonoids and acyclovir against herpesviruses in cell cultures. *Acta Microbiologica Hungarica*, *39*(2), 137–147.

Bilz, N. C., Jahn, K., Lorenz, M., Lüdtke, A., Hübschen, J. M., Geyer, H., Mankertz, A., et al., (2018). Rubella viruses shift cellular bioenergetics to a more oxidative and glycolytic phenotype with a strain-specific requirement for glutamine. *J. Virol.*, *92*(17). https://doi.org/10.1128/JVI.00934-18.

Birungi, G., Chen, S. M., Loy, B. P., Ng, M. L., & Li, S. F., (2010). Metabolomics approach for investigation of effects of Dengue virus infection using the EA.hy926 cell line. *J. Proteome Res.*, *9*(12), 6523–6534. https://doi.org/10.1021/PR100727M.

Blut, A., Krankheitserreger, U., & Bewertung, B., (2009). Influenza virus. *Transfus. Med. Hemotherapy*, *36*(1), 32. https://doi.org/10.1159/000197314.

Bose, M., Kamra, M., Mullick, R., Bhattacharya, S., Das, S., & Karande, A. A., (2017). Identification of a flavonoid isolated from plum (*Prunus domestica*) as a potent inhibitor of hepatitis C virus entry. *Sci. Rep.*, *7*(1). https://doi.org/10.1038/S41598-017-04358-5.

Calland, N., Albecka, A., Belouzard, S., Wychowski, C., Duverlie, G., Descamps, V., Hober, D., et al., (2012). (-)-Epigallocatechin-3-gallate is a new inhibitor of hepatitis C virus entry. *Hepatology, 55*(3), 720–729. https://doi.org/10.1002/HEP.24803.

Carneiro, B. M., Batista, M. N., Braga, A., Nogueira, M. L., & Rahal, P., (2016). The green tea molecule EGCG inhibits *Zika virus* entry. *Virology, 496*, 215–218. https://doi.org/10.1016/J.VIROL.2016.06.012.

Cataneo, A. H. D., Kuczera, D., Koishi, A. C., Zanluca, C., Silveira, G. F., Arruda, T. B. De., Suzukawa, A. A., et al., (2019). The citrus flavonoid naringenin impairs the *in vitro* infection of human cells by *Zika virus*. *Sci. Reports, 9, 91*(1), 1–15. https://doi.org/10.1038/s41598-019-52626-3.

Chang, C. W., Li, H. C., Hsu, C. F., Chang, C. Y., & Lo, S. Y., (2009). Increased ATP generation in the host cell is required for efficient *Vaccinia virus* production. *J. Biomed. Sci., 16*(1). https://doi.org/10.1186/1423-0127-16-80.

Chantrill, B. H., Coulthard, C. E., Dickinson, L., Inkley, G. W., Morris, W., & Pyle, A. H., (1952). The action of plant extracts on a bacteriophage of *Pseudomonas pyocyanea* and on influenza A virus. *Microbiology, 6*(1, 2), 74–84. https://doi.org/10.1099/00221287-6-1-2-74.

Chen, L., Dou, J., Su, Z., Zhou, H., Wang, H., Zhou, W., Guo, Q., & Zhou, C., (2011). Synergistic activity of baicalein with ribavirin against influenza A (H1N1) virus infections in cell culture and in mice. *Antiviral Res., 91*(3), 314–320. https://doi.org/10.1016/j.antiviral.2011.07.008.

Cheng, Z., Sun, G., Guo, W., Huang, Y., Sun, W., Zhao, F., & Hu, K., (2015). Inhibition of hepatitis B virus replication by quercetin in human hepatoma cell lines. *Virol. Sin., 30*(4), 261–268. https://doi.org/10.1007/S12250-015-3584-5.

Cherry, J. J., Rietz, A., Malinkevich, A., Liu, Y., Xie, M., Bartolowits, M., Davisson, V. J., et al., (2013). Structure based identification and characterization of flavonoids that disrupt human papillomavirus-16 E6 function. *PLoS One, 8*(12). https://doi.org/10.1371/JOURNAL.PONE.0084506.

Chiow, K. H., Phoon, M. C., Putti, T., Tan, B. K. H., & Chow, V. T., (2016). Evaluation of antiviral activities of *Houttuynia cordata Thunb.* extract, quercetin, quercetrin and cinanserin on murine coronavirus and Dengue virus infection. *Asian Pac. J. Trop. Med., 9*(1), 1–7. https://doi.org/10.1016/J.APJTM.2015.12.002.

Choi, E. J., Lee, C. H., Kim, Y. C., & Shin, O. S., (2015). Wogonin inhibits varicella-zoster (shingles) virus replication via modulation of type I interferon signaling and adenosine monophosphate-activated protein kinase activity. *J. Funct. Foods, 17*, 399–409. https://doi.org/10.1016/J.JFF.2015.05.031.

Choi, H. J., Kim, J. H., Lee, C. H., Ahn, Y. J., Song, J. H., Baek, S. H., & Kwon, D. H., (2009). Antiviral activity of quercetin 7-rhamnoside against porcine epidemic Diarrhea virus. *Antiviral Res., 81*(1), 77–81. https://doi.org/10.1016/J.ANTIVIRAL.2008.10.002.

Choudhary, S., Malik, Y. S., & Tomar, S., (2020). Identification of SARS-CoV-2 cell entry inhibitors by drug repurposing using *in Silico* structure-based virtual screening approach. *Front. Immunol., 11*, 1664. https://doi.org/10.3389/FIMMU.2020.01664.

Chowdhury, P., Sahuc, M. E., Rouillé, Y., Rivière, C., Bonneau, N., Vandeputte, A., Brodin, P., et al., (2018). Theaflavins, polyphenols of black tea, inhibit entry of hepatitis C virus in cell culture. *PLoS One, 13*(11). https://doi.org/10.1371/JOURNAL.PONE.0198226.

Chu, M., Xu, L., Zhang, M. B., Chu, Z. Y., & Wang, Y. D., (2015). Role of baicalin in anti-influenza virus A as a potent inducer of IFN-gamma. *Biomed Res. Int., 2015*. https://doi.org/10.1155/2015/263630.

Chu, Y., Lv, X., Zhang, L., Fu, X., Song, S., Su, A., Chen, D., et al., (2020). Wogonin inhibits *in vitro* herpes simplex virus type 1 and 2 infection by modulating cellular NF-KB and MAPK pathways. *BMC Microbiol.*, *20*(1). https://doi.org/10.1186/S12866-020-01916-2.

Clark, K. J., Grant, P. G., Sarr, A. B., Belakere, J. R., Swaggerty, C. L., Phillips, T. D., & Woode, G. N., (1998). An *in vitro* study of theaflavins extracted from black tea to neutralize bovine rotavirus and bovine coronavirus infections. *Vet. Microbiol.*, *63*(2–4), 147–157. https://doi.org/10.1016/S0378-1135(98)00242-9.

Cloherty, A. P. M., Olmstead, A. D., Ribeiro, C. M. S., & Jean, F., (2020). Hijacking of lipid droplets by hepatitis C, dengue and *Zika viruses*—From viral protein moonlighting to extracellular release. *Int. J. Mol. Sci.*, *21*(21), 1–18. https://doi.org/10.3390/IJMS21217901.

Cloyd, M. W. (1996). Human Retroviruses. In: Baron S, editor. Medical Microbiology. 4th edition. Galveston (TX): University of Texas Medical Branch at Galveston; Chapter 62. Available from: https://www.ncbi.nlm.nih.gov/books/NBK7934/

Codo, A. C., Davanzo, G. G., Monteiro, L. De B., De Souza, G. F., Muraro, S. P., Virgilio-da-Silva, J. V., et al., (2020). Elevated glucose levels favor SARS-CoV-2 infection and monocyte response through a HIF-1α/glycolysis-dependent axis. *Cell Metab.*, *32*(3), 437–446.e5. https://doi.org/10.1016/J.CMET.2020.07.007.

Cook, N. C., & Samman, S., (1996). Flavonoids – chemistry, metabolism, cardioprotective effects, and dietary sources. *J. Nutr. Biochem.*, *7*(2), 66–76. https://doi.org/10.1016/0955-2863(95)00168-9.

Coronavirus. https://www.who.int/health-topics/coronavirus/coronavirus#tab=tab_1 (accessed on 24 June 2023).

Cushnie, T. P., & Lamb, A. J., (2005). Antimicrobial activity of flavonoids. *Int. J. Antimicrob. Agents*, *26*(5), 343–356. https://doi.org/10.1016/J.IJANTIMICAG.2005.09.002.

Da Silva, T. F., Ferraz, A. C., Almeida, L. T., Caetano, C., Camini, F. C., Lima, R., Andrade, A., et al., (2020). Antiviral effect of silymarin against *Zika virus in vitro*. *Acta Trop.*, *211*. https://doi.org/10.1016/J.ACTATROPICA.2020.105613.

De Groot, H., & Rauen, U., (1998). Tissue injury by reactive oxygen species and the protective effects of flavonoids. *Fundam. Clin. Pharmacol.*, *12*(3), 249–255. https://doi.org/10.1111/J.1472-8206.1998.TB00951.X.

Dejani, N. N., Elshabrawy, H. A., Bezerra, F. C. Da. S. M., De Sousa, D. P., (2021). Anticoronavirus and immunomodulatory phenolic compounds: Opportunities and pharmacotherapeutic perspectives. *Biomolecules*, *11*(8), 1254. https://doi.org/10.3390/BIOM11081254.

Deng, L., Shoji, I., Ogawa, W., Kaneda, S., Soga, T., Jiang, D., Ide, Y. H., & Hotta, H., (2011). Hepatitis C virus infection promotes hepatic gluconeogenesis through an NS5A-mediated, FoxO1-dependent pathway. *J. Virol.*, *85*(17), 8556. https://doi.org/10.1128/JVI.00146-11.

Dengue and Severe Dengue. https://www.who.int/news-room/fact-sheets/detail/dengue-and-severe-dengue (accessed on 24 June 2023).

Desideri, N., Conti, C., Sestili, I., Tomao, P., Stein, M. L., & Orsi, N., (2016). *In Vitro Evaluation of the Anti-Picornavirus Activities of New Synthetic Flavonoids*, http://dx.doi.org/10.1177/095632029500600503, *6*(5), 298–306. https://doi.org/10.1177/095632029500600503.

Dial, C. N., Tate, P. M., Kicmal, T. M., & Mounce, B. C., (2019). Coxsackievirus B3 responds to polyamine depletion via enhancement of 2A and 3C protease activity. *Viruses*, *11*(5). https://doi.org/10.3390/V11050403.

Dos, S. A. E., Kuster, R. M., Yamamoto, K. A., Salles, T. S., Campos, R., De Meneses, M. D., Soares, M. R., & Ferreira, D., (2014). Quercetin and quercetin 3-O-glycosides from bauhinia longifolia (bong.) steud. show anti-Mayaro virus activity. *Parasites Vectors, 7*(1), 1–7. https://doi.org/10.1186/1756-3305-7-130.

Du, J., Chen, Z., Zhang, T., Wang, J., & Jin, Q., (2016). Inhibition of Dengue virus replication by diisopropyl chrysin-7-yl phosphate. *Science China, 59*(8), 832–838. https://doi.org/10.1007/s11427-016-5050-5.

Du, Y., Mi, Z., Xie, Y., Lu, D., Zheng, H., Sun, H., Zhang, M., & Niu, Y., (2021). Insights into the molecular basis of tick-borne encephalitis from multiplatform metabolomics. *PLoS Negl. Trop. Dis.*, *15*(3). https://doi.org/10.1371/JOURNAL.PNTD.0009172.

Eberle, R. J., Olivier, D. S., Pacca, C. C., Avilla, C., Nogueira, M. L., Amaral, M. S., Willbold, D., et al., (2021). *In vitro* study of hesperetin and hesperidin as inhibitors of *Zika* and chikungunya virus proteases. *PLoS One*, *16*(3). https://doi.org/10.1371/JOURNAL.PONE.0246319.

Fan, W., Qian, S., Qian, P., & Li, X., (2016). Antiviral activity of luteolin against Japanese encephalitis virus. *Virus Res.*, *220*, 112–116. https://doi.org/10.1016/J.VIRUSRES.2016.04.021.

Fatima, K., Mathew, S., Suhail, M., Ali, A., Damanhouri, G., Azhar, E., & Qadri, I., (2014). Docking studies of Pakistani HCV NS3 helicase: A possible antiviral drug target. *PLoS One*, *9*(9), e106339. https://doi.org/10.1371/JOURNAL.PONE.0106339.

Fatma, B., Kumar, R., Singh, V. A., Nehul, S., Sharma, R., Kesari, P., Kuhn, R. J., & Tomar, S., (2020). Alphavirus capsid protease inhibitors as potential antiviral agents for chikungunya infection. *Antiviral Res.*, *179*. https://doi.org/10.1016/J.ANTIVIRAL.2020.104808.

Fauvelle, C., Lambotin, M., Heydmann, L., Prakash, E., Bhaskaran, S., Vishwaraman, M., Baumert, T. F., & Moog, C., (2017). A cinnamon-derived procyanidin type a compound inhibits hepatitis C virus cell entry. *Hepatol. Int.*, *11*(5), 440–445. https://doi.org/10.1007/S12072-017-9809-Y.

Ferenci, P., Scherzer, T. M., Kerschner, H., Rutter, K., Beinhardt, S., Hofer, H., Schöniger-Hekele, M., et al., (2008). Silibinin is a potent antiviral agent in patients with chronic hepatitis C not responding to pegylated interferon/ribavirin therapy. *Gastroenterology*, *135*(5), 1561–1567. https://doi.org/10.1053/J.GASTRO.2008.07.072.

Ferraz, A. C., Moraes, T., Nizer, W., Santos, M. D., Tótola, A. H., Ferreira, J., Vieira-Filho, S. A., et al., (2019). Virucidal activity of proanthocyanidin against Mayaro virus. *Antiviral Res.*, *168*, 76–81. https://doi.org/10.1016/J.ANTIVIRAL.2019.05.008.

Filipe, A., & McLauchlan, J., (2015). Hepatitis C virus and lipid droplets: Finding a niche. *Trends Mol. Med.*, *21*(1), 34–42. https://doi.org/10.1016/J.MOLMED.2014.11.003.

Firpo, M. R., Mastrodomenico, V., Hawkins, G. M., Prot, M., Levillayer, L., Gallagher, T., Simon-Loriere, E., & Mounce, B. C., (2021). Targeting polyamines inhibits coronavirus infection by reducing cellular attachment and entry. *ACS Infect. Dis.*, *7*(6), 1423–1432. https://doi.org/10.1021/ACSINFECDIS.0C00491.

Fischl, W., & Bartenschlager, R., (2011). Exploitation of cellular pathways by Dengue virus. *Curr. Opin. Microbiol.*, *14*(4), 470–475. https://doi.org/10.1016/J.MIB.2011.07.012.

Fontaine, K. A., Sanchez, E. L., Camarda, R., & Lagunoff, M., (2015). Dengue virus induces and requires glycolysis for optimal replication. *J. Virol.*, *89*(4), 2358–2366. https://doi.org/10.1128/JVI.02309-14.

Forester, S. C., & Lambert, J. D., (2011). The role of antioxidant versus pro-oxidant effects of green tea polyphenols in cancer prevention. *Mol. Nutr. Food Res.*, *55*(6), 844–854. https://doi.org/10.1002/MNFR.201000641.

Frabasile, S., Koishi, A. C., Kuczera, D., Silveira, G. F., Verri, W. A. Jr., Santos, C. N. D. D., & Bordignon, J., (2017). The citrus flavanone naringenin impairs Dengue virus replication in human cells. *Sci. Rep.*, *7*. https://doi.org/10.1038/SREP41864.

Frugier, M., Florentz, C., Hosseini, M. W., Lehn, J. M., & Giegé, R., (1994). Synthetic polyamines stimulate *in vitro* transcription by T7 RNA polymerase. *Nucleic Acids Res.*, *22*(14), 2784–2790. https://doi.org/10.1093/NAR/22.14.2784.

Fukuma, I., & Cohen, S. S., (1975). Polyamines in bacteriophage R17 and its RNA. *J. Virol.*, *16*(2), 222–227. https://doi.org/10.1128/JVI.16.2.222-227.1975.

Ganesan, S., Faris, A. N., Comstock, A. T., Wang, Q., Nanua, S., Hershenson, M. B., & Sajjan, U. S., (2012). Quercetin inhibits rhinovirus replication *in vitro* and *in vivo*. *Antiviral Res.*, *94*(3), 258–271. https://doi.org/10.1016/J.ANTIVIRAL.2012.03.005.

Gibson, W., & Roizman, B., (1971). Compartmentalization of spermine and spermidine in the herpes simplex virion. *Proc. Natl. Acad. Sci. U. S. A.*, *68*(11), 2818–2821. https://doi.org/10.1073/PNAS.68.11.2818.

Goldmacher, V. S., (2005). Cell death suppression by cytomegaloviruses. *Apoptosis, 10*(2), 251–265. https://doi.org/10.1007/S10495-005-0800-Z.

Goldwasser, J., Cohen, P. Y., Lin, W., Kitsberg, D., Balaguer, P., Polyak, S. J., Chung, R. T., et al., (2011). Naringenin inhibits the assembly and long-term production of infectious hepatitis C virus particles through a PPAR-mediated mechanism. *J. Hepatol.*, *55*(5), 963–971. https://doi.org/10.1016/J.JHEP.2011.02.011.

González, M. E., Martínez-Abarca, F., & Carrasco, L., (2016). *Flavonoids: Potent Inhibitors of Poliovirus RNA Synthesis:* http://dx.doi.org/10.1177/095632029000100304, *1*(3), 203–209. https://doi.org/10.1177/095632029000100304.

Gonzalez, O., Fontanes, V., Raychaudhuri, S., Loo, R., Loo, J., Arumugaswami, V., Sun, R., et al., (2009). The heat shock protein inhibitor quercetin attenuates hepatitis C virus production. *Hepatology*, *50*(6), 1756–1764. https://doi.org/10.1002/HEP.23232.

González-Búrquez, M. D. J., González-Díaz, F. R., García-Tovar, C. G., Carrillo-Miranda, L., Soto-Zárate, C. I., Canales-Martínez, M. M., Penieres-Carrillo, J. G., et al., (2018). Comparison between *in vitro* antiviral effect of Mexican propolis and three commercial flavonoids against canine distemper virus. *Evidence-based Complement. Altern. Med.*, *2018*. https://doi.org/10.1155/2018/7092416.

Gou, H., Bian, Z., Li, Y., Cai, R., Jiang, Z., Song, S., Zhang, K., et al., (2021). Metabolomics exploration of pseudorabies virus reprogramming metabolic profiles of PK-15 cells to enhance viral replication. *Front. Cell. Infect. Microbiol.*, *10*. https://doi.org/10.3389/FCIMB.2020.599087.

Gou, H., Zhao, M., Xu, H., Yuan, J., He, W., Zhu, M., Ding, H., et al., (2017). CSFV induced mitochondrial fission and mitophagy to inhibit apoptosis. *Oncotarget*, *8*(24), 39382–39400. https://doi.org/10.18632/ONCOTARGET.17030.

Guo, Q., Zhao, L., You, Q., Yang, Y., Gu, H., Song, G., Lu, N., & Xin, J., (2007). Anti-hepatitis B virus activity of wogonin *in vitro* and *in vivo*. *Antiviral Res.*, *74*(1), 16–24. https://doi.org/10.1016/J.ANTIVIRAL.2007.01.002.

Gutiérrez-Venegas, G., Gómez-Mora, J. A., Meraz-Rodríguez, M. A., Flores-Sánchez, M. A., & Ortiz-Miranda, L. F., (2019). Effect of flavonoids on antimicrobial activity of microorganisms present in dental plaque. *Heliyon, 5*(12). https://doi.org/10.1016/J.HELIYON.2019.E03013.

Ha, S. Y., Youn, H., Song, C. S., Kang, S. C., Bae, J. J., Kim, H. T., Lee, K. M., et al., (2014). Antiviral effect of flavonol glycosides isolated from the leaf of *Zanthoxylum piperitum* on influenza virus. *J. Microbiol., 52*(4), 340–344. https://doi.org/10.1007/S12275-014-4073-5.

Hakobyan, A., Arabyan, E., Avetisyan, A., Abroyan, L., Hakobyan, L., & Zakaryan, H., (2016). Apigenin inhibits African swine fever virus infection *in vitro. Arch. Virol., 161*(12), 3445–3453. https://doi.org/10.1007/s00705-016-3061-y.

He, W., Li, L. X., Liao, Q. J., Liu, C. L., & Chen, X. L., (2011). Epigallocatechin gallate inhibits HBV DNA synthesis in a viral replication – inducible cell line. *World J. Gastroenterol., 17*(11), 1507–1514. https://doi.org/10.3748/WJG.V17.I11.1507.

Heim, K. E., Tagliaferro, A. R., & Bobilya, D. J., (2002). Flavonoid antioxidants: Chemistry, metabolism and structure-activity relationships. *J. Nutr. Biochem., 13*(10), 572–584. https://doi.org/10.1016/S0955-2863(02)00208-5.

Magnius et al., (2020). ICTV Virus Taxonomy Profile: Hepadnaviridae, Journal of General Virology, 101:571–572.

Hepatitis B – FAQs, Statistics, Data, & Guidelines | CDC https://www.cdc.gov/hepatitis/hbv/index.htm (accessed on 24 June 2023).

Hepatitis C. https://www.who.int/news-room/fact-sheets/detail/hepatitis-c (accessed on 24 June 2023).

Herker, E., Harris, C., Hernandez, C., Carpentier, A., Kaehlcke, K., Rosenberg, A. R., Farese, R. V., & Ott, M., (2010). Efficient hepatitis C virus particle formation requires diacylglycerol acyltransferase-1. *Nat. Med., 16*(11), 1295–1298. https://doi.org/10.1038/NM.2238.

Hofmann, S., Krajewski, M., Scherer, C., Scholz, V., Mordhorst, V., Truschow, P., Schöbel, A., et al., (2018). Complex lipid metabolic remodeling is required for efficient hepatitis C virus replication. *Biochim. Biophys. Acta – Mol. Cell Biol. Lipids, 1863*(9), 1041–1056. https://doi.org/10.1016/J.BBALIP.2018.06.002.

Hollman, P. C., & Katan, M. B., (1998). Bioavailability and health effects of dietary flavonols in man. *Arch. Toxicol. Suppl., 20*, 237–248. https://doi.org/10.1007/978-3-642-46856-8_21.

Hollman, P. C., Bijsman, M. N., Van, G. Y., Cnossen, E. P., De Vries, J. H., & Katan, M. B., (1999). The sugar moiety is a major determinant of the absorption of dietary flavonoid glycosides in man. *Free Radic. Res., 31*(6), 569–573. https://doi.org/10.1080/10715769900301141.

Hu, Z., Hu, J., Ren, F., Xu, H., Tan, M., Wang, Q., & Ren, J., (2020). Nobiletin, a novel inhibitor, inhibits HBsAg production and hepatitis B virus replication. *Biochem. Biophys. Res. Commun., 523*(3), 802–808. https://doi.org/10.1016/J.BBRC.2019.12.099.

Huang, H. C., Tao, M. H., Hung, T. M., Chen, J. C., Lin, Z. J., & Huang, C., (2014). (-)-epigallocatechin-3-gallate inhibits entry of hepatitis B virus into hepatocytes. *Antiviral Res., 111*, 100–111. https://doi.org/10.1016/J.ANTIVIRAL.2014.09.009.

Huang, H., Liao, D., Liang, L., Song, L., & Zhao, W., (2015). Genistein inhibits rotavirus replication and upregulates AQP4 expression in rotavirus-infected Caco-2 cells. *Arch. Virol., 160*(6), 1421–1433. https://doi.org/10.1007/S00705-015-2404-4.

Huang, H., Zhou, W., Zhu, H., Zhou, P., & Shi, X., (2017). Baicalin benefits the anti-HBV therapy via inhibiting HBV viral RNAs. *Toxicol. Appl. Pharmacol., 323*, 36–43. https://doi.org/10.1016/J.TAAP.2017.03.016.

Hulsebosch, B. M., & Mounce, B. C., (2021). Polyamine analog diethylnorspermidine restricts coxsackievirus B3 and is overcome by 2a protease mutation *in vitro. Viruses, 13*(2). https://doi.org/10.3390/V13020310.

Hung, P. Y., Ho, B. C., Lee, S. Y., Chang, S. Y., Kao, C. L., Lee, S. S., & Lee, C. N., (2015). Houttuynia cordata targets the beginning stage of herpes simplex virus infection. *PLoS One*, *10*(2), e0115475. https://doi.org/10.1371/JOURNAL.PONE.0115475.

Hutchins, W. A., & Wheeler, T. S., (1939). 17. Chalkones: A new synthesis of chrysin, apigenin, and luteolin. *J. Chem. Soc.*, 91–94. https://doi.org/10.1039/JR9390000091.

Ibrahim, M. A. A., Mohamed, E. A. R., Abdelrahman, A. H. M., Allemailem, K. S., Moustafa, M. F., Shawky, A. M., Mahzari, A., et al., (2021). Rutin and flavone analogs as prospective SARS-CoV-2 main protease inhibitors: In silico drug discovery study. *J. Mol. Graph. Model.*, *105*, 107904. https://doi.org/10.1016/J.JMGM.2021.107904.

Imanishi, N., Tuji, Y., Katada, Y., Maruhashi, M., Konosu, S., Mantani, N., Terasawa, K., & Ochiai, H., (2002). Additional inhibitory effect of tea extract on the growth of influenza a and B viruses in MDCK cells. *Microbiol. Immunol.*, *46*(7), 491–494. https://doi.org/10.1111/J.1348-0421.2002.TB02724.X.

Imran, M., Rauf, A., Abu-Izneid, T., Nadeem, M., Shariati, M. A., Khan, I. A., Imran, A., et al., (2019). Luteolin, a flavonoid, as an anticancer agent: A review. *Biomed. Pharmacother.*, *112*. https://doi.org/10.1016/J.BIOPHA.2019.108612.

Jaganath, I. B., Mullen, W., Edwards, C. A., & Crozier, A., (2006). The relative contribution of the small and large intestine to the absorption and metabolism of rutin in man. *Free Radic. Res.*, *40*(10), 1035–1046. https://doi.org/10.1080/10715760600771400.

Jang, M., Park, R., Park, Y. I., Cha, Y. E., Yamamoto, A., Lee, J. I., & Park, J., (2021). EGCG, a green tea polyphenol, inhibits human coronavirus replication *in vitro*. *Biochem. Biophys. Res. Commun.*, *547*, 23. https://doi.org/10.1016/J.BBRC.2021.02.016.

Jo, S., Kim, S., Kim, D. Y., Kim, M. S., & Shin, D. H., (2020). Flavonoids with inhibitory activity against SARS-CoV-2 3CLpro. *J. Enzyme Inhib. Med. Chem.*, *35*(1), 1539–1544. https://doi.org/10.1080/14756366.2020.1801672.

Johari, J., Kianmehr, A., Mustafa, M. R., Abubakar, S., & Zandi, K., (2012). Antiviral activity of baicalein and quercetin against the Japanese encephalitis virus. *Int. J. Mol. Sci.*, *13*(12), 16785–16795. https://doi.org/10.3390/IJMS131216785.

Kant, K., Lal, U. R., & Ghosh, M. (2016). In-Silico Discovery of Natural Lead Hits From The Genus of Arisaema Against Human Rhino Virus, in Proceedings of the 20th International Electronic Conference on Synthetic Organic Chemistry, 1–30 November 2016, MDPI: Basel, Switzerland, doi:10.3390/ecsoc-20-e004

Kaul, T. N., Middleton, E. Jr., & Ogra, P. L. (1985). Antiviral effect of flavonoids on human viruses. *J. Med. Virol.*, *15*(1), 71–79. https://doi.org/10.1002/JMV.1890150110.

Kaushik-Basu, N., Bopda-Waffo, A., Talele, T. T., Basu, A., Costa, P. R., Da Silva, A. J., Sarafianos, S. G., & Noël, F., (2008). Identification and characterization of coumestans as novel HCV NS5B polymerase inhibitors. *Nucleic Acids Res.*, *36*(5), 1482–1496. https://doi.org/10.1093/NAR/GKM1178.

Kawai, K., Tsuno, N. H., Kitayama, J., Okaji, Y., Yazawa, K., Asakage, M., Hori, N., et al., (2003). Epigallocatechin gallate, the main component of tea polyphenol, binds to CD4 and interferes with Gp120 binding. *J. Allergy Clin. Immunol.*, *112*(5), 951–957. https://doi.org/10.1016/S0091-6749(03)02007-4.

Kawser, H. M., Yeon, C. H., Hwang, J. S., Abdal, D. A., Kim, J. H., Bong, K. Y., Poo, H., & Cho, S. G., (2014). Antiviral activity of 3,4-dihydroxyflavone on influenza A virus. *J. Microbiol.*, *52*(6), 521–526. https://doi.org/10.1007/s12275-014-4212-z.

Keservani, R. K., & Sharma, A. K., (2014). Flavonoids: Emerging trends and potential health benefits. *Journal of Chinese Pharmaceutical Sciences*, *23*(12), 815.

Keservani, R. K., Kesharwani, R. K., Sharma, A. K., Vyas, N., & Chadoker, A., (2010b). Nutritional supplements: An overview. *International Journal of Current Pharmaceutical Review and Research, 1*(1), 59–75.

Keservani, R. K., Kesharwani, R. K., Vyas, N., Jain, S., Raghuvanshi, R., & Sharma, A. K., (2010a). Nutraceutical and functional food as future food: A review. *Der Pharmacia Lettre, 2*(1), 106–116.

Keservani, R. K., Sharma, A. K., & Kesharwani, R. K., (2020). *Nutraceuticals and Dietary Supplements: Applications in Health Improvement and Disease Management*. CRC Press. ISBN: 9781771888738.

Kicmal, T. M., Tate, P. M., Dial, C. N., Esin, J. J., & Mounce, B. C., (2019). Polyamine depletion abrogates enterovirus cellular attachment. *J. Virol., 93*(20). https://doi.org/10.1128/JVI.01054-19.

Kim, C. H., Kim, J. E., & Song, Y. J., (2020). Antiviral activities of quercetin and isoquercitrin against human herpesviruses. *Molecules, 25*(10). https://doi.org/10.3390/MOLECULES25102379.

Kim, N., Park, S. J., Nhiem, N. X., Song, J. H., Ko, H. J., & Kim, S. H., (2019). Cycloartane-Type triterpenoid derivatives and a flavonoid glycoside from the burs of *Castanea crenata*. *Phytochemistry, 158*, 135–141. https://doi.org/10.1016/J.PHYTOCHEM.2018.11.001.

Kim, S. J., Khan, M., Quan, J., Till, A., Subramani, S., & Siddiqui, A., (2013). Hepatitis B virus disrupts mitochondrial dynamics: Induces fission and mitophagy to attenuate apoptosis. *PLoS Pathog., 9*(12), 1–12. https://doi.org/10.1371/JOURNAL.PPAT.1003722.

Kim, S. J., Syed, G. H., Khan, M., Chiu, W. W., Sohail, M. A., Gish, R. G., & Siddiqui, A., (2014). Hepatitis C virus triggers mitochondrial fission and attenuates apoptosis to promote viral persistence. *Proc. Natl. Acad. Sci., 111*(17), 6413–6418. https://doi.org/10.1073/PNAS.1321114111.

Kolokoltsov, A. A., Adhikary, S., Garver, J., Johnson, L., Davey, R. A., & Vela, E. M., (2012). Inhibition of Lassa virus and Ebola virus infection in host cells treated with the kinase inhibitors genistein and tyrphostin. *Arch. Virol., 157*(1), 121–127. https://doi.org/10.1007/S00705-011-1115-8.

Kumar, K. A., Sharma, M., Dalal, V., Singh, V., Tomar, S., & Kumar, P., (2021). Multifunctional inhibitors of SARS-CoV-2 by MM/PBSA, essential dynamics, and molecular dynamic investigations. *J. Mol. Graph. Model., 107*, 107969. https://doi.org/10.1016/J.JMGM.2021.107969.

Kumar, R., Nehul, S., Singh, A., & Tomar, S., (2021). Identification and evaluation of antiviral potential of thymoquinone, a natural compound targeting chikungunya virus capsid protein. *Virology, 561*, 36–46. https://doi.org/10.1016/J.VIROL.2021.05.013.

Lai, Y. H., Sun, C. P., Huang, H. C., Chen, J. C., Liu, H. K., & Huang, C., (2018). Epigallocatechin gallate inhibits hepatitis B virus infection in human liver chimeric mice. *BMC Complement. Altern. Med., 18*(1). https://doi.org/10.1186/S12906-018-2316-4.

Lani, R., Hassandarvish, P., Shu, M. H., Phoon, W. H., Chu, J. J., Higgs, S., Vanlandingham, D., et al., (2016). Antiviral activity of selected flavonoids against chikungunya virus. *Antiviral Res., 133*, 50–61. https://doi.org/10.1016/J.ANTIVIRAL.2016.07.009.

Lanzer, W., & Holowczak, J. A., (1975). Polyamines in vaccinia virions and polypeptides released from viral cores by acid extraction. *J. Virol., 16*(5), 1254–1264. https://doi.org/10.1128/JVI.16.5.1254-1264.1975.

Lee, J. L., Loe, M., Lee, R., & Chu, J., (2019). Antiviral activity of pinocembrin against *Zika virus* replication. *Antiviral Res.*, *167*, 13–24. https://doi.org/10.1016/J.ANTIVIRAL.2019.04.003.

Lee, J. Y., Cortese, M., Haselmann, U., Tabata, K., Romero-Brey, I., Funaya, C., Schieber, N. L., et al., (2019). Spatiotemporal coupling of the hepatitis C virus replication cycle by creating a lipid droplet–proximal membranous replication compartment. *Cell Rep.*, *27*(12), 3602–3617.e5. https://doi.org/10.1016/J.CELREP.2019.05.063.

Lee, M., Son, M., Ryu, E., Shin, Y. S., Kim, J. G., Kang, B. W., Sung, G. H., et al., (2015). Quercetin-induced apoptosis prevents EBV infection. *Oncotarget*, *6*(14), 12603. https://doi.org/10.18632/ONCOTARGET.3687.

Lee, S., Lee, H. H., Shin, Y. S., Kang, H., & Cho, H., (2017). The anti-HSV-1 effect of quercetin is dependent on the suppression of TLR-3 in raw 264.7 cells. *Arch. Pharmacal. Res.*, *40*(5), 623–630. https://doi.org/10.1007/S12272-017-0898-X.

Lee, Y. R., Wu, S. Y., Chen, R. Y., Lin, Y. S., Yeh, T. M., & Liu, H. S., (2020). Regulation of autophagy, glucose uptake, and glycolysis under Dengue virus infection. *Kaohsiung J. Med. Sci.*, *36*(11), 911–919.

Li, B. Q., Fu, T., Dongyan, Y., Mikovits, J. A., Ruscetti, F. W., & Wang, J. M., (2000). Flavonoid baicalin inhibits HIV-1 infection at the level of viral entry. *Biochem. Biophys. Res. Commun.*, *276*(2), 534–538. https://doi.org/10.1006/BBRC.2000.3485.

Li, B., Robinson, D. H., & Birt, D. F., (1997). Evaluation of properties of apigenin and [G-3H] apigenin and analytic method development. *J. Pharm. Sci.*, *86*(6), 721–725. https://doi.org/10.1021/JS960383S.

Li, X., Zhu, Q., Cao, Q., Chen, H., & Qian, P., (2014). Japanese encephaliţis virus upregulates the expression of SOCS3 in mouse brain and raw264.7 cells. *Viruses, 6*(11), 4280–4293. https://doi.org/10.3390/V6114280.

Liao, W., Liu, X., Yang, Q., Liu, H., Liang, B., Jiang, J., Huang, J., et al., (2020). Deguelin inhibits HCV replication through suppressing cellular autophagy via down regulation of Beclin1 expression in human hepatoma cells. *Antiviral Res.*, *174*. https://doi.org/10.1016/J.ANTIVIRAL.2020.104704.

Lima, C. S., Mottin, M., De Assis, L. R., Mesquita, N., Sousa, B., Coimbra, L. D., Santos, K. B., et al., (2021). Flavonoids from pterogyne nitens as *Zika virus* NS2B-NS3 protease inhibitors. *Bioorg. Chem.*, *109*. https://doi.org/10.1016/J.BIOORG.2021.104719.

Lin, S. C., Chen, M. C., Li, S., Lin, C. C., & Wang, T. T., (2017). Antiviral activity of nobiletin against chikungunya virus *in vitro*. *Antivir. Ther.*, *22*(8), 689–697. https://doi.org/10.3851/IMP3167.

Lin, S., Liu, N., Yang, Z., Song, W., Wang, P., Chen, H., Lucio, M., et al., (2010). GC/MS-based metabolomics reveals fatty acid biosynthesis and cholesterol metabolism in cell lines infected with influenza A virus. *Talanta*, *83*(1), 262–268. https://doi.org/10.1016/J.TALANTA.2010.09.019.

Lipson, S. M., Karalis, G., Karthikeyan, L., Ozen, F. S., Gordon, R. E., Ponnala, S., Bao, J., et al., (2017). Mechanism of anti-rotavirus synergistic activity by epigallocatechin gallate and a proanthocyanidin-containing nutraceutical. *Food Environ. Virol.*, *9*(4), 434–443. https://doi.org/10.1007/S12560-017-9299-Z.

Liu, C. H., Jassey, A., Hsu, H. Y., & Lin, L. T., (2015). Antiviral activity of silymarin against chikungunya virus. *Sci. Rep.*, *5*. https://doi.org/10.1038/SREP11421.

Liu, K. C., Yang, S. L., Roberts, M. F., Elford, B. C., & Phillipson, J. D., (1992). Antimalarial activity of artemisia annua flavonoids from whole plants and cell cultures. *Plant Cell Rep.*, *11*(12), 637–640. https://doi.org/10.1007/BF00236389.

Liu, S., Li, H., Chen, L., Yang, L., Li, L., Tao, Y., Li, W., et al., (2013) (-)-epigallocatechin-3-gallate inhibition of Epstein-Barr virus spontaneous lytic infection involves ERK1/2 and PI3-K/Akt signaling in EBV-positive cells. *Carcinogenesis*, *34*(3), 627–637. https://doi.org/10.1093/CARCIN/BGS364.

Liu, Y. Y., Liang, X. D., Liu, C. C., Cheng, Y., Chen, H., Baloch, A. S., Zhang, J., et al., (2021). Fatty acid synthase is involved in classical swine fever virus replication by interaction with NS4B. *J. Virol.*, *95*(17). https://doi.org/10.1128/JVI.00781-21.

Logette, E., Lorin, C., Favreau, C., Oshurko, E., Coggan, J. S., Casalegno, F., Sy, M. F., et al., (2021). A machine-generated view of the role of blood glucose levels in the severity of COVID-19. *Front. Public Heal.*, *9*. https://doi.org/10.3389/FPUBH.2021.695139.

Loisel-Meyer, S., Swainson, L., Craveiro, M., Oburoglu, L., Mongellaz, C., Costa, C., Martinez, M., et al., (2012). Glut1-mediated glucose transport regulates HIV infection. *Proc. Natl. Acad. Sci. U. S. A.*, *109*(7), 2549–2554. https://doi.org/10.1073/PNAS.1121427109.

Lu, J. W., Hsieh, P. S., Lin, C. C., Hu, M. K., Huang, S. M., Wang, Y. M., Liang, C. Y., et al., (2017). Synergistic effects of combination treatment using EGCG and suramin against the chikungunya virus. *Biochem. Biophys. Res. Commun.*, *491*(3), 595–602. https://doi.org/10.1016/J.BBRC.2017.07.157.

Lv, X., Qiu, M., Chen, D., Zheng, N., Jin, Y., & Wu, Z., (2014). Apigenin inhibits enterovirus 71 replication through suppressing viral IRES activity and modulating cellular JNK pathway. *Antiviral Res.*, *109*(1), 30–41. https://doi.org/10.1016/J.ANTIVIRAL.2014.06.004.

Lyu, S. Y., Rhim, J. Y., & Park, W. B., (2005). Antiherpetic activities of flavonoids against herpes simplex virus type 1 (HSV-1) and type 2 (HSV-2) *in vitro*. *Arch. Pharm. Res.*, *28*(11), 1293–1301. https://doi.org/10.1007/BF02978215.

Makino, T., Shimizu, R., Kanemaru, M., Suzuki, Y., Moriwaki, M., & Mizukami, H., (2009). Enzymatically modified isoquercitrin, α-oligoglucosyl quercetin 3-O-glucoside, is absorbed more easily than other quercetin glycosides or aglycone after oral administration in rats. *Biol. Pharm. Bull.*, *32*(12), 2034–2040. https://doi.org/10.1248/BPB.32.2034.

Maleki, S. J., Crespo, J. F., & Cabanillas, B., (2019). Anti-inflammatory effects of flavonoids. *Food Chem.*, *299*. https://doi.org/10.1016/J.FOODCHEM.2019.125124.

Manaresi, E., & Gallinella, G., (2019). Advances in the development of antiviral strategies against parvovirus B19. *Viruses*, *11*(7). https://doi.org/10.3390/V11070659.

Manchester, M., & Anand, A., (2017). Metabolomics: Strategies to define the role of metabolism in virus infection and pathogenesis. *Adv. Virus Res.*, *98*, 57–81. https://doi.org/10.1016/BS.AIVIR.2017.02.001.

Mastrodomenico, V., Esin, J. J., Graham, M. L., Tate, P. M., Hawkins, G. M., Sandler, Z. J., Rademacher, D. J., et al., (2019). Polyamine depletion inhibits bunyavirus infection via generation of noninfectious interfering virions. *J. Virol.*, *93*(14), 530–549. https://doi.org/10.1128/JVI.00530-19.

Mead, J. R., & McNair, N., (2006). Antiparasitic activity of flavonoids and isoflavones against *Cryptosporidium parvum* and encephalitozoon intestinalis. *FEMS Microbiol. Lett.*, *259*(1), 153–157. https://doi.org/10.1111/J.1574-6968.2006.00263.X.

Mekky, R. Y., El-Ekiaby, N., El Sobky, S. A., Elemam, N. M., Youness, R. A., El-Sayed, M., Hamza, M. T., et al., (2019). Epigallocatechin gallate (EGCG) and MiR-548m reduce

HCV entry through repression of CD81 receptor in HCV cell models. *Arch. Virol.*, *164*(6), 1587–1595. https://doi.org/10.1007/S00705-019-04232-X.

Mhatre, S., Gurav, N., Shah, M., & Patravale, V., (2021). Entry-inhibitory role of catechins against SARS-CoV-2 and its UK variant. *Comput. Biol. Med.*, *135*, 104560. https://doi.org/10.1016/J.COMPBIOMED.2021.104560.

Mhatre, S., Naik, S., & Patravale, V., (2021). A molecular docking study of EGCG and theaflavin digallate with the druggable targets of SARS-CoV-2. *Comput. Biol. Med.*, *129*, 104137. https://doi.org/10.1016/J.COMPBIOMED.2020.104137.

Middleton, E., (1998). Effect of plant flavonoids on immune and inflammatory cell function. *Adv. Exp. Med. Biol.*, *439*, 175–182. https://doi.org/10.1007/978-1-4615-5335-9_13.

Miki, K., Nagai, T., Suzuki, K., Tsujimura, R., Koyama, K., Kinoshita, K., Furuhata, K., et al., (2007). Anti-influenza virus activity of biflavonoids. *Bioorg. Med. Chem. Lett.*, *17*(3), 772–775. https://doi.org/10.1016/J.BMCL.2006.10.075.

Miyazawa, M., & Hisama, M., (2003). Antimutagenic activity of flavonoids from *Chrysanthemum* morifolium. *Biosci. Biotechnol. Biochem.*, *67*(10), 2091–2099. https://doi.org/10.1271/BBB.67.2091.

Miyazawa, M., Okuno, Y., Fukuyama, M., Nakamura, S. I., & Kosaka, H., (1999). Antimutagenic activity of polymethoxyflavonoids from *Citrus aurantium*. *J. Agric. Food Chem.*, *47*(12), 5239–5244. https://doi.org/10.1021/JF990176O.

Miyazawa, M., Okuno, Y., Nakamura, S., & Kosaka, H., (2000). Antimutagenic activity of flavonoids from *Pogostemon cablin*. *J. Agric. Food Chem.*, *48*(3), 642–647. https://doi.org/10.1021/JF990160Y.

Miyazawa, M., Sakano, K., Nakamura, S. I., & Kosaka, H., (2001). Antimutagenic activity of isoflavone from *Pueraria lobata*. *J. Agric. Food Chem.*, *49*(1), 336–341. https://doi.org/10.1021/JF000255W.

Moghaddam, E., Teoh, B. T., Sam, S. S., Lani, R., Hassandarvish, P., Chik, Z., Yueh, A., et al., (2014). Baicalin, a metabolite of baicalein with antiviral activity against Dengue virus. *Sci. Reports*, *4*(1), 1–8. https://doi.org/10.1038/srep05452.

Mounce, B. C., Cesaro, T., Vlajnić, L., Vidiņa, A., Vallet, T., Weger-Lucarelli, J., Passoni, G., et al., (2017). Chikungunya virus overcomes polyamine depletion by mutation of NsP1 and the opal stop codon to confer enhanced replication and fitness. *J. Virol.*, *91*(15). https://doi.org/10.1128/JVI.00344-17.

Mounce, B. C., Poirier, E. Z., Passoni, G., Simon-Loriere, E., Cesaro, T., Prot, M., Stapleford, K. A., et al., (2016). Interferon-induced spermidine-spermine acetyltransferase and polyamine depletion restrict zika and chikungunya viruses. *Cell Host Microbe*, *20*(2), 167–177. https://doi.org/10.1016/J.CHOM.2016.06.011.

Mucsi, I., (1984). Combined antiviral effects of flavonoids and 5-ethyl-2'-deoxyuridine on the multiplication of herpesviruses. *Acta Virol.*, *28*(5), 395–400.

Munger, J., Bajad, S. U., Coller, H. A., Shenk, T., & Rabinowitz, J. D., (2006). Dynamics of the cellular metabolome during human cytomegalovirus infection. *PLoS Pathog.*, *2*(12), 1165–1175. https://doi.org/10.1371/JOURNAL.PPAT.0020132.

Munger, J., Bennett, B. D., Parikh, A., Feng, X. J., McArdle, J., Rabitz, H. A., Shenk, T., & Rabinowitz, J. D., (2008). Systems-level metabolic flux profiling identifies fatty acid synthesis as a target for antiviral therapy. *Nat. Biotechnol.*, *26*(10), 1179. https://doi.org/10.1038/NBT.1500.

Murayama, T., (1998). Interrelationship between human cytomegalovirus infection and chemokine. *Nihon Rinsho. Japanese Journal of Clinical Medicine*, *56*(1), 69–74.

Naqvi, A. A. T., Fatima, K., Mohammad, T., Fatima, U., Singh, I. K., Singh, A., Atif, S. M., et al., (2020). Insights into SARS-CoV-2 genome, structure, evolution, pathogenesis and therapies: Structural genomics approach. *Biochim. Biophys. Acta. Mol. Basis Dis.*, *1866*(10), 165878. https://doi.org/10.1016/J.BBADIS.2020.165878.

Narwal, M., Singh, H., Pratap, S., Malik, A., Kuhn, R. J., Kumar, P., & Tomar, S., (2018). Crystal structure of chikungunya virus NsP2 cysteine protease reveals a putative flexible loop blocking its active site. *Int. J. Biol. Macromol.*, *116*, 451–462. https://doi.org/10.1016/J.IJBIOMAC.2018.05.007.

Nayak, M. K., Agrawal, A. S., Bose, S., Naskar, S., Bhowmick, R., Chakrabarti, S., Sarkar, S., & Chawla-Sarkar, M., (2014). Antiviral activity of baicalin against influenza virus H1N1-Pdm09 is due to modulation of NS1-mediated cellular innate immune responses. *J. Antimicrob. Chemother.*, *69*(5), 1298–1310. https://doi.org/10.1093/JAC/DKT534.

Nichols, D. B., Leão, R. A., Basu, A., Chudayeu, M., De Moraes, P., Talele, T. T., Costa, P. R., & Kaushik-Basu, N., (2013). Evaluation of coumarin and neoflavone derivatives as HCV NS5B polymerase inhibitors. *Chem. Biol. Drug Des.*, *81*(5), 607–614. https://doi.org/10.1111/CBDD.12105.

Ninfali, P., Antonelli, A., Magnani, M., & Scarpa, E. S., (2020). Antiviral properties of flavonoids and delivery strategies. *Nutrients*, *12*(9), 1–19. https://doi.org/10.3390/NU12092534.

Oo, A., Rausalu, K., Merits, A., Higgs, S., Vanlandingham, D., Bakar, S. A., & Zandi, K., (2018). Deciphering the potential of baicalin as an antiviral agent for chikungunya virus infection. *Antiviral Res.*, *150*, 101–111. https://doi.org/10.1016/J.ANTIVIRAL.2017.12.012.

Pang, J. Y., Zhao, K. J., Wang, J. B., Ma, Z. J., & Xiao, X. H., (2014). Green tea polyphenol, epigallocatechin-3-gallate, possesses the antiviral activity necessary to fight against the hepatitis B virus replication *in vitro*. *J. Zhejiang Univ. Sci. B*, *15*(6), 533–539. https://doi.org/10.1631/JZUS.B1300307.

Parrish, C. R., Holmes, E. C., Morens, D. M., Park, E. C., Burke, D. S., Calisher, C. H., Laughlin, C. A., et al., (2008). Cross-species virus transmission and the emergence of new epidemic diseases. *Microbiol. Mol. Biol. Rev.*, *72*(3), 457. https://doi.org/10.1128/MMBR.00004-08.

Pauleau, A. L., Larochette, N., Giordanetto, F., Scholz, S. R., Poncet, D., Zamzami, N., Goldmacher, V. S., & Kroemer, G., (2007). Structure-function analysis of the interaction between Bax and the cytomegalovirus-encoded protein VMIA. *Oncogene*, *26*(50), 7067–7080. https://doi.org/10.1038/SJ.ONC.1210511.

Peng, M., Swarbrick, C. M. D., Chan, K. W. K., Luo, D., Zhang, W., Lai, X., Li, G., & Vasudevan, S. G., (2018). Luteolin escape mutants of Dengue virus map to PrM and NS2B and reveal viral plasticity during maturation. *Antiviral Res.*, *154*, 87–96. https://doi.org/10.1016/J.ANTIVIRAL.2018.04.013.

Peng, M., Watanabe, S., Chan, K., He, Q., Zhao, Y., Zhang, Z., Lai, X., et al., (2017). Luteolin restricts Dengue virus replication through inhibition of the proprotein convertase Furin. *Antiviral Res.*, *143*, 176–185. https://doi.org/10.1016/J.ANTIVIRAL.2017.03.026.

Pietta, P. G., (2000). Flavonoids as antioxidants. *J. Nat. Prod.*, *63*(7), 1035–1042. https://doi.org/10.1021/NP9904509.

Polansky, H., & Itzkovitz, E., (2013). Gene-Eden-VIR is antiviral: Results of a post marketing clinical study. *Pharmacol. & Pharm.*, *4*(6), 1–8. https://doi.org/10.4236/pp.2013.46a001.

Polansky, H., Itzkovitz, E., & Javaherian, A., (2016). Clinical study of gene-Eden-VIR/Novirin in genital herpes: Suppressive treatment safely decreases the duration of outbreaks in both severe and mild cases. *Clin. Transl. Med.*, *5*(1). https://doi.org/10.1186/s40169-016-0121-6.

Polansky, H., Itzkovitz, E., & Javaherian, A., (2017). Human papillomavirus (HPV): Systemic treatment with Gene-Eden-VIR/Novirin safely and effectively clears virus. *Drug Des. Devel. Ther.*, *11*, 575–583. https://doi.org/10.2147/DDDT.S123340.

Polansky, H., Javaherian, A., & Itzkovitz, E., (2018). Clinical trial of herbal treatment gene-Eden-VIR/Novirin in oral herpes. *J. Evidence-Based Integr. Med.*, *23*, 1–9. https://doi.org/10.1177/2515690X18806269.

Pollicino, T., Musolino, C., Irrera, N., Bitto, A., Lombardo, D., Timmoneri, M., Minutoli, L., et al., (2018). Flavocoxid exerts a potent antiviral effect against hepatitis B virus. *Inflamm. Res.*, *67*(1), 89–103. https://doi.org/10.1007/S00011-017-1099-2.

Prusinkiewicz, M. A., Tu, J., Dodge, M. J., MacNeil, K. M., Radko-Juettner, S., Fonseca, G. J., Pelka, P., & Mymryk, J. S., (2020). Differential effects of human adenovirus E1A protein isoforms on aerobic glycolysis in A549 human lung epithelial cells. *Viruses*, *12*(6). https://doi.org/10.3390/V12060610.

Puranik, N. V., Rani, R., Singh, V. A., Tomar, S., Puntambekar, H. M., & Srivastava, P., (2019). Evaluation of the antiviral potential of halogenated dihydrorugosaflavonoids and molecular modeling with NsP3 protein of chikungunya virus (CHIKV). *ACS Omega*, *4*(23), 20335–20345. https://doi.org/10.1021/ACSOMEGA.9B02900.

Purdy, J. G., Shenk, T., & Rabinowitz, J. D., (2015). Fatty acid elongase 7 catalyzes the lipidome remodeling essential for human cytomegalovirus replication. *Cell Rep.*, *10*(8), 1375. https://doi.org/10.1016/J.CELREP.2015.02.003.

Qaddir, I., Rasool, N., Hussain, W., & Mahmood, S., (2017). Computer-aided analysis of phytochemicals as potential Dengue virus inhibitors based on molecular docking, ADMET and DFT studies. *J. Vector Borne Dis.*, *54*(3), 255–262. https://doi.org/10.4103/0972-9062.217617.

Qadir, M. I., (2017). Role of green tea flavonoids and other related contents in cancer prevention. *Crit. Rev. Eukaryot. Gene Expr.*, *27*(2), 163–171. https://doi.org/10.1615/CRITREVEUKARYOTGENEEXPR.2017019493.

Qian, S., Fan, W., Qian, P., Zhang, D., Wei, Y., Chen, H., & Li, X., (2015). Apigenin restricts FMDV infection and inhibits viral IRES driven translational activity. *Viruses, 7*(4), 1613–1626. https://doi.org/10.3390/V7041613.

Qiu, X., Kroeker, A., He, S., Kozak, R., Audet, J., Mbikay, M., & Chrétien, M., (2016). Prophylactic efficacy of quercetin 3-β-O-D-glucoside against Ebola virus infection. *Antimicrob. Agents Chemother.*, *60*(9), 5182–5188. https://doi.org/10.1128/AAC.00307-16.

Raekiansyah, M., Buerano, C. C., Luz, M., & Morita, K., (2018). Inhibitory effect of the green tea molecule EGCG against Dengue virus infection. *Arch. Virol.*, *163*(6), 1649–1655. https://doi.org/10.1007/S00705-018-3769-Y.

Ramière, C., Rodriguez, J., Enache, L. S., Lotteau, V., André, P., & Diaz, O., (2014). Activity of hexokinase is increased by its interaction with hepatitis C virus protein NS5A. *J. Virol.*, *88*(6), 3246–3254. https://doi.org/10.1128/JVI.02862-13.

Rani, R., Singh, A., Pareek, A., & Tomar, S., (2020). *In Silico Guided Drug Repurposing to Combat SARS-CoV-2 by Targeting Mpro, the Key Virus Specific Protease*. https://doi.org/10.26434/CHEMRXIV.12030345.V1.

Reid, S. P., Shurtleff, A. C., Costantino, J. A., Tritsch, S. R., Retterer, C., Spurgers, K. B., & Bavari, S., (2014). HSPA5 is an essential host factor for Ebola virus infection. *Antiviral Res., 109*(1), 171–174. https://doi.org/10.1016/J.ANTIVIRAL.2014.07.004.

Ren, L., Zhang, W., Zhang, J., Zhang, J., Zhang, H., Zhu, Y., Meng, X., et al., (2021). Influenza A virus (H1N1) infection induces glycolysis to facilitate viral replication. *Virol. Sin.* https://doi.org/10.1007/S12250-021-00433-4.

Rendina, M., D'Amato, M., Castellaneta, A., Castellaneta, N. M., Brambilla, N., Giacovelli, G., Rovati, L., et al., (2014). Antiviral activity and safety profile of silibinin in HCV patients with advanced fibrosis after liver transplantation: A randomized clinical trial. *Transpl. Int., 27*(7), 696–704. https://doi.org/10.1111/TRI.12324.

Rizzuti, B., Grande, F., Conforti, F., Jimenez-Alesanco, A., Ceballos-Laita, L., Ortega-Alarcon, D., Vega, S., et al., (2021). Rutin is a low micromolar inhibitor of SARS-CoV-2 main protease 3CLpro: Implications for drug design of quercetin analogs. *Biomedicines, 9*(4), 375. https://doi.org/10.3390/BIOMEDICINES9040375.

Robin, V., Irurzun, A., Amoros, M., Boustie, J., & Carrasco, L., (2016). *Antipoliovirus Flavonoids from Psiadia Dentata:* http://dx.doi.org/10.1177/095632020101200503, *12*(5), 283–291. https://doi.org/10.1177/095632020101200503.

Rojas, Á., Campo, J. A. D., Clement, S., Lemasson, M., García-Valdecasas, M., Gil-Gómez, A., Ranchal, I., et al., (2016). Effect of quercetin on hepatitis C virus life cycle: From viral to host targets. *Sci. Rep., 6*. https://doi.org/10.1038/SREP31777.

Romagnolo, A. G., & Carvalho, K. I., (2021). Dengue and metabolomics in humans. *Metabolomics, 17*(3). https://doi.org/10.1007/S11306-021-01783-6.

Rotavirus Vaccination | CDC https://www.cdc.gov/vaccines/vpd/rotavirus/index.html (accessed on 24 June 2023).

Routhu, N. K., Xie, Y., Dunworth, M., Casero, R. A. Jr, Oupicky, D., & Byrareddy, S. N., (2018). Polymeric prodrugs targeting polyamine metabolism inhibit *Zika virus* replication. *Mol. Pharm., 15*(9), 4284–4295. https://doi.org/10.1021/ACS.MOLPHARMACEUT.8B00068.

Ryu, W. S., (2017). Retroviruses. *Mol. Virol. Hum. Pathog. Viruses*, 227–246. https://doi.org/10.1016/B978-0-12-800838-6.00017-5.

Ryu, Y. B., Jeong, H. J., Kim, J. H., Kim, Y. M., Park, J. Y., Kim, D., Naguyen, T. T. H., et al., (2010). Biflavonoids from *Torreya nucifera* displaying SARS-CoV 3CLpro inhibition. *Bioorg. Med. Chem., 18*(22), 7940. https://doi.org/10.1016/J.BMC.2010.09.035.

Saha, A., Acharya, B. N., Priya, R., Tripathi, N. K., Shrivastava, A., Rao, M. K., Kesari, P., et al., (2018). Development of NsP2 protease based cell free high throughput screening assay for evaluation of inhibitors against emerging chikungunya virus. *Sci. Rep., 8*(1). https://doi.org/10.1038/S41598-018-29024-2.

Samsa, M. M., Mondotte, J. A., Iglesias, N. G., Assunção-Miranda, I., Barbosa-Lima, G., Da Poian, A. T., Bozza, P. T., & Gamarnik, A. V., (2009). Dengue virus capsid protein usurps lipid droplets for viral particle formation. *PLoS Pathog., 5*(10). https://doi.org/10.1371/JOURNAL.PPAT.1000632.

Samuelsson, G., (1999). *Drugs of Natural Origin: A Textbook of Pharmacognosy* (4th rev. edn.). Apotekarsocieteten: Stockholm.

Sanchez, I., Gómez-Garibay, F., Taboada, J., & Ruiz, B. H., (2000). Antiviral effect of flavonoids on the Dengue virus. *Phytotherapy Research: An International Journal Devoted to Pharmacological and Toxicological Evaluation of Natural Product Derivatives, 14*(2), 89–92.

Sauter, D., Schwarz, S., Wang, K., Zhang, R., Sun, B., & Schwarz, W., (2014). Genistein as antiviral drug against HIV ion channel. *Planta Med.*, *80*(8, 9), 682–687. https://doi.org/10.1055/S-0034-1368583.

Scalbert, A., Morand, C., Manach, C., & Rémésy, C., (2002). Absorption and metabolism of polyphenols in the gut and impact on health. *Biomed. Pharmacother.*, *56*(6), 276–282. https://doi.org/10.1016/S0753-3322(02)00205-6.

Seelinger, G., Merfort, I., Wölfle, U., & Schempp, C. M., (2008). Anti-carcinogenic effects of the flavonoid luteolin. *Molecules*, *13*(10), 2628–2651. https://doi.org/10.3390/MOLECULES13102628.

Semple, S. J., Nobbs, S. F., Pyke, S. M., Reynolds, G. D., & Flower, R. L., (1999). Antiviral flavonoid from *Pterocaulon sphacelatum*, an Australian aboriginal medicine. *J. Ethnopharmacol.*, *68*(1–3), 283–288. https://doi.org/10.1016/S0378-8741(99)00050-1.

Seong, R. K., Kim, J. A., & Shin, O. S., (2018). Wogonin, a flavonoid isolated from *Scutellaria baicalensis*, has anti-viral activities against influenza infection via modulation of AMPK pathways. *Acta Virol.*, *62*(1), 78–85. https://doi.org/10.4149/AV_2018_109.

Serafini, M., Peluso, I., & Raguzzini, A., (2010). Flavonoids as anti-inflammatory agents. *Proc. Nutr. Soc.*, *69*(3), 273–278. https://doi.org/10.1017/S002966511000162X.

Shadrack, D. M., Deogratias, G., Kiruri, L. W., Onoka, I., Vianney, J. M., Swai, H., & Nyandoro, S. S., (2021). Luteolin: A blocker of SARS-CoV-2 cell entry based on relaxed complex scheme, molecular dynamics simulation, and metadynamics. *J. Mol. Model.*, *27*(8). https://doi.org/10.1007/S00894-021-04833-X.

Sharma, R., Fatma, B., Saha, A., Bajpai, S., Sistla, S., Dash, P. K., Parida, M., Kumar, P., & Tomar, S., (2016). Inhibition of chikungunya virus by picolinate that targets viral capsid protein. *Virology*, *498*, 265–276. https://doi.org/10.1016/J.VIROL.2016.08.029.

Sharma, R., Kesari, P., Kumar, P., & Tomar, S., (2018). Structure-function insights into chikungunya virus capsid protein: Small molecules targeting capsid hydrophobic pocket. *Virology*, *515*, 223–234. https://doi.org/10.1016/J.VIROL.2017.12.020.

Shawan, M. M. A. K., Halder, S. K., & Hasan, M. A., (2021). Luteolin and abyssinone II as potential inhibitors of SARS-CoV-2: An in silico molecular modeling approach in battling the COVID-19 outbreak. *Bull. Natl. Res. Cent.*, *45*(1), 27. https://doi.org/10.1186/S42269-020-00479-6.

Shibata, C., Ohno, M., Otsuka, M., Kishikawa, T., Goto, K., Muroyama, R., Kato, N., et al., (2014). The flavonoid apigenin inhibits hepatitis C virus replication by decreasing mature microRNA122 levels. *Virology*, *462*, *463*(1), 42–48. https://doi.org/10.1016/J.VIROL.2014.05.024.

Shoji, I., Deng, L., Matsui, C., & Hotta, H., (2015). Hepatitis C virus-induced glucose metabolic disorder. *Uirusu*, *65*(2), 263–268. https://doi.org/10.2222/JSV.65.263.

Singh, H., Mudgal, R., Narwal, M., Kaur, R., Singh, V. A., Malik, A., Chaudhary, M., & Tomar, S., (2018). Chikungunya virus inhibition by peptidomimetic inhibitors targeting virus-specific cysteine protease. *Biochimie*, *149*, 51–61. https://doi.org/10.1016/J.BIOCHI.2018.04.004.

Sithisarn, P., Michaelis, M., Schubert-Zsilavecz, M., & Cinatl, J., (2013). Differential antiviral and anti-inflammatory mechanisms of the flavonoids biochanin A and baicalein in H5N1 influenza A virus-infected cells. *Antiviral Res.*, *97*(1), 41–48. https://doi.org/10.1016/J.ANTIVIRAL.2012.10.004.

Song, J. H., & Choi, H. J., (2011). Silymarin efficacy against influenza A virus replication. *Phytomedicine*, *18*(10), 832–835. https://doi.org/10.1016/J.PHYMED.2011.01.026.

Song, L., Zhong, P., Zhu, X., Zhou, R., Gao, M., Lan, Q., Chen, J., et al., (2021). The anti-rotavirus effect of baicalin via the gluconeogenesis-related p-JNK-PDK1-AKT-SIK2 signaling pathway. *Eur. J. Pharmacol.*, *897*. https://doi.org/10.1016/J.EJPHAR.2021.173927.

Spencer, C. M., Schafer, X. L., Moorman, N. J., & Munger, J., (2011). Human cytomegalovirus induces the activity and expression of acetyl-coenzyme A carboxylase, a fatty acid biosynthetic enzyme whose inhibition attenuates viral replication. *J. Virol.*, *85*(12), 5814. https://doi.org/10.1128/JVI.02630-10.

Stermitz, F. R., Scriven, L. N., Tegos, G., & Lewis, K., (2002). Two flavonols from *Artemisa annua* which potentiate the activity of berberine and norfloxacin against a resistant strain of *Staphylococcus aureus*. *Planta Med.*, *68*(12), 1140–1141. https://doi.org/10.1055/S-2002-36347.

Sze, A., Olagnier, D., Hadj, S. B., Han, X., Tian, X. H., Xu, H. T., Yang, L., et al., (2017). Sophoraflavenone G restricts dengue and *Zika virus* infection via RNA polymerase interference. *Viruses*, *9*(10). https://doi.org/10.3390/V9100287.

Takano-Ishikawa, Y., Goto, M., & Yamaki, K., (2006). Structure-Activity relations of inhibitory effects of various flavonoids on lipopolysaccharide-induced prostaglandin E2 production in rat peritoneal macrophages: Comparison between subclasses of flavonoids. *Phytomedicine*, *13*(5), 310–317. https://doi.org/10.1016/J.PHYMED.2005.01.016.

Tang, K., He, S., Zhang, X., Guo, J., Chen, Q., Yan, F., Banadyga, L., et al., (2018). Tangeretin, an extract from citrus peels, blocks cellular entry of arenaviruses that cause viral hemorrhagic fever. *Antiviral Res.*, *160*, 87–93. https://doi.org/10.1016/J.ANTIVIRAL.2018.10.011.

Tang, W. C., Lin, R. J., Liao, C. L., & Lin, Y. L., (2014). Rab18 facilitates Dengue virus infection by targeting fatty acid synthase to sites of viral replication. *J. Virol.*, *88*(12), 6793–6804. https://doi.org/10.1128/JVI.00045-14.

Tao, J., Hu, Q., Yang, J., Li, R., Li, X., Lu, C., Chen, C., et al., (2007). *In vitro* anti-HIV and -HSV activity and safety of sodium rutin sulfate as a microbicide candidate. *Antiviral Res.*, *75*(3), 227–233. https://doi.org/10.1016/J.ANTIVIRAL.2007.03.008.

Thai, M., Graham, N. A., Braas, D., Nehil, M., Komisopoulou, E., Kurdistani, S. K., McCormick, F., et al., (2014). Adenovirus E4ORF1-induced MYC activation promotes host cell anabolic glucose metabolism and virus replication. *Cell Metab.*, *19*(4), 694–701. https://doi.org/10.1016/J.CMET.2014.03.009.

Thomas, O. J., Luisa, S. M., Isabel, S. A., Baptista, J., Helene, P. F., Vicenta, C. L., Rodolfo, C. H., & Rafael, R. H., (2019). Antiviral activity of flavonoids present in aerial parts of *Marcetia taxifolia* against hepatitis B virus, poliovirus, and herpes simplex virus *in vitro*. *EXCLI J.*, *18*. https://doi.org/10.17179/excli2019-1837.

Tripathi, P. K., Singh, J., Gaurav, N., Garg, D. K., & Patel, A. K., (2020). In-silico and biophysical investigation of biomolecular interaction between naringin and NsP2 of the chikungunya virus. *Int. J. Biol. Macromol.*, *160*, 1061–1065. https://doi.org/10.1016/J.IJBIOMAC.2020.05.165.

Vafeiadou, K., Vauzour, D., Lee, H. Y., Rodriguez-Mateos, A., Williams, R. J., & Spencer, J. P. E., (2009). The citrus flavanone naringenin inhibits inflammatory signaling in glial cells and protects against neuroinflammatory injury. *Arch. Biochem. Biophys.*, *484*(1), 100–109. https://doi.org/10.1016/j.abb.2009.01.016.

Vastag, L., Koyuncu, E., Grady, S. L., Shenk, T. E., & Rabinowitz, J. D., (2011). Divergent effects of human cytomegalovirus and herpes simplex virus-1 on cellular metabolism. *PLoS Pathog.*, *7*(7). https://doi.org/10.1371/JOURNAL.PPAT.1002124.

Veerasamy, R., & Rajak, H., (2021). QSAR studies on neuraminidase inhibitors as anti-influenza agents. *Turkish J. Pharm. Sci., 18*(2), 151. https://doi.org/10.4274/TJPS.GALENOS.2020.45556.

Vieyres, G., & Pietschmann, T., (2019). HCV pit stop at the lipid droplet: Refuel lipids and put on a lipoprotein coat before exit. *Cells, 8*(3), 233. https://doi.org/10.3390/CELLS8030233.

Pickett, B. E., Greer, D. S., Zhang, Y., Stewart, L., Zhou, L., Sun, G., Gu, Z., Kumar, S., Zaremba, S., Larsen, C. N., Jen, W., Klem, E. B., & Scheuermann, R. H. (2012). Virus pathogen database and analysis resource (ViPR): a comprehensive bioinformatics database and analysis resource for the coronavirus research community. *Viruses, 4*(11), 3209–3226. https://doi.org/10.3390/v4113209.

Vrijsen, R., Everaert, L., Van, H. L. M., Vlietinck, A. J., Vanden, B. D. A., & Boeyé, A., (1987). The poliovirus-induced shut-off of cellular protein synthesis persists in the presence of 3-methylquercetin, a flavonoid which blocks viral protein and RNA synthesis. *Antiviral Res., 7*(1), 35–42. https://doi.org/10.1016/0166-3542(87)90037-4.

Wagoner, J., Negash, A., Kane, O. J., Martinez, L. E., Nahmias, Y., Bourne, N., Owen, D. M., et al., (2010). Multiple effects of silymarin on the hepatitis C virus lifecycle. *Hepatology, 51*(6), 1912–1921. https://doi.org/10.1002/HEP.23587.

Wallace, D. C., (2012). Mitochondria and cancer. *Nat. Rev. Cancer, 12*(10), 685–698. https://doi.org/10.1038/nrc3365.

Wang, H. Q., Meng, S., Li, Z. R., Peng, Z. G., Han, Y. X., Guo, S. S., Cui, X. L., et al., (2013). The antiviral effect of 7-hydroxyisoflavone against enterovirus 71 *in vitro*. *J. Asian Nat. Prod. Res., 15*(4), 382–389. https://doi.org/10.1080/10286020.2013.770737.

Wang, J., Su, H., Zhang, T., Du, J., Cui, S., Yang, F., & Jin, Q., (2014). Inhibition of enterovirus 71 replication by 7-hydroxyflavone and diisopropyl-flavon7-Yl phosphate. *PLoS One, 9*(3). https://doi.org/10.1371/JOURNAL.PONE.0092565.

Wang, S., Zhang, J., & Ye, X., (2012). Protein kinase inhibitor flavopiridol inhibits the replication of influenza virus *in vitro*. *Wei Sheng Wu Xue Bao, 52*(9), 1137–1142.

Wang, Z. Y., Li, Y. Q., Guo, Z. W., Zhou, X. H., Lu, M. D., Xue, T. C., & Gao, B., (2020). ERK1/2-HNF4α axis is involved in epigallocatechin-3-gallate inhibition of HBV replication. *Acta Pharmacol. Sin., 41*(2), 278–285. https://doi.org/10.1038/S41401-019-0302-0.

Weber, C., Sliva, K., Von, R. C., Kümmerer, B. M., & Schnierle, B. S., (2015). The green tea catechin, epigallocatechin gallate inhibits chikungunya virus infection. *Antiviral Res., 113*, 1–3. https://doi.org/10.1016/J.ANTIVIRAL.2014.11.001.

Weng, S. C., Tsao, P. N., & Shiao, S. H., (2021). Blood glucose promotes Dengue virus infection in the mosquito aedes aegypti. *Parasites and Vectors, 14*(1). https://doi.org/10.1186/S13071-021-04877-1.

Whitley, R. J., (1996). Herpesviruses. *Dis. Swine*, 548–575.

Wong, G., He, S., Siragam, V., Bi, Y., Mbikay, M., Chretien, M., & Qiu, X., (2017). Antiviral activity of quercetin-3-β-O-D-glucoside against *Zika virus* infection. *Virol. Sin., 32*(6), 545–547. https://doi.org/10.1007/S12250-017-4057-9.

Wu, C. C., Fang, C. Y., Cheng, Y. J., Hsu, H. Y., Chou, S. P., Huang, S. Y., Tsai, C. H., & Chen, J. Y., (2017). Inhibition of Epstein-Barr virus reactivation by the flavonoid apigenin. *J. Biomed. Sci., 24*(1). https://doi.org/10.1186/S12929-016-0313-9.

Wu, W., Li, R., Li, X., He, J., Jiang, S., Liu, S., & Yang, J., (2016). Quercetin as an antiviral agent inhibits influenza A virus (IAV) entry. *Viruses, 8*(1). https://doi.org/10.3390/V8010006.

Wu, Z., Jia, J., Xu, X., Xu, M., Peng, G., Ma, J., Jiang, X., et al., (2020). Human herpesvirus 6A promotes glycolysis in infected T cells by activation of MTOR signaling. *PLOS Pathog.*, *16*(6), e1008568. https://doi.org/10.1371/JOURNAL.PPAT.1008568.

Xiao, L., Hu, Z. Y., Dong, X., Tan, Z., Li, W., Tang, M., Chen, L., et al., (2014). Targeting Epstein-Barr virus oncoprotein LMP1-mediated glycolysis sensitizes nasopharyngeal carcinoma to radiation therapy. *Oncogene*, *33*(37), 4568–4578. https://doi.org/10.1038/ONC.2014.32.

Xiao, T., Wei, Y., Cui, M., Li, X., Ruan, H., Zhang, L., Bao, J., et al., (2021). Effect of dihydromyricetin on SARS-CoV-2 viral replication and pulmonary inflammation and fibrosis. *Phytomedicine*, *91*. https://doi.org/10.1016/J.PHYMED.2021.153704.

Xu, G., Dou, J., Zhang, L., Guo, Q., & Zhou, C., (2010). Inhibitory effects of baicalein on the influenza virus *in vivo* is determined by baicalin in the serum. *Biol. Pharm. Bull.*, *33*(2), 238–243. https://doi.org/10.1248/bpb.33.238.

Xu, L., Su, W., Jin, J., Chen, J., Li, X., Zhang, X., Sun, M., et al., (2014). Identification of luteolin as enterovirus 71 and coxsackievirus A16 inhibitors through reporter viruses and cell viability-based screening. *Viruses*, *6*(7), 2778–2795. https://doi.org/10.3390/V6072778.

Xu, P., Ganaie, S. S., Wang, X., Wang, Z., Kleiboeker, S., Horton, N. C., Heier, R. F., et al., (2019). Endonuclease activity inhibition of the NS1 protein of parvovirus B19 as a novel target for antiviral drug development. *Antimicrob. Agents Chemother.*, *63*(3). https://doi.org/10.1128/AAC.01879-18.

Xu, Q., Tang, Y., & Huang, G., (2021). Innate immune responses in RNA viral infection. *Front. Med.*, *15*(3), 333–346. https://doi.org/10.1007/S11684-020-0776-7.

Yang, C. S., & Wang, H., (2011). Mechanistic issues concerning cancer prevention by tea catechins. *Mol. Nutr. Food Res.*, *55*(6), 819–831. https://doi.org/10.1002/MNFR.201100036.

Yao, L. H., Jiang, Y. M., Shi, J., Tomás-Barberán, F. A., Datta, N., Singanusong, R., & Chen, S. S., (2004). Flavonoids in food and their health benefits. *Plant Foods Hum. Nutr.*, *59*(3), 113–122. https://doi.org/10.1007/S11130-004-0049-7.

Yogev, O., Lagos, D., Enver, T., & Boshoff, C., (2014). Kaposi's sarcoma herpesvirus microRNAs induce metabolic transformation of infected cells. *PLOS Pathog.*, *10*(9), e1004400. https://doi.org/10.1371/JOURNAL.PPAT.1004400.

Zandi, K., Musall, K., Oo, A., Cao, D., Liang, B., Hassandarvish, P., Lan, S., et al., (2021). Baicalein and baicalin inhibit SARS-CoV-2 RNA-dependent-RNA polymerase. *Microorganisms*, *9*(5). https://doi.org/10.3390/MICROORGANISMS9050893.

Zandi, K., Teoh, B. T., Sam, S. S., Wong, P. F., Mustafa, M. R., & Abubakar, S., (2011). *In vitro* antiviral activity of fisetin, rutin and naringenin against dengue virus type-2. *J. Med. Plants Res.*, *5*(23), 5534–5539.

Zandi, K., Teoh, B. T., Sam, S. S., Wong, P. F., Mustafa, M. R., & AbuBakar, S., (2012). Novel antiviral activity of baicalein against Dengue virus. *BMC Complement. Altern. Med.*, *12*. https://doi.org/10.1186/1472-6882-12-214.

Zandi, K., Teoh, B. T., Sam, S. S., Wong, P. F., Mustafa, M., & Abubakar, S., (2011). Antiviral activity of four types of bioflavonoid against Dengue virus type-2. *Virol. J.*, *8*. https://doi.org/10.1186/1743-422X-8-560.

Zhang, J., Lan, Y., Li, M. Y., Lamers, M. M., Fusade-Boyer, M., Klemm, E., Thiele, C., et al., (2018). Flaviviruses exploit the lipid droplet protein AUP1 to trigger lipophagy and drive virus production. *Cell Host Microbe.*, *23*(6), 819–831.e5. https://doi.org/10.1016/J.CHOM.2018.05.005.

Zhong, L., Hu, J., Shu, W., Gao, B., & Xiong, S., (2015). Epigallocatechin-3-gallate opposes HBV-induced incomplete autophagy by enhancing lysosomal acidification, which is unfavorable for HBV replication. *Cell Death Dis.*, *6*(5). https://doi.org/10.1038/CDDIS.2015.136.

Zhou, H. Y., Gao, S. Q., Gong, Y. S., Lin, T., Tong, S., Xiong, W., Shi, C. Y., et al., (2020). Anti-HSV-1 effect of dihydromyricetin from *Ampelopsis grossedentata* via the TLR9-dependent anti-inflammatory pathway. *J. Glob. Antimicrob. Resist.*, *23*, 370–376. https://doi.org/10.1016/J.JGAR.2020.10.003.

Zika Virus. https://www.who.int/news-room/fact-sheets/detail/zika-virus (accessed on 24 June 2023).

CHAPTER 5

REGULATION OF GENE EXPRESSION BY FLAVONOIDS

POOJA SABHARWAL[1] and PRIYANKA BHARDWAJ[2]

[1]*Assistant Professor, PG Department of Sharir Rachana, CBPACS, Affiliated to GGSIP University New Delhi, India*

[2]*MD Scholar, Department of Sharir Rachana, CBPACS, New Delhi, India*

ABSTRACT

The compelling increase rate of diseases is a quest of concern for medical professionals. The gigantic burden is the focal point of all the possible etiologies for the causation of diseases. One such etiology is the Gene affected. In the past five years, numerous groundbreaking research has been concerned with epigenetics. It is now well understood that the gene itself is not responsible for the disease causation, but its expression without any change in underlying DNA sequence also has emerged as a vast field of study for research. The naturally occurring polyphenolic compounds – flavonoids are celebrated to contribute well to the treatment of various diseases and also to prevent the side effects caused by various treatment modalities. The pace at which the regulation of genes can be made "naturally" is a topic of concern for all researchers and biologists around the globe.

5.1 INTRODUCTION

Genes are the factor of distinction of inherited traits. Genes, the fundamental unit of DNA, are arranged linearly on a chromosome.

Flavonoids as Nutraceuticals. Rajesh K. Kesharwani, Deepika Saini, Raj K. Keservani, and Anil Kumar Sharma (Eds.)

The scrutinizing of human genetics can reveal all the secrets concerning human nature. It can also assist in understanding various diseases, their course, and treatment. Human genetics Ingrid has diverse, overlapping sectors, which include – classical genetics, cytogenetics, genomics, molecular genetics, developmental genetics, population genetics, clinical genetics, and genetic counseling.

5.1.1 CLASSIFICATION OF HUMAN GENETICS

5.1.1.1 CLASSICAL GENETICS

The branch of genetics that is based on exclusively over-detectable results of reproduction is referred to as Classical Genetics. It is footed on ancient discipline experiments on Mendelian inheritance by Gregor Mendel, describing the fundamental mechanisms of heredity.

5.1.1.2 CYTOGENETICS

Cytogenetics is the branch of genetics and cell biology/cytology. It is concerned with the relation of chromosomes to cell behavior, singularly during mitosis and meiosis (Ringer et al., 1968). The methodology includes karyotyping, G-banded Chromosomes Analysis, and other cytogenetic banding techniques. It also includes molecular cytogenetics, i.e., fluorescence in situ hybridization (FISH) and comparative genomic hybridization (CGH).

5.1.1.3 GENOMICS

Genomics is concerned with the anatomy, physiology, evolution, and mapping of genomes. The World Health Organization explains genomics as the goal of collectively characterizing and quantifying all of an organism's genes, their interrelationships, and effects on the organisms. Unlike genomics, genetics refers to the study of individual genes and their role in inheritance. The enzymes and messenger substances support protein production and, thus, the development of tissues, organs, and organ systems in the body. It also controls innumerable chemical reactions and transportation of signals at the intercellular level. The sequencing and analysis of genomes via high scalability DNA sequencing and bioinformatics to study and derive conclusions of entire genomes (Klug & Cummings, 2012).

5.1.1.4 MOLECULAR GENETICS

Molecular genetics, a sub-field of biology, addresses the manifestation of variation among organisms regarding structural differences or expression of DNA molecules. Molecular genetics appertain an "evaluation outlook" for the determination of the structure and/or gene function utilizing genetic screens (Waters, 2013). Molecular genetics is essentially the combined forces of numerous subdivisions of biology – Mendelian inheritance, cell biology, molecular biology, biochemistry, and biotechnology. Molecular genetics is a dynamic methodology for the linkage of mutations to genetic conditions to encourage the research for advanced treatments of genetic diseases.

5.1.1.5 DEVELOPMENTAL GENETICS

The branch of genetics that deals with the affair of growth and development of plants and animals, their regeneration, asexual reproduction, metamorphosis, and also the growth and differentiation of stem cells in the adult are referred to as Developmental Genetics.

5.1.1.6 POPULATION GENETICS

The category of evolutionary biology that studies genetic differences within and between populations is population genetics.

5.1.1.7 CLINICAL GENETICS

The branch of genetics concerning its application in the diagnosis and management of hereditary disorders in the medical field is studied under Clinical Genetics (Culver et al., 2002).

5.1.1.8 GENETIC COUNSELLING

The guidance provided to the public affected by or at risk of any genetic disorder for the medical, psychological, and familial implications of genetic contributions to disease is studied under genetic counseling. It is mandatory for the implementation of genomic medicine (Servedio et al., 2018).

We observe various health indicators reflecting the evidence presenting the connection between various diseases, including cancer, neurobehavioral disorders, autoimmune disorders, etc. This reflection is produced by proven and suspected agents, for example, tobacco, smoke, radioactive agents, etc. The etiology of human diseases from a genetic perspective is being discovered at an unrivaled pace. Epigenetics and epigenomics science have risen as climacteric in the medical world and hence require an in-depth understanding of the concepts related to evolution, stem cells, synthetic biology, species conservation, and agriculture. The flexible genomic parameters can alter the physiology of the genome under the extrinsic influence but is also entitled to stable propagation of gene activity from one cell generation to another. Epigenetics refers to the alteration in phenotype or gene expression. This may be caused by the mechanisms without affecting the underlying DNA sequence. Environment prompts trigger changes to epigenetics tags on our genome, which configure the "expression of gene." Today, several research have proven the influence of the environment affecting the expression of genes through various epigenetic mechanisms – DNA methylation, histone modifications, and microRNA expression and hence, the health status of an individual is determined from its genetic background and environmental factors.

5.2 EPIGENETICS

The alteration of genes was among the biggest challenges scientists faced for the longest time. But what came as a surprise was the change in the way of gene expression without compromising the DNA sequence. Numerous studies have been conducted over the past two decades that have established that gene function cannot be altered simply by altering the DNA sequence. Interest was overwhelming when it was clear that a crystal understanding of epigenetics and epigenomics would be pivotal for a better understanding of a wide spectrum of disorders of the respiratory system, cardiovascular system, reproductive system, and also cognitive dysfunction, autoimmune, and neurobehavioral disorders. Research on this topic has proven that several disorders and cellular behavior are linked to epigenetic mechanisms (Patch et al., 2018). Numerous forces inducing epigenetic processes are heavy metals, pesticides, smoke, polycyclic aromatic hydrocarbons, hormones, radioactive agents, viruses, and bacteria. Epigenetics generally involves changes/mechanisms that affect gene activity and expression, but the term can be made malleable to expound heritable phenotypic change at cellular or physiological grounds.

The name epigenetics includes the prefix "epi," which implies "over, outside of, around, and addition" to the traditional genetic grounds for inheritance (Rutherford, 2015). Epigenetics is the study of heritable phenotypic changes that occur without involving changes in the DNA sequence. Any procedure that can change gene activity without affecting the DNA sequence is considered epigenetics, leading to modifications transferrable to daughter cells. Methylation, acetylation, phosphorylation, ubiquitylation, and sumoylation are some of the epigenetic processes. The possibility of addition to this shall be observed with the commencement of research. DNA methylation is the best and easiest epigenetic process to examine. Where cytosine bases appear sequentially, DNA methylation refers to the addition or removal of a methyl group (CH_3). DNA methylation was originally discovered in human cancer in 1983, and since then, it has been found in a variety of health disorders.

Another notable process of epigenetics is Chromatin Modification. Chromatin complex is made of histones and DNA firmly bundled to adjust into the cell's nucleus. It can be modified by substances, for example – acetyl groups (acetylation), enzymes, microRNAs, and small interfering RNAs are all examples of acetyl groups. The chromatin structure is altered as a result of this transformation, which influences gene expression.

These processes result in several effects at the level of genes, and one such effect is Imprinting. Imprinting in maize was originally identified in 1910, and it was confirmed in mammals in 1991. When one of the two alleles of a pair of the gene is silenced by methylation or acetylation is known as Imprinting. However, if the allele expressed is damaged, it actually increases the individual's vulnerability to numerous harmful microbes, toxic agents, etc. So far, researchers have recognized 80 human genes that can be imprinted.

5.3 GENE EXPRESSION

Gene expression provides a viaduct between encoded information in a gene and a final functioning gene (for example, a protein or non-coding RNA) (ncRNA). It is a multi-level process that can be modulated at any stage and controls the quality and spatiotemporal parameters of functional protein appearance. It entails transcription, mRNA splicing, translation, and post-translational protein modification and is essential for cellular structure and function to remain normal. It is the key to developmental changes, such as the differentiation and morphogenesis of cells. The regulation of gene expression mechanism includes various physiological and pathological processes (adaptations to new environments, homeostasis maintenance, and damage recovery) which allows the delivery of functional protein to the cell required for its normal functioning. Gene expression is regulated by the activity of repressor proteins that bind to silencer regions of DNA and last for the lifetime of the cell, sometimes even numerous generations, with no changes in the underlying DNA sequence (Griesbach, 2005). The non-genetic factors cause the genes of organisms to "express themselves" apparently (Bird, 2007).

The quantitative analysis to understand the area where the target gene expression serves as an essential task in the medical field. This analysis can be achieved within a specific cell or the whole organism and helps to understand the relationship between the expression of genes and cellular/organism phenotypes. A structural gene involves two divisions of components:

- Exons are represented in the mature mRNA molecule; it affects the amino acid sequence of the protein output and code for amino acids; and
- Introns, which are the noncoding section for amino acids, are entwined from the mRNA molecule prior to translation.

A structural gene comprises of control regions:

i. **Initiation Site:** The beginning site for transcription.

ii. **A Promoter:** It serves a crucial role in regulating gene transcription but is not transcribed into mRNA. Transcription factors (TFs) adhere to particular nucleotide sequences in the promoter section and assist in the adhering of RNA polymerases.

iii. **Enhancers:** The TFs which bind to the regions and activate/increase the rate of transcribing in the end.

iv. **Silencers:** The TFs which bind to the regions and ultimately deactivate/decrease/regress the rate of transcription.

5.3.1 THE GENE EXPRESSION PROCESS

The gene expression process follows two essential stages:

1. **Transcription:** It refers to the process of RNA synthesis, which is controlled by the promoters and enhancers interaction. Various types

of RNA are produced, which include messenger RNA (mRNA) (specifies the amino acids sequence in the protein product), transfer RNA (tRNA), and ribosomal RNA (rRNA) (both play a major role in the Translation). The process of transcription involves the following steps:

i. **Initiation:** The DNA molecule unravels and disperses into a little open complex. The promoter of the template strand/sense strand/coding strand is bound by RNA polymerase. The template strand must be 3′ to 5′ since RNA synthesis proceeds in a 5′ to 3′ orientation.

ii. **Elongation:** When RNA polymerase proceeds along the template strand, it produces mRNA. RNA polymerase is a holoenzyme in prokaryotes that consists of a specific number of subunits, including a sigma transcription factor that identifies the promoter. In eukaryotes, there are three types of RNA polymerases: I, II, and III. A proofreading mechanism is visible in the process.

iii. **Termination:** The process of termination is carried out in two ways amongst the Prokaryotes. In any "XYZ"-dependent termination, the "xyz" protein is responsible for the disruption of the complex, which involves the template strand, RNA polymerase, and RNA molecule. In "XYZ"-independent termination, a loop is formed at the extremity of the RNA molecule, resulting in the detachment. In eukaryotes, the process of termination is more intricate and involves the addition of adenine nucleotides addition at the 3′ of the RNA transcript. This process is termed polyadenylation.

iv. **Processing:** Post transcription process, the RNA molecule is processed in varied ways – removal of introns and the exons are spliced together, leading to the formation of a mature mRNA molecule that has a single protein-coding sequence. The synthesis of RNA involves the rules of normal base pairing, but thymine is replaced with uracil.

2. **Translation:** In translation, mRNA is used to instruct the protein synthesis and post-translational processing of the protein. The mature mRNA is used as a standard for assembling amino acid series to allow the production of the polypeptide with a definite amino

acid sequence. The site for this in the cytoplasm is the ribosome consisting of a large and a small subunit. The process of Transcription is carried out in the following steps:

i. **Initiation:** The tiny component of the ribosome attaches to the 5′ end of the mRNA molecule. It moves in the direction of 3′. The movement continues until it encounters a begin codon (AUG) and forms a complex with the major subunit of the ribosome and an initiation tRNA molecule.

ii. **Elongation:** The number of codons on the mRNA molecule to which a tRNA molecule linked to an amino acid binds is determined. Peptidyl transferase (an enzyme) binds the amino acids together using peptide bonds. As the ribosome travels along the mRNA molecule, this process continues, resulting in a chain of amino acids.

iii. **Termination:** When the ribosomal complex reaches one or more stop codons, it is said to have terminated (UAA, UAG, UGA).

Some genes are accountable for the formation of other types of RNA which play an active role in translation, including tRNA and rRNA.

5.3.2 CONTROL OF GENE EXPRESSION

The cells of multicellular organisms differ compellingly both structure and function-wise. The simplest of all examples is when the structure and function of a neuron and lymphocyte are compared. Both vary dramatically structurally and on a function basis as well and even make it impossible to imbibe that the genome of both is the same. This is why biologists suspected that genes might be selective while expressing themselves during cell differentiation. It is well-known that the differentiation of cells depends on gene expression changes in lieu of any alterations in the nucleotide sequence of the cell's genome.

Regulation of Gene Expression refers to the biological processes that control the rate and method of gene expression. When and where genes are activated, as well as the amount of protein or RNA product produced, are determined by a series of complex interactions between genes, RNA molecules, proteins, TFs, and other components of the gene expression system. Some genes express themselves at a predictable rate, as they produce several proteins involved in metabolic functions; few genes are expressed as segments of the cell differentiation process, and few genes are expressed as a consequence of cell differentiation.

Mechanisms of gene regulation include the following:

1. **Regulating the Transcription Rate:** This is the most economical gene regulation method. The commencing step in gene expression in gene transcription. Recent research concluded that the control of transcription is a censorious regulatory step in gene expression control. In considering transcription control, the transcriptional unit is defined that specifies the initiation and the termination of transcription. This incorporates every signal required essentially for proper transcription. The mechanisms that regulate transcription factor activity:

   i. **Control of Synthesis of the Transcription Factor:** This is the footing for tissue-specific control, i.e., a prime regulatory factor/factors; only seen in the cell type that the target gene is expressed. For example, the gene albumin is transcribed in the liver, not the brain, because the necessary TFs are not present in the brain.

   ii. **Control of the DNA Binding Activity of the Factor:** The transcription factor is present in this situation, but it is not actively involved in DNA binding. For example, the steroid hormone receptors are intracellular (cytoplasmic) TFs/proteins that bind specifically to the hormone when it enters the cell. As the hormone binds, the receptor is then activated and can enter the nucleus, bind to the gene, and stimulate the process of transcription.

   iii. **Control of the Transcriptional Stimulatory Activity of the Factor:** In this instance, the protein can bind to DNA but is unable to stimulate transcription. For example, E2F transcription factor activity is responsible for the control of transcription of various genes essential for DNA replication and cell growth and is regulated by interaction with the retinoblastoma (Rb) tumor suppressor protein. When Rb binds to E2F (regulated by phosphorylation), the resulting complex can still bind to DNA, but it is inactive in stimulating transcription. That is, unphosphorylated Rb can bind to and regulate E2F, but when Rb is phosphorylated by cell cycle-regulated protein kinases, it loses the capacity to bind to E2F.

2. **Regulating the Processing of RNA Molecules:** Comprising alternative splicing for the production of several protein products from a single gene. Although there are no clear examples whereby the nuclear/cytoplasmic transport of a cellular mRNA is regulated, there are at least two instances in viral infections in which RNA transport is affected. First, adenovirus infection results in the inhibition of transport of most cellular mRNAs – a specific viral gene product is required for this to occur, and at the same time, this protein facilitates the transport of viral RNA.
3. **Regulating the Stability of mRNA Molecules:** The stability of mRNAs differs over a large radius. Some RNAs are essentially stable, with half-lives approaching the cell division duration. Other RNAs turn over very rapidly (half-lives of a few minutes). RNAs encoding cytokines, as well as early responses to mitogens, are unstable, dependent on specific sequences in the 3′ untranslated region of the RNA. The unstable nature of the mRNA as a result of the recognition of this sequence is associated with the shortening of the poly-A tail.
4. **Regulating the Rate of Translation:** Alterations of translation factors can change the translation efficiency of mRNAs. For example, phosphorylation of eIF2 inhibits its action. Cis-acting regions in the mRNA, particularly those surrounding the AUG start codon, also influence translation efficiency.

5.4 FLAVONOIDS

Flavonoids are a group of secondary metabolites with different polyphenolic structures. These are isolated from naturally occurring products – roots, bark, stem, fruits, vegetables, grains, tea, and flowers. The term ‘flavonoid’ is derived from the Latin word “flavus,” i.e., “yellow," and the majority of flavonoids found are yellow in color. Flavonoids are imperative components in nutraceutical, pharmaceutical, medicinal, and cosmetic utilization. Flavonoids are classified as flavones, flavonols, flavanones, flavanonols, flavanols or catechins, anthocyanins, and chalcones, with anti-oxidative, anti-inflammatory, anti-mutagenic, and anti-carcinogenic effects.

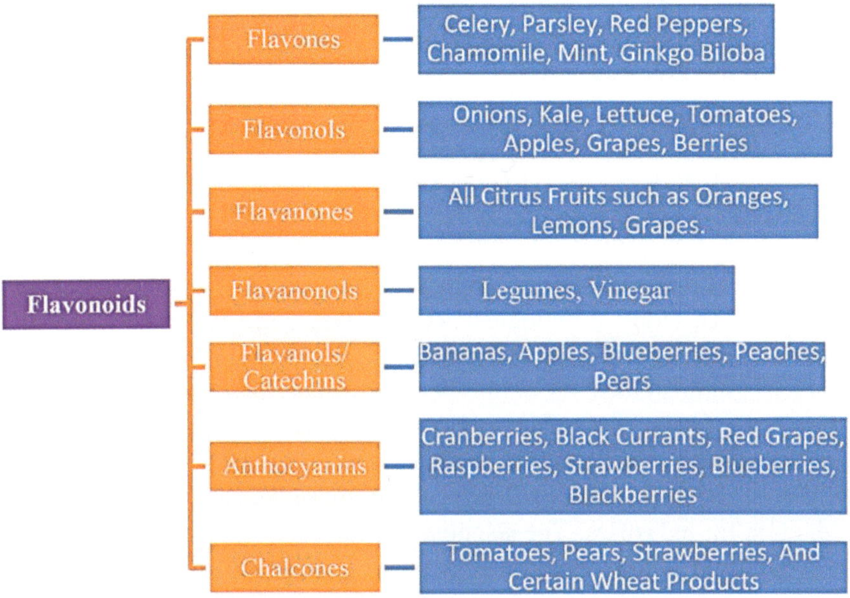

5.4.1 BIOLOGICAL ACTIVITIES OF FLAVONOIDS IN PLANTS

Flavonoids play various biological roles in plants, animals, and bacteria. Presently there are about 6,000 known flavonoids contributing to the colorful pigments of medicinal plants. Flavonoids are generated at specific places and transfer color and aroma to flowers and fruits to attract pollinators, followed by fruit dispersion to aid seed and spore germination, as well as seedling growth and development (Dupont et al., 2009).

They contribute to major functions in plants as follows:

- Protect plants against a variety of biotic and abiotic stressors, as well as act as UV filters (Hunter, 2008).
- Signal molecules, allopathic substances, phytoalexins, detoxifying agents, and antimicrobial defense compounds are all examples of signal molecules.
- Acts against frost hardness, drought resistance and may contribute to plant heat habituation and freezing tolerance (Takahashi & Ohnishi, 2004).
- Controls the growth and development of plants through their distinct action on cell wall synthesis (Samanta et al., 2011).

- Allelopathic action when there occur interactions with symbiotic mycorrhizal fungi and rhizobia, and in defense against fungal pathogens (Mathesius, 2018).

5.4.2 BIOLOGICAL ACTIVITIES OF FLAVONOIDS IN HUMANS

The working operation particulars of flavonoids are still ambiguous. However, it is renowned for a long that the derivatives of plant origin own a comprehensive scope of biological activity. The focus of current flavonoids research and development is on the identification and isolation of flavonoids and their administration on public health for benefits. Flavonoids accredit positive responses on human and animal health status, and the current intrigue concerns disease therapy and chemoprevention. Molecular docking, along with bioinformatics, is utilized to predict latent implementation and production by industries. The scientists – Dixon & Pasinetti assessed plant flavonoids and isoflavonoids and delineated their administration to agriculture and neurosciences (Dixon & Pasinetti, 2010). Kumar and Pandey looked into flavonoids' defense mechanisms against human diseases as well as their functions in plants (Kumar & Pandey, 2013) involved mechanisms (Panche et al., 2015). Panche et al. analyzed the extensive use of flavonoids as plant secondary metabolites for the treatment of Alzheimer's disease and the mechanisms involved when observing AD and recent therapy procedures (Panche et al., 2015).

5.5 GENE EXPRESSION AND FLAVONOIDS

5.5.1 INTERACTION WITH STEROID RECEPTORS

The estrogenic activity of flavonoids is customary, and its physiological implications have been concluded, too (Kurzer & Xu, 1997). The expression of endogenous estrogen-responsive genes is increased by flavonoids at the molecular level. The isoflavones genistein, daidzein, and biochanin A have been demonstrated to have strong estrogen-stimulatory effect. Flavonol, quercetin, morin, and myricetin, on the other hand, did not provide any stimulation (Miksicek, 1993). Estrogenic flavonoids bind to estrogen receptors (ERs), mimicking the hormone I76-estradiol (Miksicek, 1993; Kuiper et al., 1998). The Estrogen Receptor binding activities of flavonoids were generally lower than that of 176-estradiol, but flavonoids displayed a higher affinity for ERB compared to Era (Kuiper et al., 1998; Buslig & Manthey, 2002).

Recent research has distinguished an inhibitory effect of flavonoids-apigenin, quercetin, and fisetin, on vitamin D receptor expression in human keratinocytes (Segaert et al., 2000). The effect was observed at the level of both protein and mRNA and led to complete suppression of the vitamin D responsiveness in these cells. This observation's importance has yet to be determined.

5.5.2 REGULATION OF ANTIOXIDANT SYSTEM

Flavonoids possess chemical antioxidant properties since known. Intracellular contents of antioxidant protein thiols were also regulated by some flavonoids. Black tea polyphenols and green tea were shown to moderately induce glutathione in normal human Chang liver cells but not individual catechin components of the tea. Flavonoids inhibit cancer cell growth. At the molecular level, suppression of gene expression connected to cell proliferation has been found, although stimulation of genes linked to apoptosis has also been observed. For example, Ras proteins (p21) are G protein-like proteins involved in cell signaling. In various human cancers, mutations at Ras proto-oncogenes and the subsequent production of overactive Ras proteins have been observed (Bos, 1989). Quercetin treatment can induce cell cycle arrest and inhibit the expression of all three forms of p21 Ras, K-Ras, H-Ras, and N-Ras in human colon cancer cell lines and in primary colorectal tumors (Ranelletti et al., 2000). Similarly, the expression of oncogene N-myc was decreased by genistein in neuroblastoma cells (Brown et al., 1998), and the expression of c-myc was decreased by apigenin in human keratinocytes (Segaert et al., 2000).

The inhibition of cyclin 01 expressions was observed in prostate carcinoma cells treated with a major component of silymarin, silibinin (Agarwal, 2002), and in rat hepatic stellate cells treated with quercetin (Kawada et al., 1998). Because of the estrogenic property of genistein, as described above, the effect of genistein on cell cycle-related proteins could vary depending on the cell line chosen and the treatment condition. In the absence of serum, genistein at low concentrations actually enhanced the synthesis of cyclin 01 in human breast cancer cells mimicking the function of 17B-estradiol (Dees et al., 1997). The expression of several other tumor markers was decreased by flavonoids, but some effects could just be a consequence of cell-cycle blockage. Induction of ornithine decarboxylase by TPA was decreased by green tea as well as its catechin components in 2C5 cells (Steele et al., 2000); and by topical apigenin application to mouse skin (Wei et al., 1990). Prostate-specific antigen synthesis and secretion from prostate carcinoma cells were decreased by silibinin treatment (Agarwal, 1999). Inhibition of cyclooxygenase-2 activity has been linked to the prevention of colon carcinogenesis.

In breast carcinoma cells, flavopiridol (a synthetic flavonoid) induced cell cycle arrest and downregulated cyclin D I, which is an important protein for both cell cycle progression and neoplastic transformation (Carlson et al., 1999).

5.5.3 REGULATION OF CYTOKINE RELEASE AND CYTOKINE RESPONSE

Cellular responses are maintained by essential mediators called Cytokines. It was found that flavonoids tend to reduce the basal and induced secretion level of cytokines. In the research conducted (Tabary et al., 1999). Genistein dose-dependently decreased the production of IL-8 by human bronchial gland cells. Genistein was also responsible for decreasing the secretion of IL-2 and leukotriene B4 from lectin-stimulated human mononuclear cells (Atluru et al., 1991). In a study directed, it was observed that lipopolysaccharide (LPS)-induced TNF-a release was decreased by quercetin in the macrophage cell line RAW 264.7 (Wadsworth & Koop, 1999).

Hypoxia-induced transcription of the plasminogen activator inhibitor-I gene was blocked by genistein flavonoid in bovine aortic endothelial cells (Uchiyama et al., 2000). Under the stimulated condition, it was proposed that the inhibition of protein tyrosine kinases was done by genistein-inhibited cytokine secretion (Atluru et al., 1991; Uchiyama et al., 2000). Cell growth and differentiation is regulated by transforming growth factor B (TGFB). Quercetin was shown to increase TGFB activity in the cultured medium of ovarian cancer cells (Scambia et al., 1994); genistein, on the other hand, enhanced TGF expression only in normal human mammary epithelial cells but not in breast tumor cells (Sathyamoorthy et al., 1998). It was shown that genistein impacted the expression of TGFa, epidermal growth factor (EGF), and EGF receptor (EGFR) in the rat mammary gland in an age-dependent manner in a study assessing the effects of prepubertal genistein administration (Brown et al., 1998). The natural immunological response includes the creation of nitric oxide (NO). Flavonoids lowered NO generation, which is consistent with their immunomodulatory properties. Apigenin and kaempferol inhibited LPS-inducible nitric oxide synthase (iNOS) in macrophages, according to the findings (Liang et al., 1999). Many cytokines' signaling is maintained via the signal transducer and activator of the transcription (STAT) pathway. STAT proteins are phosphorylated on tyrosine and then translocated to the nucleus once the cytokine binds to its receptor. STAT proteins, by binding to specific sequences in the promoter region of target genes, exert their gene regulatory functions (Hill & Treisman, 1995). Many flavonoids, especially genistein, inhibit protein kinases. Therefore, flavonoids could inhibit STAT activation and thus modulate various cytokines actions.

5.6 EFFECTS ON DRUG DETOXIFICATION ENZYMES

Chemical carcinogenesis could be prevented by the enhancement of drug detoxification activity. Thus, it is observed that the drugs for cancer chemoprevention include medicinal chemicals or dietary components that can increase the activity of phase II detoxification enzymes. Contrary to this, modulators of phase I enzymes could possibly give mixed results. When examined together, flavonoids were revealed to exhibit structural-dependent effects on both phase I and phase II enzymes. Tea and its catechin components were the potent inducers of NADPH: quinone reductase and glutathione S-transferase, both of them being phase II enzymes, in human Chang liver cells (Steele et al., 2000). However, 4′-bromoflavone which is a synthetic flavonoid, induced phase I and phase II enzymes adjunctly in rat hepatoma cell culture and in rat tissues, even though the impact was remarkable for its induction of phase II enzymes (Song et al., 1999). The activation of a xenobiotic response element (XRE) and an antioxidant response element (ARE) in the quinone reductase gene mediated the induction of quinone reductase by 4′-bromoflavone (Song et al., 1999). Consistent with this observation, equol and genistein – the two isoflavones that failed to enhance glutathione S-transferase in a mouse study also could not activate XRE (Helsby et al., 1997). Genistein and equol also could not affect the phase I enzyme activity (Helsby et al., 1997), but quercetin was shown to inhibit cytochrome P-450 1Al gene expression in Hep G2 cells (Kang et al., 1999).

KEYWORDS

- **epigenetics**
- **flavonoids**
- **gene expression**
- **regulation of gene**
- **transcription**
- **translation**

REFERENCES

Atluru, S., & Atluru, D., (1991). Evidence that genistein, a protein-tyrosine kinase inhibitor, inhibits CD28 monoclonal-antibody-stimulated human T cell proliferation. *Transplantation, 51*(2), 448–450.

Bird, A., (2007). Perceptions of epigenetics. *Nature, 447*(7143), 396–398. Bibcode:2007Natur.447.396B.

Bos, J. L., (1989). Ras oncogenes in human cancer: A review. *Cancer Res., 49*(17), 4682–4689.

Brad, C., (1999). Down-regulation of cyclin D1 by transcriptional repression in MCF-7 human breast carcinoma cells induced by flavopiridol. *Cancer Res., 59*(18), 4634–4641.

Brown, E. T., & Fuller, G. M., (1998). Detection of a complex that associates with the Bbeta fibrinogen G-455-A polymorphism. *Blood, 92*(9), 3286–3293.

Buslig, B., & Manthey, J., (2002). *Flavonoids in Cell Function* (pp. 191–200). Kluwer Academic/Plenum Publishers.

Culver, K. W., & Labow, M. A., (2002). Genomics. In: Robinson, R., (ed.), *Genetics*. Macmillan Science Library. Macmillan Reference USA.

Dees, C., et al., (1997). Dietary estrogens stimulate human breast cells to enter the cell cycle. *Environment Health Perspective, 105*(3).

Dixon, R., & Pasinetti, G., (2010). Flavonoids and isoflavonoids: From plant biology to agriculture and neuroscience. *Plant Physiol., 154*, 453–457.

Dupont, C., Armant, D. R., & Brenner, C. A., (2009). Epigenetics: Definition, mechanisms and clinical perspective. *Seminars in Reproductive Medicine, 27*(5), 351–357.

Griesbach, R., (2005). Biochemistry and genetics of flower color. *Plant Breed Rev., 25*, 89–114.

Helsby, et al., (1997). The isoflavones equol and genistein do not induce xenobiotic-metabolizing enzymes in mouse and in human cells. *Xenobiotica, 27*(6).

Hill, C. S., & Treisman, R., (1995). Differential activation of c-fos promoter elements by serum, lysophosphatidic acid, G proteins and polypeptide growth factors. *EMBO J., 14*(20), 5037–5047.

Hunter, P., (2008). *What Genes Remember*. Prospect Magazine. Archived from the original on 1 May 2008.

Jorgensen, J. R., (1995). Co-suppression, flower color patterns, and metastable gene expression states. *Science, 268*, 686–691.

Kang, et al., (1999). Quercetin inhibits benzo[a]pyrene-induced DNA adducts in human hep G2 cells by altering cytochrome P-450 1A1 gene expression. *Nutrition and Cancer, 35*(2), 175–179.

Kawada, et al., (1998). Effect of antioxidants, resveratrol, quercetin, and *N*-acetylcysteine, on the functions of cultured rat hepatic stellate cells and Kupffer cells. *Hepatology, 27*(1), 1265–1274.

Klug, W. S., & Cummings, M. R., (2012). *Concepts of Genetics* (10th edn.). San Francisco: Pearson Education.

Kuiper, et al., (1998). Interaction of estrogenic chemicals and phytoestrogens with estrogen receptor beta. *Endocrinology, 139*(10), 4252–4263.

Kumar, S., & Pandey, A. K., (2013). Chemistry and biological activities of flavonoids: An overview. *Scientific World Journal, 2013*, 16275.

Kurzer, M. S., & Xu, X., (1997). Dietary phytoestrogens. *Annu. Rev. Nutr., 17*, 353–381.

Mathesius, U., (2018). Flavonoids function in plants and their interaction with other organisms. *Plants (Basel), 7*(2), 30.

Miksicek, (1993). Commonly occurring plant flavonoids have estrogenic activity. *Molecular Pharmacology, 44*(1), 37–43.

Nadine, M. B., et al., (1998). Prepubertal genistein treatment modulates TGF-α, EGF, and EGF-receptor mRNAs and proteins in the rat mammary gland. *Molecular and Cellular Endocrinology, 144*(1, 2), 149–165.

Olivier, T., et al., (1999). Genistein inhibits constitutive and inducible NFκB activation and decreases IL-8 production by human cystic fibrosis bronchial gland cells. *The American Journal of Pathology, 155*(2), 473–481.

Panche, A., Chandra, S., Diwan, A., et al., (2015). Alzheimer's and current therapeutics: A review. *Asian J. Pharm. Clin. Res., 8*, 14–19.

Patch, C., & Middleton, A., (2018). Genetic counseling in the era of genomic medicine. *British Medical Bulletin, 126*(1), 27–36.

Ranelletti, F. O., et al., (2000). Quercetin inhibits p21-RAS expression in human colon cancer cell lines and in primary colorectal tumors. *Int. J. Cancer, 85*(3), 438–445.

Rieger, R., Michaelis, A., & Green, M. M., (1968). *A Glossary of Genetics and Cytogenetics: Classical and Molecular*. New York: Springer-Verlag.

Rutherford, A., (2015). *Beware The Pseudo Gene Genies*. The Guardian.

Samanta, A., Das, G., & Das, S., (2011). Roles of flavonoids in plants. *Int. J. Pharm. Sci. Tech., 6*, 12–35.

Sathyamoorthy, N., Gilsdorf, J. S., & Wang, T. T., (1998). Differential effect of genistein on transforming growth factor beta 1 expression in normal and malignant mammary epithelial cells. *Anticancer Research, 18*(4A), 2449–2453.

Scambia, et al., (1994). Quercetin potentiates the effect of Adriamycin in a multidrug-resistant MCF-7 human breast-cancer cell line: P-glycoprotein as a possible target. *Cancer Chemotherapy and Pharmacology, 34*, 459–464.

Segaert, et al., (2000). Vitamin D receptor expression is linked to cell cycle control in normal human keratinocytes. *Biochemical and Biophysical Research Communications, 279*(1), 89–94.

Servedio, M. R., Brandvain, Y., Dhole, S., Fitzpatrick, C. L., Goldberg, E. E., Stern, C. A., Van, C. J., & Yeh, D. J., (2014). *Population Genetics – Latest Research and News*. www.nature.com. (accessed on 24 June 2023).

Song, et al., (1999). Cancer chemopreventive activity mediated by 4′-bromoflavone, a potent inducer of phase II detoxification enzymes. *Cancer Res., 59*(3), 578–585.

Takahashi, A., & Ohnishi, T., (2004). The significance of the study about the biological effects of solar ultraviolet radiation using the exposed facility on the international space station. *Biol. Sci. Space, 18*, 255–260.

Tyagi, A., Agarwal, C., & Agarwal, R., (2002). Inhibition of retinoblastoma protein (Rb) phosphorylation at serine sites and an increase in Rb-E2F complex formation by silibinin in androgen-dependent human prostate carcinoma LNCaP cells: Role in prostate cancer prevention. *Mol. Cancer Ther., 1*(7), 525–532.

Uchiyama, et al., (2000). Hypoxia induces transcription of the plasminogen activator inhibitor-1 gene through genistein-sensitive tyrosine kinase pathways in vascular endothelial cells. *Arteriosclerosis, Thrombosis, and Vascular Biology, 20*, 1155–1161.

Vernon, E. S., et al., (2000). Progress in cancer chemoprevention: Development of diet-derived chemopreventive agents. *The Journal of Nutrition, 130*(2), 467S–471S.

Vernon, E. S., Gary, J. K., Douglas, B., Charles, W. B., Rajendra, M., Donya, B., Caroline, C. S., Songyun, Z., & Sheela, S., *(2000).* Comparative chemopreventive mechanisms of green tea, black tea, and selected polyphenol extracts measured by *in vitro* bioassays. *Carcinogenesis, 21*(1), 63–67.

Wadsworth, T. L., & Koop, D. R., (1999). Effects of the wine polyphenolics quercetin and resveratrol on pro-inflammatory cytokine expression in RAW 264.7 macrophages. *Biochem. Pharmacol., 57*(8), 941–949.

Waters, K., (2013). Molecular genetics. In: Zalta, E. N., (ed.), *The Stanford Encyclopedia of Philosophy (Fall 2013 ed.)*. Metaphysics Research Lab, Stanford University. Alberts, Bruce (2014–11–18). *Molecular Biology of the Cell (6th edn.)*. New York, NY.

Wei et al., (1990). Inhibitory effect of apigenin, a plant flavonoid, on epidermal ornithine decarboxylase and skin tumor promotion in mice. *Birt. Cancer Res., 50*(3), 499–502.

Yu-Chih, L., et al., (1999). Suppression of inducible cyclooxygenase and inducible nitric oxide synthase by apigenin and related flavonoids in mouse macrophages. *Carcinogenesis, 20*(10), 1945–1952.

CHAPTER 6

FLAVONOIDS OF ASTERACEAE-PROMISING ANTI-INFLAMMATORY AGENTS

S. R. SUJA,[1] V. ASWATHY,[1] B. S. BIJUKUMAR,[2] and R. PRAKASHKUMAR[1]

[1]*Ethnomedicine and Ethnopharmacology Division, KSCSTE–Jawaharlal Nehru Tropical Botanic Garden and Research Institute, Palode, Thiruvananthapuram, Kerala, India*

[2]*Post-Graduate, Department of Zoology and Research Center, Mahatma Gandhi College, Thiruvananthapuram, Kerala, India*

ABSTRACT

One of the most important plant families is the Asteraceae, and their medicinal effects are attributed to phytochemical compounds such as polyphenols, synthetic resin acids, flavonoids, acetylenes, and triterpenes. Nowadays, there is a growing interest in natural sources, which has sparked study interest in the Asteraceae family. The majority of non-infectious diseases are developed and worsened by prolonged and persistent chronic inflammation. The prevailing therapies for many of those chronic diseases generally cause a lot of deleterious side effects, warranting the requirement of safer, less toxic, and less expensive treatment for patients. For hundreds of years, flavonoids and their preparations were used to treat numerous human diseases. Recent studies have additionally shown that flavonoids, particularly flavone derivatives, regulation of pro-inflammatory mediators such as cyclooxygenase, shown unique therapeutic activity, inducible nitric oxide synthase (iNOS),

Flavonoids as Nutraceuticals. Rajesh K. Kesharwani, Deepika Saini, Raj K. Keservani, and Anil Kumar Sharma (Eds.)

and a number of other cytokines. Because of these diverse mechanisms of action, flavonoids are thought to be affordable candidates for drug development. Understanding the involvement of inflammatory mediators in several disease conditions and, additionally, the mechanism of action of phytochemical compounds from natural sources can open the way for herbal drug development against inflammatory diseases. The chapter explored the role of flavonoids of plants of the Asteraceae family in combating several inflammatory processes underlying chronic disease conditions.

6.1 INTRODUCTION

India is one of the largest biodiversity-rich countries with an immense wealth of medicinal plants. With about 1,620 genera and 23,600 species of plants with a worldwide distribution, the Asteraceae family (Sunflower family) is one of the largest plant groups. The Asteraceae family includes a variety of well-known species like chicory, sunflower, lettuce, coreopsis, dahlias, wormwood, and other medicinal plants. The main characteristics of the family are the presence of composite flower heads and one-seeded fruits. The medicinal plants coming under the family are traditionally used against many diseases. The plants are typically found within numerous ecosystems of India because of their adaptations, like floral characteristics, fertilization, seed spreading mechanisms, etc. A large number of Asteraceae family plants have numerous medicinal applications. For more than thousands of years, some of the medicinal plants coming under the Asteraceae family have been cultivated for food and medical purposes. Asteraceae members are commonly found in subtropical regions with arid and semi-arid conditions, but they are enjoying a cosmopolitan distribution. The medicinal plants coming under the Asteraceae family exhibited therapeutic effects such as antimicrobial, anti-inflammatory, antioxidant, and hepatoprotective (Rolnik et al., 2021).

Traditional knowledge of medicine is referred to as knowledge of the medicinal property of a particular plant species and its use against some disease conditions. It has been in use for many years, and the knowledge has been passed down the generations. It has been used since time immemorial by the ancient traditional healers to improve human health. In the villages and remote areas, traditional medicine has maintained its popularity, and it is considered a primary healthcare practice at the community level (WHO). Generally, the traditional knowledge is communicated and transferred orally, and it is informal. Traditional medicine and other indigenous medicinal practices are based on the use of potential medicinal plants for treating various ailments. According to WHO, 3.5 billion people in developing and under-developing countries use herbal medicaments for primary health care. Traditional healers use different plant parts to make crude medicines and are used for treating diseases such as wounds, inflammation, fever, cough, kidney problems, etc. Recently erosion of traditional knowledge among the people was serious due to many factors such as the lack of interest of the young generation in gaining knowledge, agricultural expansion, Orally transferred knowledge without any documentation, species unavailability, and modern education influence, the unwillingness of the traditional healers to disclose their secret knowledge to others. The indiscriminate use of root plants for the preparation of traditional medicine could also be considered as a threat.

Asteraceae plants have been used to cure various diseases from ancient times. Studies showed that they exhibit analgesic, antimicrobial, antiviral, antioxidant, anti-proliferative, anti-inflammatory, and Vasodilatory activities. These plants are an inevitable part of traditional treatments such as dermatological problems, wounds, and associated inflammations, etc. The pharmacological effects of Asteraceae plant species are due to the presence of a wide range of phytochemical compounds, including alkaloids, flavonoids, polyphenolic compounds, phenolic acids, etc.

6.2 INFLAMMATION

Inflammation refers to the physiological response to different cellular, vascular, and pathological injuries. When the body reacts to inflammation, it starts to activate the inflammatory cells. The activation of inflammatory cells such as neutrophils, basophils, and eosinophils, and mononuclear cells like monocytes and macrophages leads to the release of many inflammatory mediators. Inflammation is characterized by pain, redness on the inflamed area, heat or burning sensation, and swelling. Inflammation results increase in the blood flow to the injured area or infected tissue. It results in the redness and warmth of the inflamed area. Some of the inflammatory mediators cause fluid to leak into the tissues of the affected area. This fluid leakage results in swelling.

Microorganisms or tissue damage can induce inflammatory responses in body tissues. It will activate the release of PAMPs (pathogen-associated molecular patterns) and DAMPs (damage-associated molecular patterns).

The antigen-specific and nonspecific immune cells, like dendritic cells (DCs) and macrophages, can recognize these molecules through various receptors found on their cell surface, like pattern recognition receptors (PRRs). When get activated, these immune cells start to produce chemo-attractant molecules, which are controlled by the transcription factor NF-κB. The transcription factor NF-κB is highly significant in the regulation of the expression of inflammatory enzymes like COX and other pro-inflammatory cytokines, which makes it one of the most significant transcription factors (TFs) during the inflammatory process and pain. Cytokines and chemokines produced by these immune cells and formyl-peptide (fMLP) released by dying cells activate vascular endothelial cells. This activation provides a gradient of signals to guide neutrophils precisely to the inflamed area following a spatial, temporal, and hierarchic cascade of mediators (Catherine et al., 1998).

Non-steroidal anti-inflammatory drugs or NSAIDs are commonly used for treating inflammatory diseases and other health problems associated with inflammation. The common side effects of NSAIDs include renal problems, stomach ulcers, vomiting, headache, nausea, blurred vision, dizziness, and other allergic reactions. In this scenario, plant-based natural anti-inflammatory drugs have a significant position in the treatment of inflammatory conditions. When compared to conventional medicines, plant-based drugs and other herbal remedies have fewer side effects. Natural drugs can be carefully selected as a novel therapeutic agent for managing inflammatory ailments. The phytochemical constituents present in medicinal plants have many therapeutic properties, including anti-inflammatory, wound healing, anti-arthritic analgesics, etc. Among these bioactive phytochemicals' flavonoids have a significant position. They occur in many foods like vegetables and fruits, and medicinal plants, and they are the most active constituent, having the ability to reduce and cure inflammation.

6.3 FLAVONOIDS

Flavonoids are an important class of plant constituents like polyphenolic compounds with great structural diversity found in plants and commonly consumed in diets. These secondary metabolites are the most important plant pigments which impart flower coloration and produce red/blue- or yellow-colored petals to attract pollinator animals. Flavonoids are the major plant pigments which are synthesized from phenylalanine. They display bright colors known from flower petals; they can emit brilliant fluorescence when get excited by UV and are present in green plant cells. The flavonoid compounds are used in taxonomical classification, especially in chemotaxonomy. They have an important role in the regulation of plant growth by inhibiting exocytosis of the auxin and indole acetic acid, and they influence other biological activities of cells in various ways. Some of the flavonoid compounds showed antimicrobial effects and inhibited viral enzymes like reverse transcriptase and protease, and destroyed some pathogenic protozoans (Havsteen, 2002).

The other functions of flavonoids in plants include plant development regulation, pigmentation, protection from damage due to UV exposure, roles in plant defense mechanisms, and signaling. Many flavonoids act as bioactive compounds that interact with nucleic acids or proteins and show different kinds of pharmacological properties such as antimicrobial, insecticidal, antifungal, etc. They act as symbionts, allelochemicals, and antimicrobial and anti-herbivory factors in plants. Flavonoids are one of the major dietary components in many fruits and vegetables. Because of their common occurrence in the human diet, many flavonoids present in medicinal plants are used for controlling inflammatory responses and tumor development. Many studies have shown that flavonoids exhibit biological and pharmacological activities, including antioxidant, cytotoxic, anticancer, antiviral, antibacterial, anti-inflammatory, anti-allergic, antithrombotic, cardioprotective, hepatoprotective, neuroprotective, antimalarial, anti-leishmanial, antitrypanosomal, and antimonial properties (Erica et al., 2017). Aglycones, glycosides, and methylated derivatives are all examples of flavonoids. The basic structure of the flavonoid compound is the aglycone six-member ring condensed with the benzene ring and is either α-pyrone (flavonols and flavanones) or its dihydroderivative (flavonols and flavanones).

Flavonoids constitute one of the most important classes of bioactive compounds in higher medicinal plants. Flavonoids are divided into six subclasses based on their chemical structure, such as flavones, flavanones, isoflavones, flavanols, neoflavonoids, flavan-3-ols, chalcones, and anthocyanidins. Flavones are present in the leaves of many plants, flowers, and fruits. Luteolin and apigenin are the most studied flavones.

Flavones can be found as glucosides in flowers, plants, and fruits. The hydroxyl group in the fifth position of the A ring is found in the majority of flavones found in plants and fruits, While hydroxylation in other sites, most notably in the seventh position of the A ring or the 3′ and 4′ positions of the B ring, varies depending on taxonomic categorization (Panche et al., 2016).

6.3.1 FLAVONOLS

Another important class of flavonoids is flavanols, generally found in citrus fruits like lemons, oranges, and grapes. Flavanones are present in all citrus fruits having the characteristic bitter taste of the peel and juice. Fruits like lemons, oranges, and grapes are rich in flavanones, and major phytoconstituents are naringenin, hesperitin, etc. Flavanol compounds like quercetin, rutin, kaempferol, myricetin, fisetin, silymarin, and isorhamnetin are present in foods such as saffron, apple, onions, kale, lettuce, etc. Flavan-3-ols, also known as dihydroflavonols, include compounds such as epicatechin, catechin, gallocatechin, epicatechin gallate, epigallocatechin, epigallocatechin gallate (EGCG) and procyanidin. The flavan-3-ol compound is commonly present in black and green tea and fruits like bananas, pears, etc.

6.3.2 ISOFLAVONOIDS

Isoflavonoids are a subclass of flavonoids with a small distribution in the plant kingdom. They are dominantly present in leguminous plants such as soya beans. During plant-microbe interactions and plant defense mechanisms, they serve as a precursor for the synthesis of phytoalexins.

6.3.3 NEOFLAVONOIDS

The chemical structure of neoflavonoids includes a 4-phenylchromen backbone without hydroxyl group substitution at the second position. The first neoflavone compound isolated from natural sources is calophyllolide, from the seeds of *Calophyllum inophyllum.*

6.3.4 FLAVANOLS, FLAVAN-3-OLS, OR CATECHINS

Flavanols are a diversified and multi-substituted subgroup of flavanones and 3-hydroxy derivatives. They are also called flavan-3-ols, as the third position of the C ring is occupied by the hydroxyl group. There is no double bond between the second and third positions. Fruits like apples, blueberries, peaches, bananas, and pears are high in flavanols.

6.3.5 ANTHOCYANINS

Anthocyanins are pigments which impart various colors to flowers and fruits of plants. The most common and abundant anthocyanins present in plants include cyanidin, malvidin, delphinidin, peonidin, and pelargonidin. They are the most studied anthocyanins (Choy et al., 2019). The outer layer of fruits such as grapes, strawberries, raspberries, and cranberries are rich in anthocyanin content.

6.3.6 CHALCONES

Chalcones are another major subclass of flavonoids, which are distinguished by the absence of the basic flavonoid skeleton's 'ring C.' Open-chain flavonoids are another name for them. The most common chalcones found in plants are phloridzin, arbutin, phloretin, and chalconaringenin. Chalcones are present in fruits like tomatoes, strawberries, pears, bearberries, and certain wheat products. Nowadays, chalcones and their derivatives are getting considerable attention because of their various biological and nutritional benefits.

Recently flavonoids are considered as important compounds in a number of nutraceutical products (Keservani et al., 2010a, b), pharmaceutical, medicinal, and cosmetic applications. They provide a wide spectrum of health benefits and disease prevention. This is due to their therapeutic efficacies like anti-inflammatory, anti-oxidative, anti-carcinogenic, and anti-mutagenic properties and their capacity to control and influence key cellular enzyme functions. They are strong inhibitors of inflammatory enzymes like cyclo-oxygenase (COX), xanthine oxidase (XO), phosphoinositide 3-kinase, and lipoxygenase. Flavonoids lowered the cardiovascular mortality rate and prevented CHD. The mechanism by which flavonoids work is still a mystery (Pancheetal, 2016; Keservani & Sharma, 2014).

6.4 ANTI-INFLAMMATORY MECHANISM OF FLAVONOIDS

Many recent studies have shown that the daily intake of flavonoids will help to reduce the risk of cardiovascular problems (Keservani et al., 2016). The antithrombotic, anti-ischemic, and antioxidant properties of flavonoids (Keservani et al., 2020) can give a protective effect against coronary heart problems. Flavonoids work in three ways to minimize the risk of coronary heart disease. They can improve the coronary vasodilation, platelets in the blood are less likely to clot, and low-density lipoproteins are less likely to oxidize. The flavonoids found in citrus plants have significant anti-inflammatory properties. The compounds inhibited the synthesis and biological activities, including signal transduction pathways of different pro-inflammatory mediators (Benavente et al., 2008).

Inflammation is the complex biological response of the immune system against some harmful stimuli, such as foreign particles, antigens, pathogens, damaged cells, and other irritants. It is a protective mechanism of the body to remove the injurious stimuli. The inflammatory response also initiates the healing process for the tissue. The inflammatory response involves the action of many enzymes, such as phosphodiesterase, phospholipase A2, lipoxygenases, COX, protein tyrosine kinases, and protein kinase C. These enzymes are vital in the activation of cells, such as endothelial cells, which are involved in generating inflammatory responses. Cyclooxygenase (COX) is an enzyme that plays a crucial role as an inflammatory mediator and is involved in the release of arachidonic acid, a precursor for the biosynthesis of eicosanoids like prostaglandins and prostacyclin. COX enzyme catalyzes the conversion reaction of arachidonic acid into prostaglandins and thromboxanes; COX-2 is an inducible enzyme that is only expressed after some inflammatory stimuli, whereas COX-1 is responsible for the supply of prostaglandins for preserving the integrity of the stomach mucosa and provide appropriate vascular homeostasis (Walker et al., 2000). An inflammatory reaction begins with the generation and release of arachidonic acid. Prostaglandins and nitric oxide (NO) biosynthesis play a role in the inflammatory response, and isoforms of inducible nitric oxide synthase (iNOS) and cyclooxygenase (COX-2) produce a significant amount of these mediators (Permender et al., 2009). The anti-inflammatory effect of most flavonoids, notably flavone derivatives, is mediated by modification of proinflammatory gene expression, such as cyclooxygenase-2, iNOS, and numerous key cytokines.

6.5 MEDICINAL USES OF ASTERACEAE PLANTS

Plants of the Asteraceae family feature hairy, fragrant leaves and flat clusters of flowers with a head inflorescence and terminal position; some are popular garden plants with a variety of colorful flowers (asters, chrysanthemum, cosmos, dahlia, marigolds, and zinnias). The majority of the Asteraceae family members are medicinal plants which have therapeutic properties.

Achillea, Carthamus, Chromolaena, Emilia, and Pluchea are among them (Achika et al., 2018).

Ethnopharmacological studies have been performed in different countries from the ancient time. The Asteraceae family of medicinal plants is used to treat wounds and cuts; they are highly efficient wound-healing agents. Information and evidence obtained from the traditional medicinal practices have the potential to open new ways for the development of novel and highly efficient therapies based on the medicinal properties of these plants (Alexander et al., 2018). Asteraceae family plants have a significant role in traditional medicinal practices. They are used to cure wounds, cuts, dermatological problems, inflammation, etc. The uses of medicinal plants from the Asteraceae family in dermatological diseases are listed in Table 6.1. Asteraceae is a widely distributed family, and most of the plants coming under this family share a similar chemical composition for the compounds: for example, a natural polysaccharide with strong prebiotic properties is found in almost all Asteraceae members. They are anti-inflammatory, antioxidant, wound healing, diuretic, and antimicrobial properties. Their potent pharmacological effects and medicinal properties, flavonoids, a s is attributed to the presence of their wide range of phytochemical constituents, including phenolic acids, polyphenols, acetylenes, and triterpenes.

All parts of the plant *Tagetes erectus* L. (Asteraceae) are of medicinal value and used to treat many diseases, including pain, inflammation, tumors, and various gastrointestinal diseases. It also has therapeutic uses in traditional medicine. Studies showed that the leaves of *T. erectus* showed antinociceptive and anti-inflammatory activities (Shinde et al., 2009).

Porophyllum ruderale (Jacq.) Cass. is a common weed widespread in subtropical America and *Conyza bonariensis* (L.) Cronq. is a prolific agricultural weed, and both of the species are used in folk medicine for treating various ailments. *Conyza bonariensis* finds popular use in the treatment of diarrhea and hemorrhoids. Species of this genus are used, among other purposes, for treating diabetes, malaria, and gastrointestinal inflammation (Souza, 2003).

Eclipta alba is an herb coming under the Asteraceae family, commonly known as Bringharaj according to the Ayurvedic system of medicine. It is used as a liver tonic to rejuvenate and for good hair. *Eclipta prostrata* is used for athlete foot eczema. It is also used for anti-venom treatment against snake bites. It is used as a scalp tonic for hair care (Bruneton et al., 2007). The leaf decoction of *Eclipta* is used by the Rakhain tribal healers of Chittagong Division, Bangladesh for treating Malaria.

Helianthus annuus Linn. belongs to the family Asteraceae, having many folkloric uses and ethnomedicinal practices in India, Pakistan, and Southeastern Nigeria. The plant is used against the conditions of inflammation, pyrexia, worm infestation, diabetes mellitus, and other stomach ailments (Samuel et al., 2019). Senecio *serratuloides* (*Asteraceae*) is a traditional herbal drug used against skin wounds in South Africa. The plant contains compounds like hepatotoxic alkaloids, which may cause the death of herbivorous animals. The plant is traditionally used to cure wounds and cuts (Gould et al., 2015).

Aspilia africana is a plant that grows in Africa (Pers.) C. D. Adams is a less toxic plant that has been used in African traditional medicine for generations to treat wounds and other related conditions, including stopping bleeding from severed arteries. They are commonly called wild sunflower, one of the highly valued wound healing plants throughout its distribution range and beyond this unique wound healing plant, commonly known as "hemorrhage plant" due to its significant ability to stop serious bleeding. *A. africana* is reported to be crucial in the treatment and management of many serious diseases and disorders, including headaches, corneal opacities, rheumatic arthritis, stomach problems, cough, gonorrhea, pains, and tuberculosis. The leaf decoction is taken as a tonic by the women immediately after delivery. *A. africana* plant is known to possess significant antimicrobial, antimalarial, and anti-inflammatory properties (Richard et al., 2019).

Elaphantopus scaber Linn. (Asteraceae) is distributed in the moist and deciduous forests of central Western Ghats region. The plant is popularly known as Elephant foot in English. In the Indian system of medicine, the medicinal attribution of the plant has been known for a long time. As per the traditional claims, the roots of this plant are used as antipyretic, cardiotonic, and diuretic. The whole plant material is macerated and applied on the surface of wounds to cure. The aqueous extract of the plant is applied externally to cure eczema and ulcers (Singh et al., 2005).

Wedelia chinensis belonging to the family Asteraceae, is often infused with hot water and is often used as herbal tea and also as an anti-inflammatory tonic or medicinal food (Lin, 2014). *Tridax procumbens* is commonly used to treat anti-inflammatory conditions and as an analgesic agent. The crude decoction prepared from leaves of *Tridax procumbens* and *Andrographis paniculata* (*Acanthaceae*) is used by some tribal healers for the cure of malaria fever.

Artemisia absinthium, commonly known as wormwood, is a perennial herb. Because of its strong odor, it is frequently used with biological pesticides. The plant showed medicinal properties like diuretic, balsamic, digestive, and depurative. It's also utilized as a supplement in the treatment of leukemia. The aerial part of the plant shows snake anti-venom activity. *Acmella oleracea* is a source plant of the compound spilanthol, which belongs to N-acrylamides, exhibits a diuretic effect, and is used in oral healthcare frequently used as a toothpaste additive. *Achillea kellalensis* flowers contain flavonoids, and their topical application can hasten wound healing. The extract of *Achillea millefolium* possesses estrogenic properties due to the presence of phytoestrogens like luteolin and apigenin. They have a stronger binding affinity to estrogen receptors (ERs) than estradiol. *Calendula officinalisis* exhibited wound-healing effects and antibacterial and antiviral properties (Rolnik et al., 2021).

Ageratum conyzoides (L.) (Asteraceae), herbal infusion of which is given for gastrointestinal ailments such as diarrhea, dysentery, and intestinal colic with flatulence. This plant is used as a worm medicine in Cameroon. The crushed fresh fruits of *Xanthium strumarium* L. are used by the coastal tribes of Odisha state, India for the treatment of filariasis (unpublished observations). The plant is used against chronic malaria by the tribes of Bannu district, Pakistan.

6.6 ETHNOMEDICINALLY IMPORTANT PLANTS FROM ASTERACEAE

TABLE 6.1 Application of Traditional Medicine

Sl. No.	Botanical Name	Local Name	Applications in Traditional Medicine
1.	*Acanthospermum hispida* DC.	Kadalmullu	Juice of leaf and young stem is used against wounds and skin scratches with 2% calcium carbonate solution.
2.	*Agreratum conyzoides* L.	Nayitulasi	Leaf juice is used on skin scars and Leprosy.
3.	*Artemisia absinthum* L.	Vilayeti	The whole plant paste is an antiseptic and used as a detergent agent for skin diseases.
4.	*Artemisia pallenswall*	Davan	Leaf juice is applied against wounds and cuts of skin.
5.	*Calendula officinalis*	Giragola	An ointment prepared from the flowers of Corolla is used on skin ailments. Leaf paste is being used on external ulcers and on open sores.

TABLE 6.1 *(Continued)*

Sl. No.	Botanical Name	Local Name	Applications in Traditional Medicine
6.	*Carthumus tinctorius* L.	Kusabhe	Leaf paste with a few drops of Pongamia seed oil is used on skin eruptions and eczema.
7.	*Chrysanthemum coronarium* L.	Jamanthi	Leaf juice is used for the burning sensation of skin diseases externally.
8.	*Chrysanthemum indicum* L.	Jamanthi	The paste prepared with flower sepals and a few drops of lemon juice is applied on scars on the facial cheeks externally to remove pimples.
9.	*Eclinops echinatus* Roxb.	Kantalu	Root paste is most useful to heal skin wounds within 3–5 days.
10.	*Eclipta alba* (L.)	Brungaraja	Leaf paste is applied twice a day and found to be most useful for eczema and ringworms.
11.	*Eclipta prostrata* L.	Kesharaja	The paste prepared from the whole plant part is useful for skin cuts and wounds.
12.	*Helianthus annus* L.	Surya kanthi	Young Leaves tender-made bandages on wound swellings, and a leaf decoction is used for washing wounds.
13.	*Tridax procumbens* L.	Chiravanaku	Leaf juice and powder are applied on cuts and injured parts of the skin.
14.	*Vernonia cinerae* (L.) Less.	Poovamkurunthal	The whole plant paste is used to cure eczema and ringworms and is also used for skin eruptions.

6.7 FLAVONOIDS IN ASTERACEAE

The medicinal plants coming under the family Asteraceae are widely used in traditional medicine and Ayurveda due to the presence of secondary metabolites such as alkaloids, flavonoids, phenolic compounds, terpenoids, and coumarins. Phenolic compounds and flavonoids are widely distributed in the plants of the Asteraceae family, and some of them have the ability to inhibit parasites. Flavonoids are classified into several subgroups, like flavone, flavonol, flavanone, dihydro-flavonol, flavan-3-ol, flavan-3,4-diol, chalcone, aurone, and anthocyanidine. Some plant flavonoids exist in bioflavonoids and glycosidic forms.

6.8 COMMON FLAVONOIDS IN ASTERACEAE FAMILY

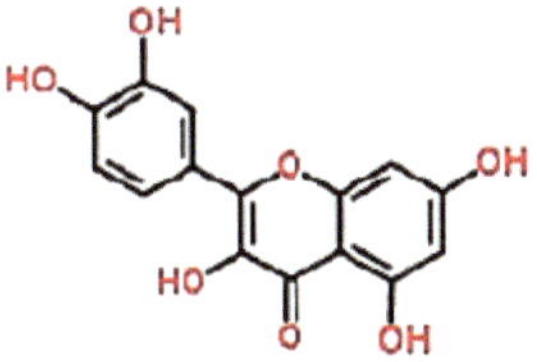

(a) quercetin

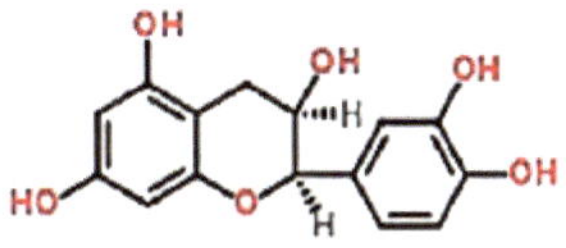

(b) (-)-epicatechin

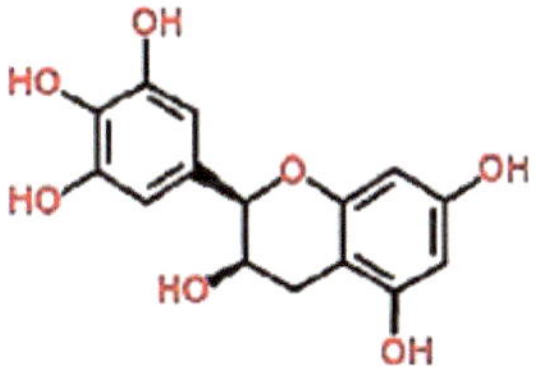

(c) (-) epigallocatechin

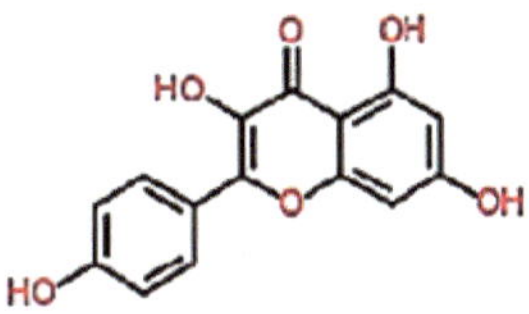

(d) Kaempferol

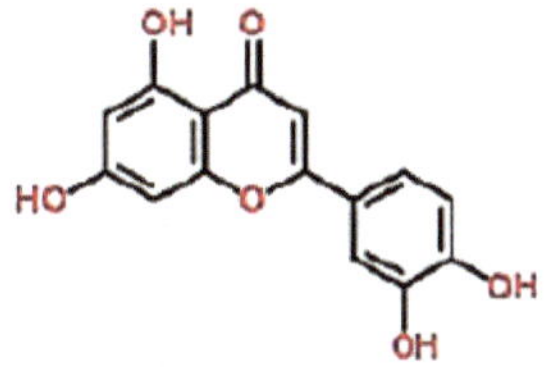

(e) Luteolin

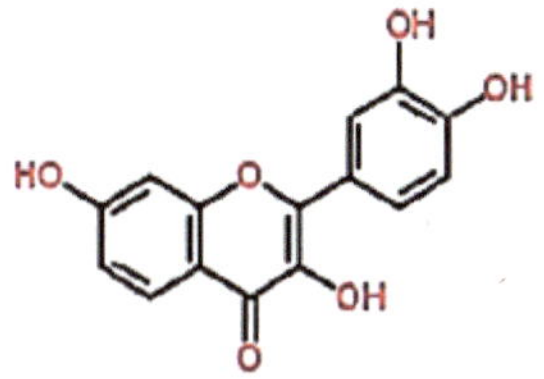

(f) Fisetin

Kristó et al. (2002) analyzed the antioxidant flavonoids by capillary electrophoresis (CE) from the plants *Solidago gigantean*, *Taraxacum officinale* Wiggers and Webers (Asteraceae). *Solidago gigantea* was investigated because it has diuretic, spasmolytic, and wound-healing effects. *Taraxacum officinale* was investigated because; it has good diuretic and choleretic activity. The study identified quercetin-3-*O*-β-rutinoside (rutin), quercetin-3-*O*-β-D-glucoside (iso-quercitrin), and the most prevalent component in *Solidago gigantea* is chlorogenic acid and apigenin-7-*O*-β-glucoside, luteolin-7-*O*-β-glucoside, and chlorogenic acid in *Taraxacum officinale* and discovered that quercetin-3-*O*-α-rhamnoside (quercitin) and quercetin-3-*O*-β-galactoside (hyperoside) were absentin *Solidago gigantea* and quercitin from *Morus nigra*. Quercetin is a flavonoid found in fruits, vegetables, and medicinal plants. It has very potent and unique biological activities that help to improve mental and physical performance and helps to reduce infection-related health problems. These effects of the compound are responsible for the potential health benefits and disease resistance.

Quercetin and isoquercetin are known for their anti-inflammatory activity. The anti-inflammatory property of quercetin and iso-quercetin investigated were reported in an experimental murine allergic asthma model in mice. Quercetin and isoquercitrin are particularly effective in eosinophilic inflammation suppressors, and they can be good against allergies, according to the findings (Rogerio et al., 2007). Quercetin also inhibits NLRP3 inflammasome formation by affecting TXNIP (Al-Saeedi et al., 2021). iNOS, cyclooxygenase-2 (COX-2), and reactive C-protein (CRP) expression levels may be regulated by kaempferol and quercetin. In a human hepatocyte-derived cell line, they can also promote alterations in the nuclear factor kappa B (NFB) pathway (Garca et al., 2007).

Sphaeranthus indicus is an aromatic, potent medicinal herb in the Asteraceae family and has been studied for its potential health-promoting properties, mainly for its anti-inflammatory activity. In traditional medicine, this plant is used to treat glandular swelling, urethral discharge, and jaundice in the neck. This plant is rich with a number of phytoconstituents such as β-sitosterol, lupeol, β-daucosterol, 7-hydroxy sitosterol, erythrodiol-3-palmitate, quercetagetin, erythrodiol, gallic acid, syringic acid, quercetin, quercetagetin-7-methyl ether, quercetagetin-7-O-glucoside and tagetone (Shetty et al., 2015). Gallic acid, quercetin, and syringic acids are phenolic compounds which are known for their significant anti-inflammatory activity. The anti-inflammatory potential of gallic acid was studied in zymosan-induced acute food pad swelling in mice. The analysis of the observation made from the study showed that the o-dihydroxy group of gallic acid is responsible for the inhibitory and anti-inflammatory activity *in vitro* (Kroes et al., 1992).

Tagetes erecta is an ornamental as well as a medicinal herb in the Asteraceae family used for digestive tract problems, including poor appetite, gas, stomach pain, intestinal worms, and dysentery. It is also used for coughs, colds, mumps, inflammations, and sore eye. It has a rich quantity of flavonoids, and its high therapeutic value is related to the high yield of flavonoids which is of effective medicinal value (Devika et al., 2014).

Sunflower, *Helianthus annuus* is the most common member of the Asteraceae family. It is grown all over the world for the seeds, which are used to obtain sunflower oil. Anti-inflammatory, antipyretic/antifebrile/febrifuge, and other therapeutic qualities are abundant in *Helianthus annuus.* Helianthus seeds and sprouts are high in flavonoids and phenolic acids, which have been found in sunflower seeds and sprouts and have been linked to medicinal activity. Sunflower seeds and sprouts have flavonoid levels of 25 and 45 mg/g quercetin, respectively. The anti-inflammatory effect of sunflower seed oil and sprouts is due to the presence of the principal kind of flavonoids.

Eclipta prostrata or (syn. E. Alba), (Asteraceae) is widely distributed in tropical and subtropical regions and is used as an anti-inflammatory, wound healing drug. Many studies showed that the stem of *E. prostrata* was effective against hyperlipidemia, hypercholesterolemia, hepatic disorders, and spleen enlargement. The stems of this plant have been used to cure anemia, asthma, and tuberculosis, while the leaves have been used to treat skin problems like inflammation and scalp infection. The quantitative and qualitative analysis of ethanolic and aqueous extracts of *E. prostrate* identified the presence of flavonoids and phenolics. The presence of flavonoids in *E. prostrata* was thought to be responsible for the anti-inflammatory and free radical scavenging properties (Wipawan et al., 2014). *Tridax procumbens* is used by the local natives against body pain, rheumatic pains of joints, and gastric problems. The plant is enriched with therapeutically important flavonoid compounds. The leaves of *T. procumbens* are highly medicinal due to the presence of bioactive molecules. They possess a high amount of flavonoids, alkaloids, hydroxyl cinnamates, tannins, and phytosterols, a moderate amount of lignans and benzoic acid derivatives, and very low carotenoid contents. Around 39 known alkaloids, 23 known flavonoids were identified in the leaves. The flavonoid content in the leaves is high (6477.706 g/kg). Around 23 known flavonoid compounds were detected in the leaves, consisting of (–)-epicatechin (12.538%), kaempferol (17.593%), (+)-catechin (7.968%), biochanin (8.202%), naringenin (6.254%), apigenin (7.875%), daidzein (6.124%), butein (4.376%), quercetin (4.418%), robinetin (3.895%), baicalein (3.047%), nobiletin (2.986%), (–)-epigallocatechin (2.806%), genistein (2.444%), (+)-gallocatechin (2.338%), ellagic acid (1.818%), luteolin (1.707%), myricetin (1.406%), baicalin (0.735%), isorhamnetin (0.498%), (–)-epigallocatechin-3-gallate (0.404%), silymarin (0.359%) and (–)-epicatechin-3-gallate (0.207%). The rich flavonoid content is responsible for the significant anti-inflammatory potential of Tridax (Catherine et al., 2008).

Chrysanthemum morifolium is an Asteraceae decorative and medicinal herb used to treat hypertension, type 2 diabetes, fever, cold, headache, dizziness, and edema. It is commonly used against dermatological problems such as skin irritations, infections, and inflammations on skin. Studies on *Chrysanthemum* varieties, such as C. *morifolium* and *C. indicum*, showed that the plants contain 21 flavones and flavonols. The flavonoids are responsible for the flower color of Chrysanthemum and their medicinal activities, including anti-inflammatory activity.

Eupatorium species are a commonly used wound-healing medicinal herb. It is widely used by the rural communities in India in wound-healing treatments. The squeezed plant extract is directly applied to the wounds and cuts for the instant stopping of blood and inflammation related to wounds. *The dihydroflavonols were obtained from chloroform extracts of Eupatorium capillifolium (Lam.) Small and Eupatorium perfoliatum L. (2R,3R)-7-methoxy-3,5,4′-trihydroxyflavanone),* (2*R*,3*R*)-3,4′-dihydroxy-5,7-dimethoxyflavanone and either (2*R*,3*R*)-3,7-dimethoxy-3,5,4′-trihydroxy- or (2*R*,3*R*)-7,4′-dimethoxy-3,5,3′-trihydroxyflavanone. The presence of flavonoids imparts the medicinal properties to the plant (Herz et al., 1972).

6.9 MECHANISM OF ACTION OF MAJOR FLAVONOIDS FROM ASTERACEAE (TABLE 6.2)

TABLE 6.2 Mechanism of Action of Major Flavonoids from Asteraceae

Sl. No.	Flavonoids	Mechanism of Action
1.	Quercetin	• Inhibition of NO production and expression of iNOS protein. • Inhibition of the activities of both cyclooxygenase and lipoxygenase.
2.	Luteolin	• Inhibiting the upregulation of THP-1 adhesion and VCAM-1 expression. • Inhibiting the activity of the NF-κB.
3.	Epicatechin	• Attenuating the activation of the NFκB signaling pathway. • Activation of tissue levels of cytokines (TNFα) and chemokines (MCP).
4.	Epigallocatechin	• Inhibit inflammation mediated by various cell types, such as vascular endothelial cells, immune cells, and fibroblasts.
5.	Kaempferol	• Inhibit LPS-induced NF-κB p65 and I-κB phosphorylation. • Inhibit cyclooxygenase enzymes and prevents the inflammatory process.

6.10 FLAVONOID COMPOUNDS IN ASTERACEAE FAMILY PLANTS

Flavonoids are said to be one of the most promising natural anti-inflammatory medication possibilities. Fruits, vegetables, and medicinal plants parts are rich in different types of bioactive flavonoid content. These secondary metabolites impart various properties of these plant colors and many nutritional as well as the therapeutic parts. The plants coming under the Asteraceae family are used in traditional medicinal systems and in home remedies, and also as anti-inflammatory agents. They are highly recommended in the treatment of dermatological diseases and also in chronic inflammations. Flavonoids are also potent antioxidants with the potential of free radical scavenging activity. Plant flavonoids show high anti-inflammatory potential both *in vivo* and *in vitro*. The molecular mechanisms of action involved in the anti-inflammatory effects of flavonoid compounds include inhibition of pro-inflammatory enzymes viz., NF-B, and activating protein-1 (AP-1) suppression and activation of phase II antioxidant detoxification enzymes, mitogen-activated protein kinase (MAPK), protein kinase C, and nuclear factor-erythroid 2-related factor 2 (Mauro et al., 2010). Through key modulation of signaling pathways, flavonoids reduce the generation of reactive oxygen species (ROS) and down-regulate many inflammatory mediators, resulting in anti-inflammatory effects. Recent studies on the biological activity of flavonoids demonstrated that flavonoids inhibited TFs or regulatory enzymes important for controlling mediators involved in inflammation. Flavonoids are good antioxidants with the ability to decrease tissue damage or fibrosis. The biological actions expressed by flavonoids reflect their diverse modes of action in inflammation (Permender et al., 2009).

Many plant flavonoids are involved in the inhibition of prostaglandin formation. Flavonoids like quercetin inhibit the COX pathway. Quercetin is one of the common flavonoids found in Asteraceae plants. It is found to be a strong inhibitor of both 5-LOX and COX-2 enzymes involved in eicosanoids production from arachidonic acid. *In vitro* investigations revealed that quercetin suppresses both cyclooxygenase and lipoxygenase activities, as well as NO generation and iNOS protein expression. Flavonoids may have an accumulative effect because of their binding to platelet membranes. Luteolin is another important flavonoid discovered in Asteraceae plants. It inhibits inflammatory response through the inhibition of the upregulation of THP-1 adhesion and VCAM-1 expression-inhibiting the activity of the NF-Kappa B. Many studies have demonstrated the antioxidant capacity of extracts (of roots, stems, bark, leaves, flowers, fruits, and seeds) *in vivo* and *in vitro*. So, the Asteraceae species are one of the highest possible natural antioxidants (Silvia et al., 2015). On human endothelial cells, hydroxy flavones and flavanols have been shown to inhibit the production of cytokine-induced ICAM-1, VCAM-1, and E-selectin. Apigenin, one of the primary flavones, inhibited adhesion protein upregulation at the transcriptional level and had a dose- and time-dependent influence on adhesion protein expression. Apigenin inhibited alpha-induced prostaglandin,

IL-1 synthesis, and TNF-alpha-induced IL-6 and IL-8 release. The studies showed that hydroxy flavones can act as inhibitors of cytokine-induced gene expression (Gerritsen et al., 1995).

The inhibitory effect on the inflammatory cells, like mast cells, appears to surpass any other clinically available compound. The ability of flavonoids to affect the activities of numerous inflammatory mediators suggests that they have the potential to influence inflammation. This could lead to the creation of new pharmacological anti-inflammatory drugs and new insights into the regulation of the inflammatory process. The flavonoids possess excellent anti-inflammatory effects and can serve as potent anti-cancer phytocompound that exerts its activity through several mechanisms of action like inactivation of carcinogens, cell cycle arrest triggering, apoptosis induction, and angiogenesis inhibition (Rashida et al., 2009). The medicinal plants coming under the Asteraceae family are promising in the treatment of acute to chronic inflammatory problems. The flavonoid content in the plants is responsible for their anti-inflammatory potential. Nowadays, there is an increasing interest in green therapy based on natural remedies for the treatment of many diseases, including chronic inflammation. Asteraceae plants could make a positive impact on human health through their antioxidant and anti-inflammatory activities.

6.11 CONCLUSION

The research on flavonoids from natural sources gains significance as flavonoids are widely distributed in medicinal plants with a wide range of therapeutic effects and are also an important component of the human diet. In the present chapter, we discussed various therapeutic effects, especially the anti-inflammatory properties, mechanism of action, and ethnopharmacological uses of flavonoid-rich plants of Asteraceae. The structural and functional relationships of the flavonoids are the epitome of major biological activities, and they act in various interrelated signaling pathways. Flavonoids of Asteraceae thus promise therapeutic options to provide a site-specific application to identify novel flavonoid-based therapies to treat inflammation by understanding the structure-activity relationship.

KEYWORDS

- **anthocyanins**
- **anti-inflammation**
- **Asteraceae**
- **chalcones**
- **flavonoids**
- **hydroxyl flavones**
- **neoflavonoids**
- **NSAIDS**
- **polyphenols**

REFERENCES

Achika, J., Arthur, D., Gerald, I., & Adedayo, A., (2014). A review of the phytoconstituents and related medicinal properties of plants in the *Asteraceae* family. *IOSR J. Appl. Chem., 7*, 1–8.

Alexsander, R., Carvalho, J., Roseana, M., Mariela, A. M., Cristiane, S. S., Figueiredo, S., Zagmignan, A., et al., (2018). Use of some *Asteraceae* plants for the treatment of wounds: From ethnopharmacological studies to scientific evidence. *Front. Pharmacol.* https://doi.org/10.3389/fphar.2018.00784.

Ali-Saeedi, B., Mohammad, R., & Mahmoudian, S., (2021). Anti-inflammatory potential of quercetin in COVID-19 treatment. *J. Inflamm., 18*(3).

Benavente-Garcia, O., & Castillo, J., (2008). Update on uses and properties of citrus flavonoids: New findings in anticancer, cardiovascular, and anti-inflammatory activity. *J. Agric. Food Chem., 56*, 6185–6205.

Bruneton, J., & Halton, C. K., (2007). *Pharmacognosy Phytochemistry Medicinal Plants* (pp. 98–112). Lavoisier Publishers.

Catherine, C., Ikewuchi, J., Ikewuchi, C., Mercy, O., Ifeanacho, De Vries, J. H., Hollman, P. C., et al., (1998). Plasma concentrations and urinary excretion of the antioxidant flavonols quercetin and kaempferol as biomarkers for dietary intake. *Am. J. Clin. Nutr., 68*, 60–65.

Choy, K. W., Murugan, D., Leong, J., Xin, F., Abas, R., Alias, A., & Mustafa, M. R., (2019). Flavonoids as natural anti-inflammatory agents targeting nuclear factor-kappa B (NFκB) signaling in cardiovascular diseases: A mini-review. *Front. Pharmacol.*

Devika, R., & Koilpillai, J., (2014). *In vitro* quantification study of flavonoids from *Tagetes erect. Asian J. Biotech., 6*, 21–24.

García-Mediavilla, M. V., Crespo, I., Collado, P. S., Esteller, A., Sánchez-Campos, S., Tuñón, M. J., & González-Gallego, J., (2007). The anti-inflammatory flavones quercetin and kaempferol cause inhibition of inducible nitric oxide synthase, cyclooxygenase-2, and reactive C-protein, and down-regulation of the nuclear factor kappa B pathway in Chang liver cells. *Eur. J. Pharmacol.*, *557*, 221–229.

Gerritsen, m. e., carley, w. W., Ranges, G. E., Shen, C. P., Phan, S. A., Ligon, G. F., & Perry, C. A., (1995). Flavonoids inhibit cytokine-induced endothelial cell adhesion protein gene expression. *Am. J. Pathol.*, *147*, 278–292.

Gould, A. N., Penny, C. B., Patel, C. C., & Candy, G. P., (2015). Enhanced cutaneous wound healing by *Senecio serratuloides* (*Asteraceae/Compositae*) in a pig model. *South African J. Bot.*, *100*, 63–68.

Guo, S., Ge, Y., & Jom, K. N., (2017). A review of phytochemistry, metabolite changes, and medicinal uses of the common sunflower seed and sprouts (*Helianthus annuus* L.). *Chem. Central J.*, *11*(95).

Havsteen, B., (2002). The biochemistry and medical significance of the flavonoids. *Pharmacol. Ther.*, *96*, 67–202.

Herz, W., Bhat, G. S., & Srinivasan, A., (1972). Dihydroflavonols and other flavonoids of *Eupatorium* species. *Phytochem.*, *11*(9), 2859–2863.

Keservani, R. K., & Sharma, A. K., (2014). Flavonoids: Emerging trends and potential health benefits. *Journal of Chinese Pharmaceutical Sciences*, *23*(12), 815.

Keservani, R. K., Kesharwani, R. K., Sharma, A. K., Vyas, N., & Chadoker, A., (2010b). Nutritional supplements: An overview. *International Journal of Current Pharmaceutical Review and Research, 1*(1), 59–75.

Keservani, R. K., Kesharwani, R. K., Vyas, N., Jain, S., Raghuvanshi, R., & Sharma, A. K., (2010a). Nutraceutical and functional food as future food: A review. *Der Pharmacia Lettre*, *2*(1), 106–116.

Keservani, R. K., Sharma, A. K., & Kesharwani, R. K., (2016a). Nutraceutical and functional foods for cardiovascular health. In: *Food Process Engineering* (pp. 257–278, 291–312). Apple Academic Press. ISBN: 9781771884020.

Keservani, R. K., Sharma, A. K., & Kesharwani, R. K. (2020). *Nutraceuticals and Dietary Supplements: Applications in Health Improvement and Disease Management*. CRC Press. ISBN: 9781771888738.

Kristó, S. T., Ganzler, K., Apáti, P., Szőke, É., & Kéry, Á., (2002). Analysis of antioxidant flavonoids from *Asteraceae* and *Moraceae* plants by capillary electrophoresis. *Chromatographia.*, *56*, S121–S126.

Kroes, B. H., Van, D. B. A. J., Quarles, V. U. H. C., Van, D., H., & Labadie, R. P., (1992). Anti-inflammatory activity of gallic acid. *Planta Med.*, *58*(6), 499–504.

Mauro, S., Ilaria, P., & Anna, R., (**2010**). *Flavonoids as Anti-inflammatory Agents*. Published online by Cambridge University Press.

Panche, N., Diwan, A. D., & Chandra, S. R. (2016). Flavonoids: An overview. *J. Nutritional Sci.*, *5*, 1–15.

Permender, R., Hema, C., Rathee, S., Rathee, D., Kumar, V., & Kohli, K., (2009). Mechanism of action of flavonoids as anti-inflammatory agents: A review. *Inflammation & Allergy – Drug Targets*, *8*, 229–235.

Rashida, G., Raina, B., & Zafar, K. (2019). Potential role of flavonoids in treating chronic inflammatory diseases with a special focus on the anti-inflammatory activity of apigenin. *Antioxid.*, *8*. doi: 10.3390.

Richard, K., Motlalepula, G. M., & Kang, Y., (2019). The wound healing potential of *Aspilia africana* (Pers.) C.D. Adams (*Asteraceae*). *Evidence-Based Compl. Alter. Med.*

Rogerio, A. P., Kanashiro, A., Fontanari, C., Da Silva, E. V. G., Lucisano-Valim, Y. M., Soares, E. G., & Faccioli, L. H., (2007). Anti-inflammatory activity of quercetin and isoquercitrin in experimental murine allergic asthma. *Inflamm. Res.*, *56*(10), 402–408.

Rolnik, A., & Olas, B., (2021). The plants of the *Asteraceae* Family as agents in the protection of human health. *Int. J. Mol. Sci.*

Shetty, L. J., Sakr, F. M., Al-Obaidy, K., Patel, M. J., & Shareef, H. (2015). A brief review on medicinal plant *Tagetes erecta* Linn A. *J. Applied Pharmaceut. Sci.*, *5*, 91–95.

Shinde, N. V., Kanase, K. G., Shilimkar, V. C., Undale, V. R., & Bhosale, A. V., (2009). Antinociceptive and anti-inflammatory effects of solvent extracts of *Tagetes erectus* Linn *(Asteraceae)*. *Tropical J. Pharmaceut. Res.*, *8*(4), 325–329.

Sílvia, M. F., João, C. M., Barreira, M., & Beatriz, P. P., (2015). *Asteraceae* species with most prominent bioactivity and their potential applications: A review. *Indus. Crops Produ.*, *76*, 604–615.

Singh, S. D. J., Krishna, V., Mankani, K. L., Manjunatha, B. K., Vidya, S. M., & Manohara, Y. N., (2005). The wound-healing activity of the leaf extract and deoxyelephatopin isolated from *Elaphantopus scaber* Linn. *Indian J. Pharmacol.*

Walker, E., Pacold, M., & Perisic, O., (2000). Structural determinations of phosphoinositide 3-kinase inhibition by wortmannin, LY294002, quercetin, myricetin, and staurosporine. *Mol. Cell.*, *6*, 909–919.

Wipawan, P., Sunee, C., & Yingmanee, T., (2014). Antioxidant activity, phenolic compound content, and phytochemical constituents of *Eclipta prostrate* (Linn.) Linn. *Chiang Mai J. Sci.*, *41*(3), 568–576.

CHAPTER 7

BIOACTIVE FLAVONOIDS FROM NATURAL SOURCES: POTENTIAL IMMUNE-BOOSTERS

S. R. SUJA,[1] N. M. KRISHNAKUMAR,[2] B. S. BIJUKUMAR,[3] and R. PRAKASHKUMAR[1]

[1]*Ethnomedicine and Ethnopharmacology Division, KSCSTE–Jawaharlal Nehru Tropical Botanic Garden and Research Institute, Palode, Thiruvananthapuram, Kerala, India*

[2]*Department of Biosciences, Rajagiri College of Social Sciences, Kalamassery, Kochi, Ernakulam, Kerala, India*

[3]*Post-Graduate, Department of Zoology and Research Center, Mahatma Gandhi College, Thiruvananthapuram, Kerala, India*

ABSTRACT

Recently there has been an upsurge of interest in the therapeutic potential of medicinal plants, which might be due to their phenolic compounds, specifically flavonoids. Flavonoids are a group of secondary metabolites having different phenolic structures and are found in vegetables, fruits, stems, flowers, bark, grains, and roots of plants. These natural compounds exhibit various biological activities like antioxidant, anti-mitogenic, anti-inflammatory, anticancer, and immunomodulatory effects complied with their capacity to modulate cellular enzyme functions. Among dietary factors, flavonoids have great potential as diet-derived immune-modulatory chemopreventive agents that might be of importance to several cancers. They regulate immunity by interfering with the regulation of immune

Flavonoids as Nutraceuticals. Rajesh K. Kesharwani, Deepika Saini, Raj K. Keservani, and Anil Kumar Sharma (Eds.)

cells, synthesis, and secretion of various proinflammatory cytokines and gene expression. Flavonoids and their derivatives modulate various transcriptional factors, differentiation, proliferation, and activation of immune cells. The versatile health benefits of flavonoids in various epidemiologic studies will provide newer insights and will certainly lead to a new era of flavonoid-based pharmaceutical agents for the treatment of many infectious and degenerative diseases. The present review focuses on the immuno-enhancing potential of various bioactive flavonoids isolated from natural sources and their mechanism of action.

7.1 INTRODUCTION

Flavonoids are an important group of naturally occurring compounds in plants belonging to a class known as secondary metabolites with polyphenolic structure (Keservani & Sharma, 2014; Keservani et al., 2010a). They play an important role in plant growth, reproduction, microbial infections, and mechanical damage. These compounds are derivatives of 2-phenylbenzopyran or 3-phenylbenzopyran, and they are present in vegetables, fruits, and some beverages with various health-promoting properties and a key component in various medicinal, pharmaceutical, nutraceuticals, and cosmeceuticals (Panche et al., 2016). Bioactive flavonoids are those which are extracted from dietary sources possessing a particular biological activity. These metabolites, produced by the combined biosynthesis of the Shikimic acid and acetic acid malonate pathway, accumulated in plant cell vacuole (Castellano et al., 2013). Flavonoids exhibited various biological activities such as antioxidant, immunomodulatory, anticancer, and anti-inflammatory effects interacting with different cellular enzymes. They are low-molecular-weight phenolic compounds based on a 15-carbon skeleton and have structural diversity that arises from methoxylation, hydroxylation, and glycosylation patterns of ring substitution (Amic et al., 2007).

Various *in vitro* and *in vivo* studies have revealed that flavonoids possess immunomodulatory effects, which means both immunostimulatory and immunosuppressive effects. It has been reported that activated immune cells such as macrophages, mast cells, T and B lymphocytes, eosinophils, neutrophils, and basophils are susceptible to the modulatory activity of flavonoids. These immune cells are influenced by particular bioactive flavonoids (Middleton, 1998).

7.2 CLASSIFICATION OF FLAVONOIDS

The classification of flavonoids is based on the oxidation state of the central carbon atom with a general backbone of C_6-C_3-C_6. There are several subgroups of flavonoids, such as flavones, flavonols, isoflavones, and chalcones. The major sources of flavonoids are dietary materials like apples, onions, leafy vegetables, cherries, citrus, soybean, berries, tea, etc. (Dewick, 2001). Flavonoids can be classified based on their structural properties (Keservani & Sharma, 2014). The basic structure of flavonoid is the flavan nucleus, which contains two benzene rings and an oxygen-containing pyran ring, with different oxidation levels in the carbon ring of the basic 4-isoflavonoid nucleus. The individual compounds differed in the substitution patterns of rings A and B. They can be classified into eight major groups: flavones, flavonols, isoflavones, flavan-3-ols, flavanonols, anthocyanidins, chalcones, and flavanones (Panche et al., 2016). Catechin, epicatechin, gallocatechin, and epicatechin-3-gallate are examples of flavans. Dihydrokaempferol and taxifolin are flavanols. Examples of flavanones are hesperetin, naringenin, homoeriodictyol, and eriodictyol. Luteolin, apegenin, quercetin, myricetin, kaempferol, and furanoflavanol are examples of anthoxanthins. The compounds such as cyanidin, malvidin, delphinidin, peonidin, and petunidin are anthocyanidins (Keservani et al., 2010b, 2020).

7.2.1 *FLAVONES*

Flavones are compounds having double bonds between C-2 and C-3 and a ketone in position 4 of the carbon ring (Chirumbolo, 2010). The differences and various changes in the simple structure among the compounds are the primary cause for the significant consequences in the pharmacological and therapeutic importance of the compounds. They are commonly present in fruits and leaves of chrysin, apigenin, wogonin, luteolin, baicalin, and tangeritin (Nile et al., 2018). Apigenin is present in chamomile, parsley, artichokes, celery, mint, *Ginkgo biloba* and oregano (Shankar et al., 2017). The flavone Chrysin or 5,7-dihydroxyflavone is present in various extracts, such as propolis, honey, and blue passion flowers, with several beneficial effects (Wang et al., 2018). Wogonin (5,7-dihydroxy-8-methoxyflavone), a naturally occurring flavonoid isolated from the root extract of *Scutellaria baicalensis* has been conventionally used in the treatment of inflammatory diseases (Khan et al., 2017) The flavone tangeritin is the compound found in tangerine and other citrus fruits having anti-tumor antioxidant, cytostatic, and anti-diabetic activities.

Apigenin

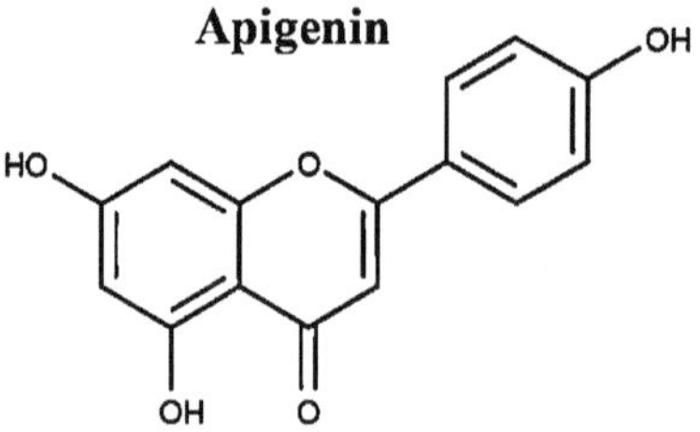

Luteolin

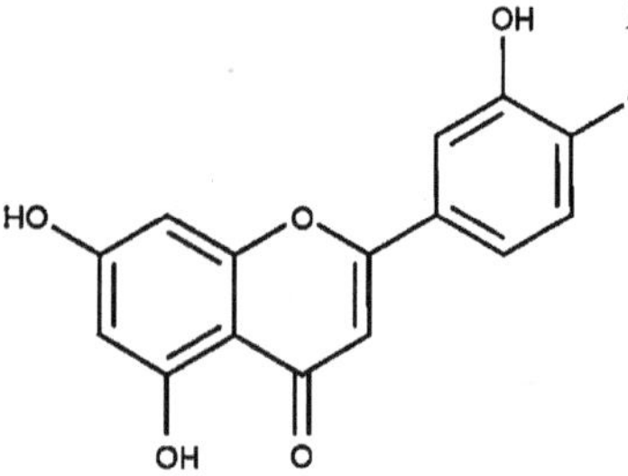

7.2.2 *FLAVONOLS*

Flavonols have the 3-hydroxy flavone backbone and hydroxyl group in the C-3 position. Quercetin, kaempferol, fisetin, and myricetin are the most prevalent plant flavonols in fruits, vegetables, and herbs. Flavonols are the building blocks of proanthocyanins (Panche et al., 2016).

Kaempferol

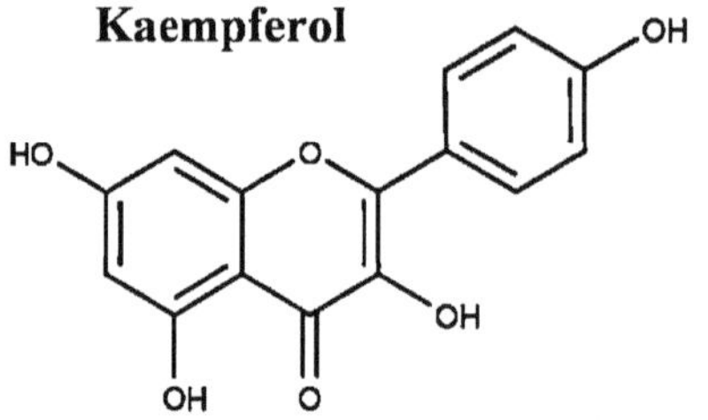

Quercetin

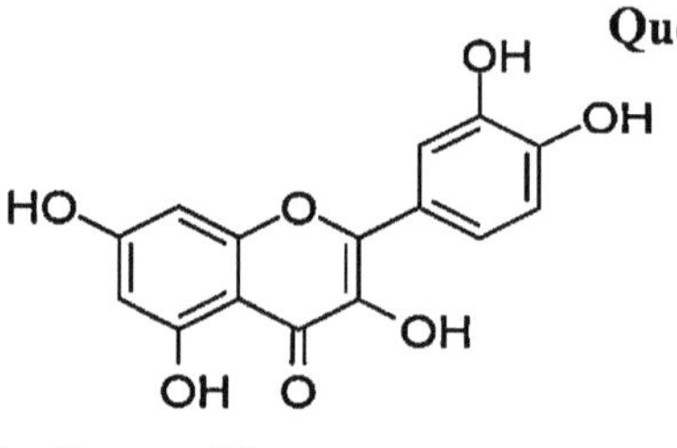

7.2.3 Isoflavonoids

The flavones and isoflavones are isomers with a heterocyclic ring attached in the C-2 and C-3 positions. The principal source of isoflavonoids like daidzein and genistein includes the plants of *Leguminosae* family (Tandon & Das, 2018).

Daidzein

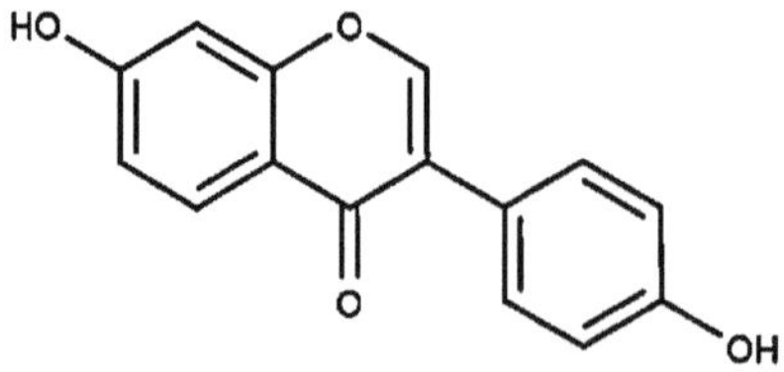

Genistein

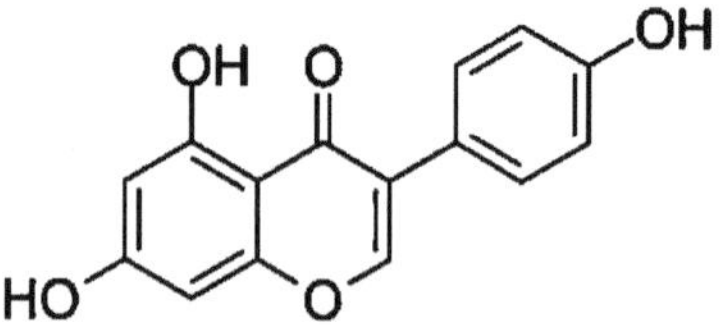

7.2.4 *FLAVANOLS, FLAVAN-3-OLS, OR CATECHINS*

Flavanols and flavan-3-ols with a hydroxyl group bound to the third position of the carbon ring, and there is no double bond between the second and third positions (Panche et al., 2016). The polyphenolic flavonoids like epicatechin, epigallocatechin, epicatechin-3-gallate, and epigallocatechin-3-gallate are present in Green tea made from the unfermented leaves of *Camellia sinensis* (Chakrawarti et al., 2016).

Epicatechin

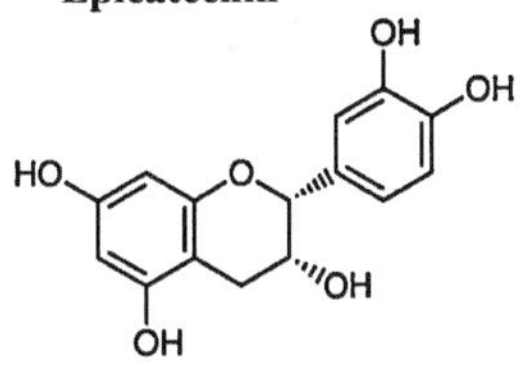

Epigallocatechin

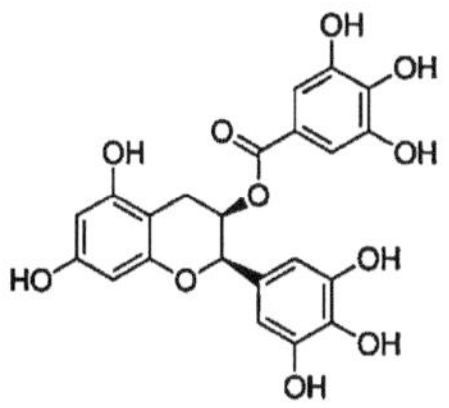

Epicatechin-3-gallate

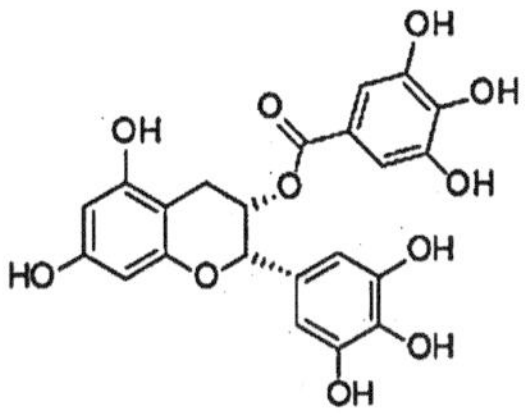

7.2.5 *FLAVANONOLS*

In flavonols, the hydroxyl group is situated in the third carbon, and the oxygen atom is attached to the fourth carbon of the ring (Kumar & Pandey, 2013). The phytocompounds silibinin, astilbin, and taxifolin are examples of flavanonols (Hua et al., 2018).

Astilbin

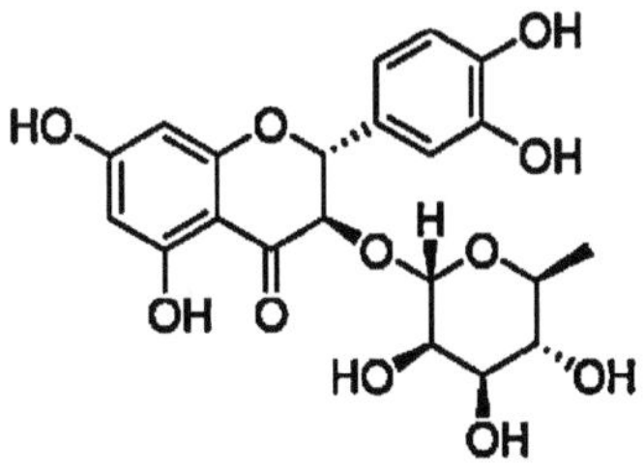

Silibinin

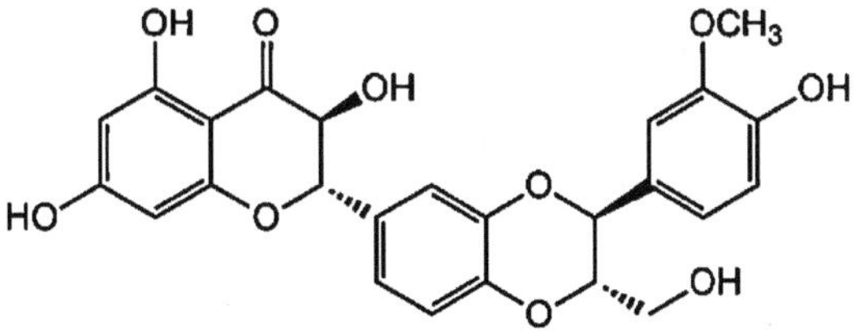

7.2.6 *ANTHOCYANINS*

Anthocyanins are the pigments responsible for most of the blue, purple, and red-colored flowers, fruits, and vegetables (D'Archivio et al., 2007).

Anthocyanidins are water-soluble aglycon forms of anthocyanins. Proanthocyanidins (PACs) are mainly present in grapes. There are multiple forms of flavonoids, dimers, or oligomers of catechin and epicatechin and their gallic acid esters. A condensed tannin or polymeric flavanol is known as PAC. PAC-type compounds are namely malvidin, pelargonidin, peonidin, cyaniding, delphinidin, and petunidin (Stalikas et al., 2007).

Cyanidin

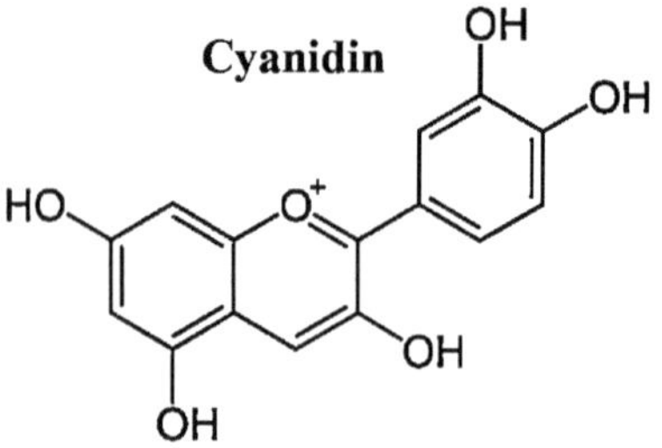

Delphinidin

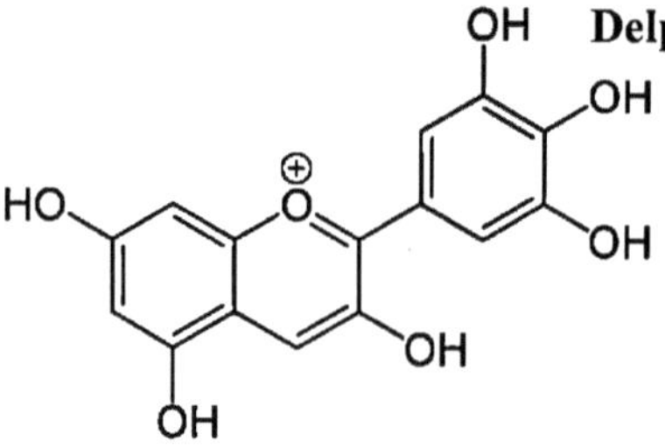

7.2.7 *CHALCONES*

Chalcones are important compounds in the flavonoid biosynthesis because it represents the precursor for a broad range of flavonoids (Raffa et al., 2017). Chalcones consist of phloridzin, arbutin, phloretin, and chalconaringenin. Major examples of Chalcones occur in significant amounts in tomatoes, pears, strawberries, bearberries, and certain wheat products. As previously mentioned, chalcones have a broad spectrum of biological properties, such as antioxidant, antimicrobial, and anti-inflammatory activities. Isoliquiritigenin, also known as 2′,4,4′-trihydroxychalcone or 6′-deoxychalcone, is a member of the class of compounds known as 2′-hydroxychalcones (Sahu et al., 2012).

Phloretin

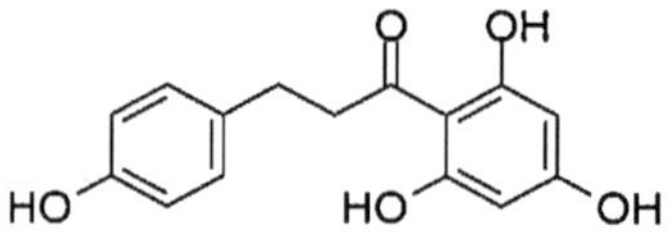

Chalconaringenin

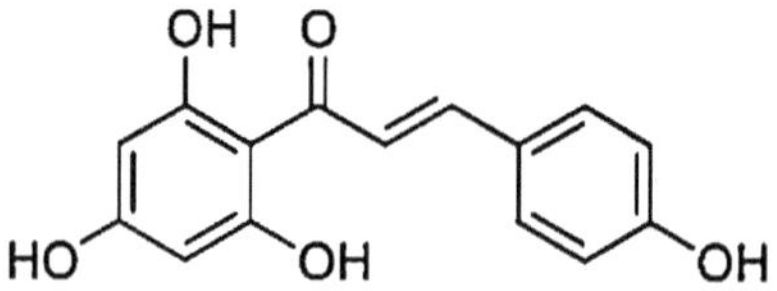

7.2.8 FLAVANONES

Flavanones, also known as dihydroxyflavones, have a bitter taste. The juice and peel of citrus fruits contain flavanones. Examples of flavanones are naringin, naringenin, and hesperidin. They differ from flavone in structure doesn't have instaurations between C-2 and C3. These compounds exhibited pharmacological properties such as anti-inflammatory, antioxidant, anti-hypercholesterolemic effects (Panche et al., 2016).

Naringin

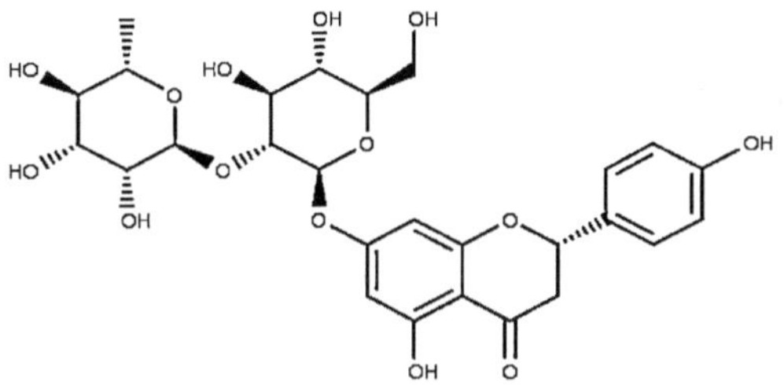

Naringenin

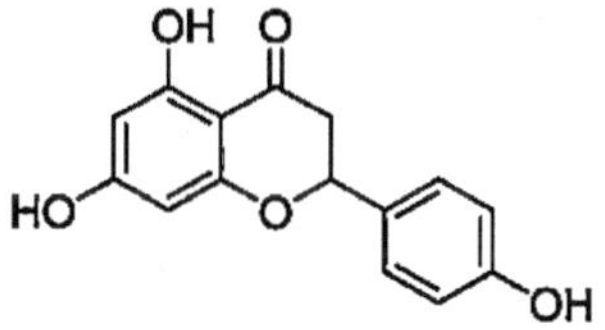

7.2.9 BIS AND BIOFLAVONOIDS

Bis and Bioflavonoids joined in a symmetrical or unsymmetrical manner, where the moieties are linked by a C-C or C-O-C bond (Mercader & Pomilio, 2013). When the units are doubled, they are called bis-flavonoids, but when the structural units are not the same, they are called biflavonoids. The common examples are amentoflavone (dimer of two apigenins), ginkgetin, moreloflavone, isoginkgetine, hinokiflavone, robustaflavone, and ochnaflavone. They have a basic structural framework of 4,5,7-trihydroxyflavon flavonoid (Silva et al., 2017).

Amentoflavone

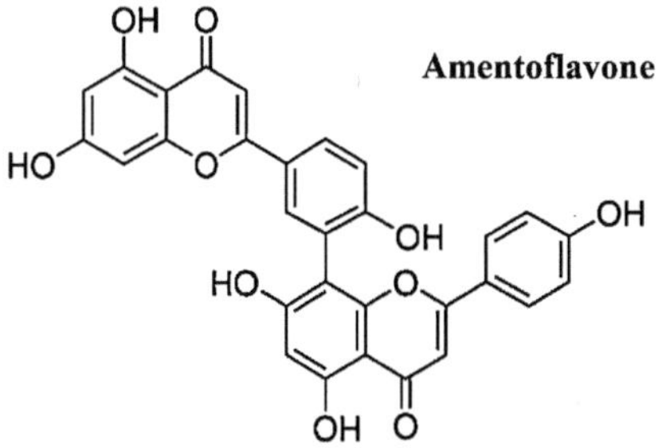

Ginkgetin

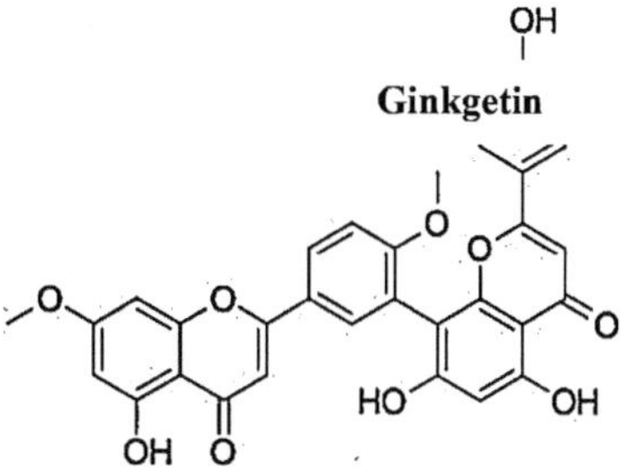

7.3 IMMUNOSTIMULATORY EFFECTS OF FLAVONOIDS OF PLANT ORIGIN

Immune boosters enhance the nonspecific humoral and cellular immune responses. The immunomodulatory activity is related to the structural characterization of phytosterols. The sugar moiety of the phytophagous forms the key point for different types of immune responses (Cherng et al., 2008). The size and molecular weights of polyphenols influence the immunomodulatory responses, particularly of T-cell cytokine secretion. The polyphenols interact with immune system components and cells like lymphocytes, dendritic cells (DCs), neutrophils, and macrophages. The stimulation of the immune system is regulated by T-lymphocytes by TH-1 ($CD4^+$), causing an increase in the protective effects producing activated macrophages and cytokines. Macrophages play an important role and act as antigen-presenting cells in the generation of immune response and trigger phagocytosis.

Masad et al. (2021) reported that quercetin, one of the flavonoid compounds present in honey, differentially activated Th1 cells providing a crucial mechanism for antitumor activity. Sutoyoto et al. (2018) reported the immunostimulant activity of kaempferol isolated from acetone extract of *Pityrogramma calomelanos* (Silver fern) in a carbon clearance assay. The phagocytic index of the compound-treated group was high compared to the normal control group. Kaempferol increased the percentage of $CD8^+$ cells. There was an increase in B-cells and lymphocytic T-cell proliferation, the release of cytokines such as IL-4, TNF-α, and IFN-γ. The compound stimulated the phagocytic property of macrophages, lysosomal enzyme activity, and NO release by macrophages. Talmale et al. (2014) studied the immunostimulatory effect of flavonoids extracted from *Zizyphus mauritiana* stem bark. It stimulated lysosomal degranulation, phagocytic index, and proliferation of lymphocytic splenocytes. Abdelsalam et al. (2017) evaluated the immune stimulant effect of flavonoids isolated *Alcea rosea* on hepatocellular carcinoma HepG2 cell line.

Flavonoids exhibited immune booster activity by stimulating mononuclear cells to secrete TNF-α, IL-1β, and IFN-γ. The flavonoids consisting of phyllantidine, quercetin, isoquercetin, and astragaline isolated from the roots and leaves of *Phyllanthus niruri* water extract increased peripheral blood mononuclear cell (PBMC) proliferation, increased the release of NO from macrophages, improved phagocytic index and increased antibody response (Muthulakshmi et al., 2016). Flavonoids such as quercetin, guajaverine, and gallocatechin extracted from the leaves of *Psidium guajava* stimulated humoral and cellular immune response in mice. It showed increased expression of the IL-8 gene, indicating immune-boosting action (Laily et al., 2015). Nair et al. (2002) carried out the evaluation of flavonoid quercetin in regulating Th-1 and Th-2 cytokine gene expression. The compound significantly induced the gene expression and production of Th-1 derived IFN-γ by normal PBMC. Quercetin treatment increased the phenotypic expression of IFN-γ cells suggesting the immunopotentiating effects mediated through the induction of Th-1 derived IFN-γ.

Jung et al. (2012) reported the efficacy of quercetin on impaired immune function in irradiation-induced inflammatory mice. Irradiation exposed animals showed diminished splenocyte proliferation. The treatment with quercetin (10 and 40 mg/kg) significantly enhanced splenocyte proliferation after 30 days and exhibited a protective effect against irradiation-induced inflammation. Quercetin exhibited *in vivo* immunostimulatory activity in ovalbumin-immunized Balb/c mice with increased ovalbumin-specific serum IgG antibody titers. The compound stimulated Th-2 immune response and increased $CD11c^+$ dendritic cell infiltration in the peritoneal cavity. The expression of GATA-3, Tbx-21, and Oct-2 proteins enhanced in mice splenocytes treated with quercetin, suggesting the immunostimulatory and adjuvant effect of quercetin (Singh et al., 2017). The *in vivo* adjuvant effect of kaempferol was studied by Singh et al. (2018) with ovalbumin antigen in Balb/c mice. The kaempferol-treated group showed a significant increase in IgG, IgG1, and IgG2a antibody titers. The increased expression of Tbx21 and GATA-3 transcription factors (TFs) in splenocytes supported the stimulation of Th1 and Th2 immune response in the treated group. Kaempferol treatment also increased the infiltration of $CD11c^+$, $MHCII^+$ DCs suggesting the adjuvant effect and immunostimulatory activity.

Oral administration (12.5–100 mg/kg) of the flavonoid daidzein stimulated IgM and IgG titer *in vivo* and dose-dependently increased delayed-type hypersensitivity reaction implicating the immune stimulatory effect of the compound in Balb/c mice (Dhiman et al., 2013). Epigallocatechin-3-gallate, an active ingredient of green tea, was found to induce T-cell activation, proliferation, differentiation, and production of cytokines, and it prevented and ameliorated T-cell-mediated autoimmune diseases (Pae & Wu, 2013). Epicatechin isolated from the chloroform extract of *Rhododendron spiciferum* significantly stimulated splenocyte proliferation when treated with concanavalin-A. It enhanced the cytotoxicity of NK cells and the phagocytic function of macrophages. The compound significantly increased Th1 cytokines (Liu et al., 2015).

Some of the flavonoids, like ugonin U, a flavonoid isolated from *Helminthostachys zeylanica*, induced activation of NLRP3 (NOD-, LRR-, and pyrin domain-containing protein-3) inflammasome. It also induced caspase-1 activation and IL-1β secretion in human monocytes pre-stimulated with lipopolysaccharide (LPS) (Chen et al., 2017). Flavonoids chrysin and hesperetin significantly increased the activity of NK cells in K562 cells (Sassi et al., 2018). Luteolin (1.5 and 10 μM) and apigenin (5, 10, and 25 μM) increased NK cell activity in mouse splenocytes. Naringenin (25 μM) stimulated the expression of ULBP-1, ULBP-2, and MIC-A/B in Raji cells (Kilani-Jaziri et al., 2016). The treatment of flavonoids on various *in vivo* models of tumors in mice stimulated NK cell activity and increased the survival time of animals (Kilani-Jaziri et al., 2016). Quercetin significantly increased the NK cell activity in a mouse leukemia model *in vivo* (Bae et al., 2010). Genistein increased the activity of cytotoxic T-cells and NK cells, secretion of IFN-γ, and activation of STAT-1 and STAT-4 (Guo et al., 2007). PAC significantly elevated the T-cell differentiation to $CD8^+$ cells in mice splenic cells (Guo et al., 2007). Flavonoids apigenin, icariin, and gallotannin differentially stimulated CD8 cells along with decreasing tumor growth by apoptosis. The flavonoids quercetin and bilclain showed modulation of B-cell proliferation and antibody production (Zhang et al., 2018).

Curcumin can act as a potent therapeutic adjuvant for DCs-related acute and chronic disorders, and it is very efficient at antigen capture via mannose receptor-mediated endocytosis (Kim et al., 2005). Activation of cell-mediated immune response is one of the most studied properties of plant flavonoids and polyphenolic compounds. A polyphenol oenothein B extracted from *Epilobium angustifolium* activated myeloid cells and stimulated innate lymphocytes and natural killer cells resulting in the increased expression of CD25 and CD69 (Ramstead et al., 2012). It enhanced IFN-γ production by human and bovine T-cells and NK cells (Ramstead et al., 2015). The administration of biflavone di-C-glucoside, 6,6″-di-C-beta-D-glucopyranoside-methylene-(8,8″)-biapigenin (0.25 mg/kg) resulted in the stimulation of both humoral and cell-mediated zero response in chicks (Abd-Alla et al., 2009). Flavonoid daidzein potentiated the proliferation of splenocyte cultures activated with LPS or concanavalin A and the secretion of IL-2 and IL-3 (Wang et al., 1997). Genistein increased the number of splenic B-cells, T-cells, cytotoxic T-cells, and macrophages (Guo et al., 2002). Epigallocatechin gallate, gallic acid, and tannic acid enhanced the mitogenic activity of B-lymphocytes (Hu et al., 1992).

The tea polyphenols stimulated the proliferation and activation of T-lymphocytes, indicated by the increase in the $CD4^+/CD8^+$ ratio (Deng et al., 2010). Green tea and its bioactive constituent epigallocatechin 3-gallate improved the symptoms and reduced the pathology in some experimental animal models of autoimmune disorders. The treated group showed a significant increase in the number and frequencies of regulatory T-cells in the spleen and lymph nodes (Pae et al., 2012). The activation of mononuclear cells and increase in the phagocytic response are induced by some flavonoids and phenolic compounds such as daidzein (20 and 40 mg/kg) (Zhang et al., 1997), procyanidin A_1 (Liu et al., 2010), procyanidin C_1 (62.5 μg/mL), procyanidin dimer B_2 (Sung et al., 2013), kaempferitrin (25 μM) (Del Carmen Juarez-Vazquez et al., 2013), oenothein B (Schepetkin et al., 2009), geraniin and isocorilagin (Liu et al., 2012) via influencing MAPK (Mitogen Activated Protein Kinase) and NF-κB signaling pathways. The humoral immune response can be quantified by the serum levels of specific immunoglobulins by antibody titer assays. Green tea, a rich source of polyphenols, exhibited a stimulatory effect on IgM and IgG-mediated humoral immune response (Khan et al., 2016). Polyphenol-rich pomegranate extract increased IgG response (Oliveira et al., 2010). Immunoglobulin synthesis was induced by cyanidol (Daniel et al., 1986) and daidzein (Zhang et al., 1997).

The flavonoid compounds bilclain, astilbin, and quercetin showed an impact on B-lymphocytes, their modulation, proliferation, and antibody production. In B-cell leukemia, wogonin (Lin et al., 2013), chrysin (Xu et al., 2019), quercetin (Spagnuolo et al., 2012), and epigallocatechin gallate (EGCG) (Huang et al., 2013) decreased cell burden and induced differentiation on B-cell leukemia (Kawai et al., 2011).

Flavonoids are one of the natural sources of immune boosters and anti-retroviral for AIDS therapy because of their potent anti-HIV activity and low toxicity (Saravanan et al., 2015). Epigallocatechin inhibited protease kinetics and post-adsorption entry of the virus (Yamaguchi et al., 2002). Robustaflavone and hinokiflavone isolated from *Garcinia multiflora* and *Rhus succedanea* showed anti-HIV activity against polymerase HIV1 reverse transcriptase enzyme. Quercetin 3-o-(2-galloyl) a-L-arabinopyranose, wikstrol B, pterocarpans, xanthohumol, 2-methoxy-3-methyl-4,6-dihydroxy-5-(3′-hydroxy) cin-namoyl benzaldehyde, lawinal, luteolin, hydroxyl panduratin A, taxifolin, aromadendrin, apigenin 7-0-beta-D-(4′-caffeoyl) glucuronide, formosanatin C, 5-hydroxy-7-methoxy flavone, baicalin, chrysin, and kaempferol are some of the flavonoids exhibiting potent activity against human immunodeficiency virus (HIV) (Saravanan et al., 2015). Toll-like receptors are important in the development of innate immune response, and their activation stimulates the secretion of inflammatory cytokines. Modulation of toll-like receptors is of pharmaceutical importance. Quercetin can effectively act as a toll-like receptor modulator and is used for the treatment of HIV (Chin et al., 2018). The flavonoid compound reduced viral replication and enhanced the immune response (Thompson, 2016).

Nanotechnology use in medicine is spreading rapidly in the drug development field of developed countries. Various nanomedicines with different therapeutic effects are developed by pharmaceutical researchers by using the active principle from the natural products and synthetic materials. There is a rapid growth in the field of herbal nanomedicine development due to the increased use of natural products for the treatment of various ailments. The factors such as the ability to cross the cell membranes like the blood-brain barrier, gain access to cells, and translocate around the body through blood and lymph make nanomedicine a promising tool for drug delivery to the site of action (Rao et al., 2009).

The nanochemical technology uses various nanocomposites and nanoparticles for biosensing flavonoids and their delivery as drugs. The flavonoid quercetin and its derivatives, such as myricetin and rutin, are used for sensing and delivery. The flavonoid-based hybrid nanocomposites exhibited significant immunomodulatory activity (Parhi et al., 2020). The nanoparticle-encapsulated drugs exhibited a higher rate of proliferation of B-lymphocytes, T-lymphocytes, and NK cells. Immunotherapeutic nanoparticles become an effective cancer treatment therapy. Gold nanoparticle-immobilized (GNP) silymarin and GNP-luteolin had antitumor effects (Dykman & Khlebtsov, 2019). Some flavonoids isolated from medicinal plants, such as quercetin and rutin, are used as reducing or capping agents in the green synthesis of nanoparticles (Selvakesavan & Franklin, 2021). The polyphenols extracted from *Phyllanthus niruri* were formulated into polymeric nanoparticles and evaluated for immunomodulatory activity. The results of the study showed that the nanoparticles were significantly effective compared to the unformulated extract, as evidenced by the increased phagocytic index *in vitro* (Pratiwi et al., 2019).

7.4 FLAVONOIDS AS CHEMOPREVENTIVE AGENTS IN CANCER TARGETED THERAPY

Natural products rich in flavonoids from fruits and other sources are potent chemo-preventive agents in cancer-targeted therapy. Naringin, naringenin, hesperetin, hesperidin, nobiletin, and tangerine are some examples, and these compounds possess inhibition of the growth of certain types of cancers by various mechanisms such as cell cycle modulation, apoptosis induction, anti-angiogenic property, etc. The antioxidant effects of some Citrus flavonoids like hesperidin, tangerine, and nobiletin have been reported. These compounds have the ability to interact with and stabilize free radicals and protect the cells and DNA from damage (Yi et al., 2008). Flavonoids also suppress the process of carcinogenesis by various mechanisms. Flavonoids such as nobiletin blocked carcinogenesis by suppressing c-Myc expression. Flavonoids like tangerine showed the lowest IC_{50} value in COLO205 human colon carcinoma cells and HL-60 human promyelocytic leukemia cells by triggering apoptosis (Pan et al., 2002).

The pretreatment of flavonoids myricetin, apigenin, and hesperidin increased humoral antibody production levels, macrophage phagocytosis, antioxidant marker enzymes, natural killer cell cytotoxicity, and splenic lymphocyte proliferation in cyclophosphamide-induced myeloid-suppressed animals. The compounds decreased proinflammatory cytokine levels, reduced lipid peroxidation, and decreased tissue damage in the bone marrow and spleen, indicating the immunostimulatory effects of the compounds. They can be very useful during cancer chemotherapy to reduce the side effects (Berkoz et al., 2021).

Polyphenolic compounds isolated from propolis stimulated immune response in mice bearing Ehrlich Ascites Carcinoma (EAC) tumor. The increase in ascetic fluid induced by EAC was markedly reduced after the treatment, and the survival time was prolonged. There was a dose-dependent increase in B-cells, cytotoxic T-cell, and NK cell activity, and the treatment increased the functional activity of macrophages, indicating increased host resistance against tumor cells (Orsolic et al., 2005).

Kaempferol augments antioxidant defense against free radicals and modulates a number of key elements in cellular signal transduction pathways linked to inflammation, tumor angiogenesis, metastasis, and apoptosis. Kaempferol-mediated MAPK activation prevented DNA damage leading to cell transformation and cancer development. The compound directly binds to the RSK2 protein and paralyzes the RSK2 protein, a key suppressor of apoptosis. Kaempferol binds to Src at its ATP site preventing its skin cancer-promoting activity (Chen & Chen, 2013). Kaempferol reduced CDK1 levels in human breast cancer cell line MDA-MB-453 and inhibited proliferation by disrupting the G2 checkpoint of the cell cycle (Luo et al., 2009). The compound impaired cancer angiogenesis through the inhibition of VEGF (vascular endothelial growth factor) in human cancer cell lines MDA-MB-453 and acted as VEGF antagonist (Luo et al., 2012). Kaempferol significantly inhibited MMP-3 protein activity in a dose-dependent manner and disrupted cancer metastasis in the MDA-MB-231 cell line. It prevented the *in vitro* migration of the cancer cell lines (Phromnoi et al., 2009).

Many of the flavonoids stimulated the cytotoxic activity of NK cells against various tumor cells. The administration of flavone-8-acetic acid in a mouse renal cancer model dose-dependently stimulated the activity of NK cells in the liver, spleen, lungs, and peritoneum (Kilani-Jaziri et al., 2016). The oral treatment with genistein (20 mg/kg) for four weeks increased the resistance of adult female B6C3fl mice to *in vivo* carcinogenesis induced in mouse melanoma tumor cells by B16F10 or DMBA (7,12-dimethyl benz (a) anthracene) (Lin et al., 2012). The flavonoids isolated from the roots and leaves of *Eleutherine palmifolia* stimulated the immune system by the increased activity of T-lymphocytes and macrophages and stimulated the secretion of IgG and IgM. The compounds stimulated the human mononuclear cells to secrete cytokines like TNF-α, IL-1β, and IFN-γ to prevent the progression of leukemia (Liao et al., 2015).

7.5 FLAVONOIDS AS A PROMISING SAFE THERAPEUTIC AGENTS AGAINST COVID-19 PANDEMIC

A novel coronavirus outbreak was reported as a severe acute respiratory disease syndrome coronavirus-2 (SARS-CoV-2) in 2019, and WHO announced it as an unexpected pandemic outbreak of a new virus from the beta coronavirus family. The infected people experienced mild to moderate respiratory illness. However, some groups become seriously ill and require medical attention. Elderly people and immune-compromised individuals may lead to severe pneumonia associated with systemic and strong inflammation implicated s airway damage, acute respiratory distress syndrome (ARDS), and multi-organ failure, and subsequently, it may become fatal (Tay et al., 2020). According to WHO, appropriate nutrition and a well-balanced diet are important to be healthier with strong immune systems and lower the risk of infectious diseases like COVID-19. Some countries have developed vaccines against the COVID-19 pandemic. Natural products are considered an adjuvant treatment for SARS-CoV-2 infection because they are safe, cheap, widely available, and without any undesirable side effects.

Flavonoids exhibited potent inhibitory effects against COVID-19, the current pandemic outbreak caused by SARS-CoV-2. Flavonoids can bind to the essential viral targets required for virus entry and replication and inhibiting various inflammatory cytokines. SARS-CoV-2 entry to host cells is mainly characterized by the interaction with viral spike protein and cellular angiotensin-converting enzyme-2 (ACE-2) and serine protease TMPRSS2. The mechanism involves over-expression of PAK1, which mediates Corona virus-induced immune system suppression, lung inflammation, and fibrosis.

Flavonoids such as naringenin, hesperidin, neohesperidin, naringin, apigenin, luteolin, cyaniding, catechin, quercetin, kaempferol, myricetin, caflanone, genistein, and linebacker were identified with potent inhibitory activity against SARS-CoV-2 by *in silico* study (Alzaabi et al., 2021). Quercetin showed significant inhibition against SARS-CoV M^{pro} (IC_{50}: 73 μM) and exhibited immunostimulatory activity when co-administered with vitamin C. The synergistic effect can be effectively employed for prophylaxis and high-risk population (Colunga et al., 2020). The enzymatic activity of SARS-CoV M^{pro} was effectively blocked by rhoifolin, herbcetin, and pectolinarin flavonoids (Jo et al., 2020). Flavonoids such as quercetin, luteolin, baicalin, hesperetin, gallocatechin gallate (GCG), EGCG, mentoflavone, scutellarein, and papyriflavonil A were evaluated *in vitro* for the inhibitory effect on key proteins PL^{pro}, $3CL^{pro}$, NTPase/helicase involved in infective cycle of corona virus. The molecular docking-based screening and *in vitro* studies using the recombinant proteins revealed that (–)-epicatechin 3-o-(3′-o-methyl) gallate for TMPRSS2 and oroxylin A glycoside and baicalein for furin bind and inhibit their respective proteases blocking virus propagation (Russo et al., 2020).

Liskova et al. (2021) evaluated the effect of flavonoids against the SARS-CoV-2-induced 'inflammatory storm' caused by uncontrolled systemic inflammatory responses. The results revealed that flavonoids can effectively modulate inflammatory signals associated with SARS-CoV-2. The *in-silico* study showed that equivir, caflanone, myricetin, linebacker, and hesperetin could bind with high-affinity spike protein, protease, and helicase sites on the ACE2 receptor used by the virus to infect cells. Caflanone inhibited the virus's key factors like ABL-2 and cathepsin-L cytokines (Ngwa et al., 2020). The *in vitro* and *in silico* studies evaluated the effect of quercetin on various stages of the virus entry and replication cycles, such as $3CL^{pro}$, PL^{pro}, and NTPase/helicase. The molecular docking and simulation studies revealed that quercetin-3-o-rhamnoside exhibited the highest binding affinity to spike protein and proteases on the virus (Cherrak et al., 2020). Clinical trials with the use of polyphenols such as cannabidiol, curcumin, resveratrol, epigallocatechin-3-gallate, and quercetin have been approved to combat COVID-19 in the light of *in vitro* and *in vivo* studies (Bernini & Velotti, 2021). The flavonoids present in propolis have adjuvant activity enhancing IL-4, IFN-γ, and IgG in serum (Berretta et al., 2020).

The flavonoid hesperidin present in *Citrus* fruits binds to the key proteins of SARS-CoV-2. The computational analysis showed that the flavonoid compound has low binding energy with both the spike protein and the main protease of the coronavirus responsible for its replication, and it can exhibit an effective antiviral effect (Bellavite & Donzelli, 2020). The flavonoid quercetin has immune booster effects as indicated by the expression of several vital genes and secretion of Th-1 derived IFN-γ and down-regulating Th-2 derived IL-4. Quercetin-3-β-galactoside inhibited *in vitro* a protease essential for the replication of SARS-CoV-2 and blocked middle east respiratory syndrome (MERS)-CoV-3CL^{pro} enzymatic activity. It also inhibited SARS-CoV-3CL^{pro} proteolytic activity (Jo et al., 2020).

Catechin showed high binding affinity towards papain-like proteinase, which is involved in RNA SARS-CoV-2 replication (Wu et al., 2020). Kaempferol glycosides exhibited an antiviral effect against the 3a channel protein of SARS-CoV, and the compound effectively reduced the proteolytic activity of MERS-CoV by its potential to occupy S1 and S2 sites of MERS-CoV-3CL^{pro} (Zhang et al., 2020). Another flavonoid apigenin was reported to inhibit SARS-CoV-3CL^{pro} proteolytic activity, and it suppressed the activity of SARS-CoV-3CL^{pro}. The flavonoid compound chrysin inhibited the interaction between the spike protein of SARS-CoV and Angiotensin Converting Enzyme ACE-2 (Wu et al., 2020). Hesperitin showed a dose-dependent suppression of the cleavage activity of 3c-like proteases of SARS-CoV in the *in vitro* assay. It is also effective in inhibiting ACE2. A recent docking study reported that cis-p-coumaric acid interferes with SARS-CoV-2 attachment to the host cell, and it can also act as an inhibitor of endoribonuclease Nsp15 encoded by MERS-CoV (Elfiky, 2020). Some docking studies have reported that polydatin is effective against COVID-19 main protease (Mpro) inhibitor (Adem et al., 2020).

7.6 FUTURE PROSPECTS

Bioactive flavonoids isolated from different medicinal plants are proven to be effective and competitive candidates for the therapy of several disorders.

They can act as potent immunomodulators. The compounds have been proven to inhibit various proinflammatory cytokines and different signal transduction pathways involved in the production of inflammatory reactions. Though pharmacokinetic and pharmacodynamic research has been carried out, the study of flavonoids is complex because of the heterogeneity of different molecular structures and the lack of sufficient data on their bioavailability. The data on the consequences of long-term consumption of flavonoids are less. More molecular docking studies have to be carried out to identify the novel, potential flavonoid molecules for their usage in the treatment of various diseases. More research is needed for the clinical trials addressing the dosage and combination of flavonoids to substantiate the therapeutic benefits. It has been reported that the generation of nanoparticles carrying flavonoid molecules can enhance the immune response, and further research is warranted in this regard. The results of various *in silico*, *in vitro, and in vivo* studies and clinical trials revealed that flavonoids have been shown to exhibit significant inhibitory effects against critical viral targets required for the entry and replication of SARS-CoV-2. Flavonoids can act as potent immune boosters, and they can also play a crucial role in the prevention and treatment of SARS-CoV-2.

7.7 CONCLUSION

Flavonoids are biologically active substances present in fruits, vegetables which constitute a major part of our diet. Nowadays, they are gaining more attention because of their usefulness and important roles in the mechanism of the pathophysiology of many diseases, including immunodeficiency diseases and cancer. It is the need of the time to explore more potential benefits of flavonoids in the field of immunomodulation, cancer, and other lifestyle diseases. There remains some ambiguity in the pharmacokinetic profile of flavonoids with certain functional groups and which is important to understand the bioavailability. Therefore, it is important to elucidate the biological fate and cellular metabolism of flavonoids. The mechanism of action at the molecular level and the structure-activity-pharmacokinetic relationship of flavonoids should be thoroughly studied. Flavonoids possess a remarkable spectrum of biological activities and have beneficial effects on health. Most of the chemotherapeutic agents adversely affect the normal cells, and they also develop multidrug resistance. Since flavonoids are non-toxic, there are endless possibilities for the development of synthetic analogs of flavonoids against various chronic diseases. Detailed studies are warranted in this regard to understand the molecular mechanism and signaling pathway of many known flavonoids for the drug development process.

KEYWORDS

- **anthocyanins**
- **bioactive flavonoids**
- **cytokines**
- **flavanones**
- **flavonoids**
- **gene expression**
- **immune cells**
- **immunomodulation**

REFERENCES

Abd-Alla, H., Moharram, F., Gaara, A., & El-Safty, M., (2009). Phytoconstituents of *Jatropha curcas* L. leaves and their immunomodulatory activity on humoral and cell-mediated immune response in chicks. *Z Naturforsch C., 64*(7, 8), 495–501.

Abdelsalam, N. A., Ghazy, N. M., Sallam, S. M., Radwan, M. M., Wanas, A. S., Elsohly, M. A., El-Demellawy, M. A., et al., (2017). Flavonoids of *Alcea rosea* L. and their immune-stimulant, antioxidant and cytotoxic activities on hepatocellular carcinoma HepG-2 cell line. *Nat. Prod. Res., 32*(6), 702–706. http://doi.org/10.1080/14780419.2017.1332602.

Adem, S., Eyupoglu, V., Sarfraz, I., Rasul, A., & Ali, M., (2020). *Identification of Potent COVID-19 Main Protease (Mpro) Inhibitors from Natural Polyphenols: An In-Silico Strategy Unveils a Hope Against CORONA*. Preprints, 2020030333. doi: 10.20944/preprints202003.0333.v1.

Alzaabi, M. M., Hamdy, R., Ashmawy, N. S., Hamoda, A. M., Alkhayat, F., Khademi, N. N., Al Joud, S. M. A., et al., (2021). Flavonoids are promising safe therapy against COVID-19. *Phytochem. Rev., 22*, 1–22. doi: 10.1007/s11101-021-09759-z .

Amić, D., Davidović-Amić, D., Beslo, D., Rastija, V., Lucić, B., & Trinajstić, N., (2007). SAR and QSAR of the antioxidant activity of flavonoids. *Curr. Med. Chem., 14*(7), 827–845.

Bae, J. H., Kim, J. Y., Kim, M. J., Chang, S. H., Park, Y. S., Son, C. H., Park, S. J., et al., (2010). Quercetin enhances susceptibility to NK cell-mediated lysis of tumor cells through induction of NKG2D ligands and suppression of HSP70. *J. Immunother., 33*(4), 391–401.

Bellavite, P., & Donzelli, A., (2020). Hesperidin and SARS-Cov-2: New light on the healthy function of *Citrus* fruits. *Antioxidants, 9*, 742. doi: 10/3390/antiox9080742.

Berköz, M., Yalın, S., Özkan, Y. F., Hunt, A. Ö., Krośniak, M., Francik, R., Yunusoğlu, O., et al., (2021). Protective effect of myricetin, apigenin, and hesperidin pre-treatments on cyclophosphamide-induced immunosuppression. *Immunopharmacol. Immunotoxicol.*, *43*(3), 353–369.

Bernini, R., & Velotti, F., (2021). Natural polyphenols as immunomodulators to rescue immune response homeostasis: Quercetin as a research model against severe COVID-19. *Molecules*, *26*, 5803, https://doi.org/103390/molecules26195803.

Berretta, A. A., Duarte, S. M. A., Condor, C. J. M., & De, J. D., (2020). Propolis and its potential against SARS-CoV-2 infection mechanisms and COVID-19 disease. *Biomed. Pharmacother.*, *131*, 110622.

Castellano, G., González-Santander, J. L., Lara, A., & Torrens, F., (2013). Classification of flavonoid compounds by using entropy of information theory. *Phytochem.*, *93*, 182–191.

Chakrawarti, L., Agrawal, R., Dang, S., Gupta, S., & Gabrani, R. (2016). Therapeutic effects of EGCG: A patent review. *Expert Opin. Ther. Pat.*, *26*(8), 907–916.

Chen, A. Y., & Chen, Y. C., (2013). A review of the dietary flavonoid, kaempferol on human health and cancer chemoprevention. *Food Chem.*, *138*, 2099–2107.

Chen, C. Y., Yang, C. H., Tsai, Y. F., Liaw, C. C., Chang, W. Y., & Hwang, T. L., (2017). Stimulates NLRP3 inflammasome activation and enhances inflammasome-mediated pathogen clearance. *Redox Biol.*, *11*, 263–274.

Cherng, J. M., Chiang, W., & Chiang, L. C., (2008). Immunomodulatory activities of common vegetables and spices of Umbelliferae and its related coumarins and flavonoids. *Food Chem.*, *106*, 944–950.

Cherrak, S. A., Merzouk, H., & Soulimane, N. M., (2020). Potential bioactive glycosylated flavonoids as SARS-CoV-2 main protease inhibitors: A molecular docking and simulation studies. *Plos One*, *15*(10), e0240653.

Chin, G., Mackman, R. L., Mish, M. R., & Zablocki, J. (2018). *Toll-like Receptor Modulator Compounds*. US2018086755.

Chirumbolo, S., (2010). The role of quercetin, flavonols, and flavones in modulating inflammatory cell function. *Inflamm. Allergy Drug Targets*, *9*(4), 263–285.

Colunga, B. R. M. L., Berrill, M., & Catravas, J. D., (2020). Quercetin and vitamin C: An experimental, synergistic therapy for the prevention and treatment of SARS-CoV-2 related diseases (COVID-19). *Front Immunol.*, *11*, 1451. doi: 10.3389/fimmu.2020.01451.

D'Archivio, M., Filesi, C., Di Benedetto, R., Gargiulo, R., Giovannini, C., & Masella, R., (2007). Polyphenols, dietary sources, and bioavailability. *Ann. Ist. SuperSanità.*, *43*(4), 348–361.

Daniel, P., Falcioni, F., Berg, A. U., & Berg, P., (1986). Influence of cianidanol on specific and non-specific immune mechanisms. *Methods Find Exp. Clin. Pharmacol.*, *8*(3), 139–145.

Del-Carmen J. V. M., Alonso-Castro, A. J., & García-Carrancá, A., (2013). Kaempferitrin induces immunostimulatory effects *in vitro*. *J. Ethnopharmacol.*, *148*(1), 337–340.

Deng, Q., Xu, J., Yu, B., He, J., Zhang, K., Ding, X., & Chen, D., (2010). Effect of dietary tea polyphenols on growth performance and cell-mediated immune response of post-weaning piglets under oxidative stress. *Arch. Anim. Nutr.*, *64*(1), 12–21. doi: 10.1080/17450390903169138.

Dewick, P. M., (2001). The shikimate pathway: Aromatic amino acids and phenylpropanoids. In: Dewick, P. M., (ed.), *Medicinal Natural Products: A Biosynthetic Approach* (pp. 137–186). John Wiley: Chichester.

Dhiman, S., Sharma, N., Sisodia, S. S., & Rajora, M., (2013). Evaluation of the immunomodulatory activity of daidzein in Balb/C mice using a panel of *in vivo* assays. *Int. J. Pharmaceut. Erud.*, *2*(4), 12–21.

Dykman, L. A., & Khlebtsov, N. G., (2019). Gold nanoparticles in chemo, immune and combined therapy: Review. *Biomed. Optics Expr.*, *10*(7), 3152–3182.

Elfiky, A. A., (2020). SARS-CoV-2 RNA dependent RNA polymerase (RdRp) targeting: An *in silico* perspective. *J. Biomol. Struct. Dyn.*, *39*(9), 1–9.

Guo, T. L., Chi, R. P., Hernandez, D. M., Auttachoat, W., & Zheng, J. F., (2007). Decreased 7, 12-dimethylbenz[a]anthracene-induced carcinogenesis coincides with the induction of antitumor immunities in adult female B6C3F1 mice pretreated with genistein. *Carcinogenesis.*, *28*(12), 2560–2566.

Guo, T. L., White, K. L. Jr., Brown, R. D., Delclos, K. B., Newbold, R. R., Weis, C., Germolec, D. R., & McCay, J. A., (2002). Genistein modulates splenic natural killer cell activity, antibody-forming cell response, and phenotypic marker expression in F(0) and F(1) generations of Sprague Dawley rats. *Toxicol. Appl. Pharmacol.*, *181*(3), 219–227.

Hu, Z. Q., Toda, M., Okubo, S., Hara, Y., & Shimamura, T., (1992). Mitogenic activity of (-) epigallocatechin gallate on B-cells and investigation of its structure-function relationship. *Int. J. Immunopharmacol.*, *14*(8), 1399–1407.

Hua, S., Zhang, Z., Liu, J., Dong, L., Huang, J., Lin, D., & Fu, X., (2018). Ethnomedicine, phytochemistry and pharmacology of *Smilax glabra*: An important traditional Chinese medicine. *Am. J. Chin. Med.*, *46*(2), 261–297.

Huang, A. C., Cheng, H. Y., Lin, T. S., Chen, W. H., Lin, J. H., & Lin, J. J., (2013). Epigallocatechin gallate (EGCG) influences a murine WEHI-3 leukemia model *in vivo* through enhancing phagocytosis of macrophages and populations of T- and B-cells. *In Vivo*, *27*(5), 627–634.

Huang, F., Li, Y., Leung, E. H., Liu, X., Liu, K., Wang, Q., Lan, Y., et al., (2020). A review of therapeutic agents and Chinese herbal medicines against SARS-COV-2 (COVID-19). *Pharmacol. Res.*, *158*, 104929. doi: 10.1016/j.phrs.2020.104929.

Jo, S., Kim, S., & Shin, D. H. (2020). Inhibition of SARS-CoV 3CL protease by flavonoids. *J. Enzyme Inhib. Med. Chem.*, *35*, 145–151. doi: 10.1080/14756366.2019.1690480.

Jung, J. H., Kang, J. I., & Kim, H. S., (2012). Effect of quercetin on impaired immune function in mice exposed to inflammation. *Nutr. Res. Pract.*, *6*(4), 301–307.

Kawai, K., Tsuno, N. H., Kitayama, J., Sunami, E., Takahashi, K., & Nagawa, H., (2011). Catechin inhibits adhesion and migration of peripheral blood B- cells by blocking CD11b. *Immunopharmacol. Immunotoxicol.*, *33*(2), 391–397.

Keservani, R. K., & Sharma, A. K., (2014). Flavonoids: Emerging trends and potential health benefits. *Journal of Chinese Pharmaceutical Sciences*, *23*(12), 815.

Keservani, R. K., Kesharwani, R. K., Sharma, A. K., Vyas, N., & Chadoker, A., (2010b). Nutritional supplements: An overview. *International Journal of Current Pharmaceutical Review and Research, 1*(1), 59–75.

Keservani, R. K., Kesharwani, R. K., Vyas, N., Jain, S., Raghuvanshi, R., & Sharma, A. K., (2010a). Nutraceutical and functional food as future food: A review. *Der Pharmacia Lettre*, *2*(1), 106–116.

Keservani, R. K., Sharma, A. K., & Kesharwani, R. K. (2020). *Nutraceuticals and Dietary Supplements: Applications in Health Improvement and Disease Management*. CRC Press. ISBN: 9781771888738.

Khan, A., Ali, A., Santercole, V., Paglietti, B., Rubino, S., Urooj, S., Farooqui, K., & Farooqui, A., (2016). *Camellia sinensis* Mediated enhancement of humoral immunity to particulate and nonparticulate antigens. *Phytother. Res.*, *30*, 41–48.

Khan, N. M., Haseeb, A., Ansari, M. Y., & Haqqi, T. M., (2017). A wogonin-rich fraction of *Scutellaria baicalensis* root extract exerts chondroprotective effects by suppressing IL-1β-induced activation of AP-1 in human OA chondrocytes. *Sci. Rep.*, *7*, 43789.

Kilani-Jaziri, S., Mustapha, N., Mokdad-Bzeouich, I., El-Gueder, D., Ghedira, K., & Ghedira-Chekir, L., (2016). Flavones induce immunomodulatory and anti-inflammatory effects by activating cellular anti-oxidant activity: A structure-activity relationship study. *Tumor Biol.*, *37*(5), 6571–6579.

Kim, G., Kim, K., Lee, S., Yoon, M., Lee, H., Moon, D., Lee, C., Ahn, S., & Park, Y., (2005). Curcumin inhibits immunostimulatory function of dendritic cells: MAPKs and translocation of NF-kappa B as potential targets. *J. Immunol.*, *174*(12), 8116–8124. doi: 10.4049/jimmunol.174.12.8116.

Kumar, S., & Pandey, A. K., (2013). Chemistry and biological activities of flavonoids: An overview. *Sci. World J.*, *2013*, Article ID: 162750, https://doi.org/10.1155/2013/162750.

Laily, N., Kusumaningtyas, R. W., Sukarti, L., & Rini, M. R. D. K., (2015). The potency of *Psidium guajava* (L.) leaves as a functional immunostimulatory ingredient. *Procedia Chem.*, *14*, 301–307.

Liao, D. Y., Chai, Y. C., Wang, S. H., Chen, C. W., & Tsai, M. S., (2015). Antioxidant activities and contents of flavonoids and phenolic acids of *Talinum triangulare* extracts and their immunomodulatory effects. *J. Food Drug. Anal.*, *23*, 294–302.

Lin, C. C., Lin, J. J., Wu, P. P., Lu, C. C., Chiang, J. H., & Kuo, C. L., (2013). Wogonin, a natural and biologically-active flavonoid, influences a murine WEHI-3 leukemia model *in vivo* through enhancing populations of T- and B-cells. *In Vivo.*, *27*(6), 733–738.

Lin, C. C., Yu, C. S., Yang, J. S., Lu, C. C., Chiang, J. H., Lin, J. P., Kuo, C. L., & Chung, J. G., (2012). Chrysin, a natural and biologically active flavonoid, influences a murine leukemia model *in vivo* through enhancing populations of T-and B-cells and promoting macrophage phagocytosis and NK cell cytotoxicity. *In vivo.*, *26*(4), 665–670.

Liskova, A., Koklesova, M. S. L., Samuel, S. M., Zhai, K., AlIshaq, R. K., Abotaleb, M., Nosal, V., et al., (2021). Flavonoids against the SARS-CoV-2 induced inflammatory storm. *Biomed. Pharmacother.*, *138*, 111430.

Liu, S., Lin, C., Hung, S., Chou, J., Chi, C., & Fu, S., (2010). Fisetin inhibits lipopolysaccharide-induced macrophage activation and dendritic cell maturation. *J. Agric. Food Chem.*, *58*(20), 10831–10839.

Liu, X., Zhao, M., Wua, K., Chai, X., Yu, H., Tao, Z., & Wang, J., (2012). Immunomodulatory and anticancer activities of phenolics from emblica fruit (*Phyllanthus emblica* L.). *Food Chem.*, *131*, 685–690.

Liu, Y., Wang, C., Dong, X., Cheng, O., & Zhou, T., (2015). Immunomodulatory effects of epicatechin (2β-o-7, 4β-8)-ent epicatechin isolated from *Rhododendron spiciferum in vitro*. *Immunopharmacol. Immunotoxicol.*, *37*(6), 1–8.

Luo, H., Rankin, G. O., Juliano, N., Jiang, B. H., & Chen, Y. C., (2012). Kaempferol inhibits VEGF expression and *in vitro* angiogenesis through a novel ERK-NFκB-cMyc-p21 pathway. *Food Chem.*, *130*(2), 321–328.

Luo, H., Rankin, G. O., Liu, L., Daddysman, M. K., Jiang, B. H., & Chen, Y. C., (2009). Kaempferol inhibits angiogenesis and VEGF expression through both HIF-dependent and independent pathways in human ovarian cancer cells. *Nutr. Cancer*, *61*(4), 554–563.

Masad, R. J., Haneefa, S. M., Mohamed, Y. A., Al-Sbiei, A., Bashir, G., Cabezudo, M. J. F., & Al-Ramadi, B. K., (2021). The immunomodulatory effects of honey and associated flavonoids in cancer. *Nutrients, 13*(4), 1269, https://doi.org/10.3390/nu13041269.

Mercader, A. G., & Pomilio, A. B., (2013). Naturally-occurring dimers of flavonoids as anticarcinogens: Anticancer agents. *Med. Chem., 13*(8), 1217–1235.

Middleton, E., (1998). Effect of plant flavonoids on immune and inflammatory cell function. In: Manthey, J. A., & Buslig, B. S., (eds.), *Flavonoids in the Living System: Advances in Experimental Medicine and Biology* (p. 439). Springer: Boston, MA. https://doi.org/10.1007/978-1-4615-5335-9_13.

Muthulakshmi, M., Subramani, P. A., & Michael, R. D. (2016). Immunostimulatory effect of the aqueous leaf extract of *Phyllanthus niruri* on the specific and nonspecific immune responses of *Oreochromis mossambicus. Iran J. Vet. Res., 17*(3), 200–202.

Nair, M. P. N., Kandaswami, C., Mahajan, S., Chadh, K. C., Chwda, R., Nair, H., Kumar, N., et al., (2002). The flavonoid quercetin differentially regulates Th-1 (IFN-γ) and Th-2 (IL-4) cytokine gene expression by normal peripheral blood mononuclear cells. *Biochim. Biophys. Acta, 1953,* 29–36.

Ngwa, W., Kumar, R., Thompson, D., Lyerly, W., Moore, R., Reid, T. E., Lowe, H., & Toyang, N. (2020). The potential of flavonoid-inspired phytomedicines against COVID-19. *Molecules., 25*(11), 2707. https://doi.org/10.3390/molecules25112707.

Nile, S. H., Keum, Y. S., Nile, A. S., Jalde, S. S., & Patel, R. V., (2018). Antioxidant, anti-inflammatory, and enzyme inhibitory activity of natural plant flavonoids and their synthesized derivatives. *J. Biochem. Mol. Toxicol., 32*(1), e22002. doi: 10.1002/jbt.22002.

Oliveira, R., Narciso, C., Bisinotto, R., Perdomo, M., Ballou, M., Dreher, M., & Santos, J., (2010). Effects of feeding polyphenols from pomegranate extract on health, growth, nutrient digestion, and immunocompetence of calves. *J. Dairy Sci., 93*(9), 4280–4291.

Orsolic, N., Terzic, S., Sver, L., & Basic, I., (2005). Polyphenolic compounds from propolis modulate immune responses and increase host resistance to tumor cells. *Food Agric. Immunol., 16*(3), 165–179.

Pae, M., & Wu, D., (2013). Immunomodulating effects of epigallocatechin-3-gallate from green tea: Mechanisms and applications. *Food Funct., 4*(9), 1287–1303.

Pae, M., Ren, Z., Meydani, M., Shang, F., Smith, D., Meydani, S., & Wu, D., (2012). Dietary supplementation with a high dose of epigallocatechin-3-gallate promotes inflammatory response in mice. *J. Nutr. Biochem., 23*(6), 526–531. doi: 10.1016/j.jnutbio.2011.02.006.

Pan, M. H., Chen, W. J., Shiau, S. Y. L., Ho, C. T., & Lin, J. K., (2002). Tangeretin-induced cell cycle G1 arrest through inhibiting cyclin-dependent kinases 2 and 4 activities as well as elevating Cdk inhibitor p21 and p27 in human colorectal carcinoma cells. *Carcinogenesis, 23,* 1677–1684.

Panche, A. N., Diwan, A. D., & Chandra, S. R. (2016). Flavonoids: An overview. *J. Nat. Sci., 5,* 1–15. doi: 10.1017/jns.2016.41.

Parhi, B., Bhratiya, D., & Swain, S. K., (2020). Application of quercetin flavonoid based hybrid nanocomposites: A review. *Saudi Pharmaceut. J., 28,* 1719–1732.

Phromnoi, K., Yodkeeree, S., Anuchapreeda, S., & Limtrakul, P., (2009). Inhibition of MMP-3 activity and invasion of the MDA-MB-231 human invasive breast carcinoma cell line by bioflavonoids. *Acta Pharmacol. Sin. , 30,* 1169–1176.

Pratiwi, G., Martien, R., & Murwanti, R., (2019). Chitosan nanoparticle as a delivery system for polyphenols from meniran extract (*Phyllanthus niruri* L.): Formulation, optimization and immunomodulatory activity. *Int. J. Appl. Pharmaceut., 11*(2), https://doi.org/10.22159/ijap.2019v11i2.29999.

Raffa, D., Maggio, B., Raimondi, M. V., Plescia, F., & Daidone, G., (2017). Recent discoveries of anticancer flavonoids. *Eur. J. Med. Chem., 142*, 213–228.

Ramstead, A., Schepetkin, I., Quinn, M., & Jutila, M., (2012). Oenothein B, A cyclic dimeric ellagitannin isolated from *Epilobium angustifolium*, enhances IFNγ production by lymphocytes. *PLoS One, 7*(11), e50546.

Ramstead, A., Schepetkin, I., Todd, K., Loeffelholz, J., Berardinelli, J., Quinn, M., & Jutila, M., (2015). Aging influences the response of T cells to stimulation by the ellagitannin, oenothein B. *Int. Immunopharmacol., 26*(2), 367–377.

Rao, K. S., Ghorpade, A., & Labhasetwar, V., (2009). Targeting anti-HIV drugs to the CNS. *Expert Opin. Drug Deliv., 6*, 771–784.

Russo, M., Moccia, S., Spagnuolo, C., Tedesco, I., & Russo, G. L., (2020). Roles of flavonoids against coronavirus infection. *Chem. Biol. Interact., 328*, 109211. doi: 10.1016/j.cbi.2020.109211.

Sahu, N. K., Balbhadra, S. S., Choudhary, J., & Kohli, D. V., (2012). Exploring the pharmacological significance of chalcone scaffold: A review. *Curr. Med. Chem., 19*(2), 209–225.

Saravanan, D., Thirumalai, D., & Asharani, I. V., (2015). Anti-HIV flavonoids from natural products: A systematic review. *Int. J. Res. Pharm. Sci., 6*(3), 248–255.

Sassi, A., Maatouk, M., El-Gueder, D., Bzéouich, I. M., Abdelkefi-Ben, H. S., & Jemni-Yacoub, S., (2018). Chrysin, a natural and biologically active flavonoid, suppresses tumor growth of mouse B16F10 melanoma cells: *In vitro* and *in vivo* study. *Chem. Biol. Interact., 283*, 10–19.

Schepetkin, I., Kirpotina, L., Jakiw, L., Khlebnikov, A., Blaskovich, C., Jutila, M., & Quinn, M., (2009). Immunomodulatory activity of oenothein B isolated from *Epilobium angustifolium. J. Immunol., 183*(10), 6754–6766.

Selvakesvan, R. K., & Franklin, G., (2021). Prospective applications of nanoparticles green synthesized using medicinal plant extracts as novel nanomedicines. *Nanotechnol. Sci. Appl., 14*, 179–195.

Shankar, E., Goel, A., Gupta, K., & Gupta, S. (2017). Plant flavone apigenin: An emerging anticancer agent. *Curr. Pharmacol. Rep., 3*(6), 423–446.

Silva, V., Dos Santos, M. H., & Viegas, C., (2017). Biological and chemical aspects of natural bioflavonoids from plants: A brief review. *Mini-Rev. Med. Chem., 17*, 834–862.

Singh, D., Tanwar, H., Das, S., Ganju, G., & Singh, S. B. A., (2018). Novel *in vivo* adjuvant activity of kaempferol enhanced Tbx-21, GATA-3 expression and peritoneal $CD11c^+$ $MHCII^+$ dendritic cell infiltration. *Immunopharmacol. Immunotoxicol.* https://doi.org/10.1080/08923973.2018.1434794.

Singh, D., Tanwar, H., Jayashankar, B., Sharma, J., Murthy, S., Chanda, S., Singh, S. B., & Ganju, L., (2017). Quercetin exhibits adjuvant activity by enhancing Th2 immune response in ovalbumin immunized mice. *Biomed. Pharmacother., 90*, 354–360.

Spagnuolo, C., Russo, M., Bilotto, S., Tedesco, I., Laratta, B., & Russo, G. L., (2012). Dietary polyphenols in cancer prevention: The example of the flavonoid quercetin in leukemia. *Ann. NY Acad. Sci., 1259*, 95–103.

Stalikas, C., (2007). Extraction, separation, and detection methods for phenolic acids and flavonoids. *J. Sep. Sci., 30*, 3268–3295.

Sung, N. Y., Yang, M. S., Song, D. S., Byun, E. B., Kim, J. K., Park, J. H., Song, B. S., et al., (2013). The procyanidin trimer C1 induces macrophage activation via NF-κB and MAPK pathways, leading to Th1 polarization in murine splenocytes. *Eur. J. Pharmacol., 714*(1–3), 218–228.

Sutoyo, S., Ismono, I., & Mitarlis, D., (2018). Immunostimulant activity of flavonoid isolated from the acetone extract of silver fern (*Pityrogramma calomelanos*). *Adv. Eng. Res., 171*, 86–88.

Talmale, S., Bhujade, A., & Patil, M., (2014). Immunostimulatory activity of flavonoids isolated from stem bark of *Ziziphus mauritiana*. *Int. J. Innovative Res. Sci. Eng. Technol., 3*(7), 14285–14296.

Tandon, V., & Das, B., (2018). Genistein: Is the multifarious botanical a natural anthelmintic too? *J. Parasit. Dis., 42*(2), 151–161.

Tay, M. Z., Poh, C. M., Rénia, L., MacAry, P. A., & Ng, L. F. P., (2020). The trinity of COVID-19: Immunity, inflammation and intervention. *Nat. Rev. Immunol., 20*(6), 363–374.

Thompson, D., (2016). *Method and Composition for Preventing and Treating Viral Infections.* US20160367517.

Wang, J., Wang, H., Sun, K., Wang, X., Pan, H., Zhu, J., Ji, X., & Li, X., (2018). Chrysin suppresses proliferation, migration, and invasion in glioblastoma cell lines via mediating the ERK/Nrf2 signaling pathway. *Drug Des. Devel. Ther., 12*, 721–733.

Wang, W., Higuchi, C. M., & Zhang, R., (1997). Individual and combinatory effects of soy isoflavones on the *in vitro* potentiation of lymphocyte activation. *Nutr. Cancer., 29*(1), 29–34.

Wu, C., Liu, Y., Yang, Y., Zhang, P., Zhong, W., Wang, Y., Wang, Q., et al., (2020). Analysis of therapeutic targets for SARS-CoV-2 and discovery of potential drugs by computational methods. *Acta Pharm. Sin. B, 10*(5), 766–788.

Xu, M., Shi, H., & Liu, D., (2019). Chrysin protects against renal ischemia reperfusion-induced tubular cell apoptosis and inflammation in mice. *Exp. Ther. Med., 17*(3), 2256–2262.

Yamaguchi, K., Honda, M., Ikigai, H., Hara, Y., & Shimamura, T., (2002). Inhibitory effects of (-) epigallocatechin gallate on the life cycle of human immunodeficiency virus type 1 (HIV-1). *Antiviral Res., 53*, 19–34.

Yi, Z., Yu, Y., Liang, Y., & Zeng, B., (2008). *In vitro* antioxidant and antimicrobial activities of the extract of *Pericarpium citri* (reticulatae) of a new *citrus* cultivar and its main flavonoids. *Food Sci. Tech., 41*, 597–603.

Zhang, R., Li, Y., & Wang, W., (1997). Enhancement of immune function in mice fed high doses of soy daidzein. *Nutr. Cancer, 29*(1), 24–28.

Zhang, T., He, Y., Xu, W., Ma, A., Yang, Y., & Xu, K. F. (2020). Clinical trials for the treatment of Corona Virus Disease 2019 (COVID-19): A rapid response to urgent need. *Sci. China Life. Sci., 63*(5), 774–776.

Zhang, X., Kang, Z., Li, Q., Zhang, J., Cheng, S., & Chang, H. (2018). Antigen-adjuvant effects of icariin in enhancing tumor-specific immunity in mastocytoma-bearing DBA/2J mice. *Biomed. Pharmacother., 99*, 810–816.

CHAPTER 8

ROLE OF FLAVONOIDS AS ANTI-INFLAMMATORY AGENTS

PARUL SAINI

Alumni, John Curtin School of Medical Research, Australian National University, Canberra, Australia

ABSTRACT

Over the past few years, inflammation has been recognized as a significant risk factor for a diverse range of human diseases. Acute inflammation is a short-term and self-limiting process that makes it easy for the host defenses to return the body to homeostasis. On the other hand, chronic inflammation is being shown to be increasingly involved in predisposition to a pathological progression of chronic illnesses, including cardiovascular diseases (CVDs), diabetes, neurodegenerative diseases, obesity, asthma, and even cancer. As a result, treatment of chronic inflammatory disease is under active investigation, and there is an immediate need to find new and safe anti-inflammatory compounds. Flavonoids are natural substances that normally occur in a diet and have been reported to play a significant role in managing various chronic inflammatory disorders. This chapter contains the current knowledge of mechanisms involved in the anti-inflammatory activities of flavonoids and the implications of these effects on protection against various chronic inflammatory diseases.

8.1 INTRODUCTION

Inflammation is a coordinated biological process induced by tissue injury or microbial pathogen infection. A significant trigger of inflammation is

Flavonoids as Nutraceuticals. Rajesh K. Kesharwani, Deepika Saini, Raj K. Keservani, and Anil Kumar Sharma (Eds.)

the recognition of microbial pathogens by specific receptors of the innate immune system, which is required for the induction and establishment of inflammatory settings (Figure 8.1). The initiation of inflammation at the site of damage involves the migration of immune cells from blood vessels and the release of mediators, resulting in the recruitment of inflammatory cells and the release of proinflammatory cytokines, reactive nitrogen species (RNS), and reactive oxygen species (ROS) to destroy foreign pathogens, resolving infection and repairing damaged tissues (Pan et al., 2009; Medzhitov et al., 2008). Thus, the ideal inflammatory response is essential for the defense of the host. However, in certain chronic infections or inflammatory disorders, inflammation causes more damage to the host than the microbial pathogen.

The immune system and inflammation are intimately coordinated to produce an effective host defense response. Indeed, an overactive innate immune response can result in chronic inflammation or chronic infection due to inefficient control of the inflammatory responses. Current steroidal anti-inflammatory drugs and non-steroidal anti-inflammatory drugs for inflammation are adequate for treating acute inflammation but are not entirely successful in curing chronic inflammatory disorders and have unforeseen side effects (Li et al., 2018; Pahwa & Jialal, 2018). Hence, there is an urgent requirement to find new and safer anti-inflammatory drugs. Traditionally, natural compounds, such as plant extracts, have been used to treat various disorders, including chronic inflammatory disorders. Flavonoids are among the active constituents of these extracts that have a diverse spectrum of biological properties, including antimicrobial, antiviral, anti-cancer, anti-thrombogenic, and anti-inflammatory (Pan et al., 2009; Garcia-Lafuente et al., 2009). The anti-inflammatory activity of flavonoids has long been used in Chinese medicine by applying crude plant extracts. Further, there is

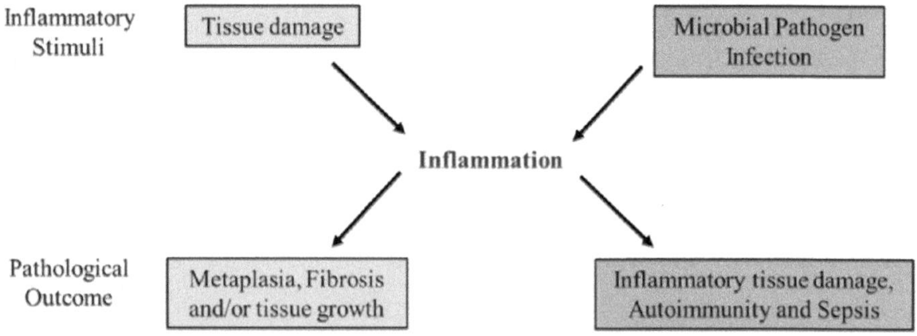

FIGURE 8.1 Causes of inflammation and associated pathological outcomes. Based on the inflammatory stimuli, inflammatory stimuli can have different pathological outcomes.

significant scientific evidence based on *in vitro* and *in vivo* models of inflammation that supports the anti-inflammatory activity of a variety of flavonoid molecules (Aswad et al., 2018; Attiq et al., 2018; Azab et al., 2016).

Inflammation is actively involved in the development of many chronic illnesses, including obesity, CVDs, neurodegenerative diseases, diabetes, and cancer (Libby, 2008; Calder et al., 2007). A significant number of epidemiological studies have indicated that an increase in the consumption of flavonoid-rich food decreases the incidences of the above-mentioned chronic illnesses attributed to the anti-inflammatory activity of flavonoids (Garcia-Lafuente et al., 2009; Mennen et al., 2004). Therefore, the study of the anti-inflammatory activity of flavonoids is vital for establishing anti-inflammatory mechanisms and the development of safe anti-inflammatory drugs that may be useful in treating these chronic illnesses.

In this chapter, we review the anti-inflammatory activities of flavonoids and the potential mechanisms associated with them. We also summarize the role of inflammation in the development of four critical chronic illnesses. The potential function of flavonoids in preventing and treating these diseases, on account of anti-inflammatory activity, is also reviewed.

8.2 FLAVANOIDS: A NOVEL COMPOUND FOR THE TREATMENT OF INFLAMMATION

Flavonoids are secondary plant metabolites present in vegetables, fruits, spices, legumes, herbs, flowers, stems, and nuts. Flavonoids are a subclass of polyphenols and are characterized by two or more aromatic rings, each having a minimum of one aromatic hydroxyl and joined with a heterocyclic pyran (Pan et al., 2009). Based on the connection of the aromatic ring to the heterocyclic ring, the functional groups on the heterocyclic rings, and oxidation state, flavonoids are classified into six different subclasses (Panche et al., 2016; Kumar et al., 2013; Keservani et al., 2010a, b, 2020; Keservani & Sharma, 2014). Table 8.1 shows the names, dietary sources, and prominent examples of each subclass.

Previous investigations have repeatedly proven the anti-inflammatory activity of flavonoid molecules from each subclass in acute and chronic inflammation through *in vivo* and *in vitro* models (Knekt et al., 2002). The molecular and biochemical mechanisms and the cell signaling pathways used by flavonoids for modulating inflammatory processes that cause or exacerbate chronic illnesses are described in Table 8.1.

TABLE 8.1 Flavonoid Subclasses, Dietary Sources, and Important Examples

Flavanoid Subclass	Dietary Source	Example
Flavanols	Apples, grapes, pears, teas	Catechin, epicatechin, gallocatechin
Flavanones	Citrus fruits, e.g., lemon, orange	Eriodictyol, hesperetin, naringenin, hesperidin
Flavones	Spices and herbs, e.g., Parsley, thyme	Apigenin, luteolin, tangeretin
Isoflavones	Legumes and derived products, e.g., soybean, tofu	Biochanin A, daidzein, genistein
Flavonols	ubiquitous in nearly all foods	Isorhamnetin, kaempferol, quercetinrutin, morin
Flavanonols	–	Aromadendrin, engeletin, taxifolin
Anthocyanidins	Blue, purple, and red berries, red wine	Cyanidin, delphinidin, pelargonidin

8.3 ANTI-INFLAMMATORY ACTIVITY OF FLAVONOIDS: MOLECULAR AND BIOCHEMICAL MECHANISMS

The anti-inflammatory activities of flavonoids are exhibited through various mechanisms, such as inhibition of transcription factors (TFs) and regulatory enzymes that play a vital role in controlling mediators of inflammation. Also, Flavonoids are potent antioxidants, thereby scavenging free radicals and reducing their formation. Consequently, flavonoids profoundly affect numerous immune cells and immune mechanisms essential for inflammation (Table 8.2).

TABLE 8.2 Anti-Inflammatory Properties and Mode of Action of Flavonoids

Property of Flavonoids	Mode of Action
Inhibition of regulatory enzymes	Inhibition of phosphodiesterases, protein kinases, and transcription factors.
Antioxidants	Inhibition of free radicals, scavenger function.
Inhibition of prostanoid synthesis	Inhibition of COX, LOX.
Inhibition of histamine release	Inhibition of histamine release in the late phase of allergic inflammatory reactions.
Impact on immune cells	Inhibition of cell activation, maturation, and signaling pathways.

8.4 INHIBITION OF PHOSPHODIESTERASES

Inhibition of phosphodiesterases holds a particular therapeutic significance in chronic as well as allergic inflammatory processes. In addition, suppression of phosphodiesters is an essential activity associated with numerous medicinal plants, and in many traditional medicines, flavonoids have been linked with this inhibitory activity (Kusano et al., 1991). Studies have shown potent inhibition of cyclic adenosine monophosphate (cAMP) phosphodiesterases by different flavonoid molecules, flavones aglycones, and C-glycosyl, biflavanoes from *Ginkgo biloba*, and four flavonoids from licorice (Dehmlow et al., 1996; Saponara et al., 1998). cAMP is a secondary messenger important for regulating various cellular functions during inflammation, and increased levels of cAMP are associated with anti-inflammatory activities. The cAMP phosphodiesterases degrade cAMP to maintain normal levels. The inhibitory functions of flavonoids on cAMP phosphodiesterases involve inhibition of cAMP degradation, prolonging cAMP signaling, and consequently promoting anti-inflammatory functions linked with high levels of cAMP (Wahlang et al., 2018; Guo et al., 2018).

8.5 INHIBITION OF PROTEIN KINASES AND TRANSCRIPTION FACTORS (TFS)

Cell activation during inflammation requires different kinases (e.g., protein kinase C, protein tyrosine kinase, and phosphatidyl kinase) for signal transduction. Based on this, flavonoids can target multiple protein kinases as part of multiple signal transduction cascades (Hou & Kumamoto, 2010). For example, previous studies have reported the inhibition of kinases such as protein kinase C, phosphoinositol kinase, tyrosine kinase, or cyclin-dependent kinase-4 phosphatidylinositol kinase by different types of flavonoids (Lolli et al., 2012; Yokoyama et al., 2015). Further, inhibition of a particular protein kinase by more than one subclass of flavonoid molecule has also been reported. For instance, phosphatidylinositol-3 kinases are inhibited by flavones apigenin, myricetin, quercetin, luteolin, and fisetin, as well as by isoflavone or obol (Agullo et al., 1998).

Flavanoids can also regulate protein kinases by inhibiting TFs like nuclear factor kappa B (NF-κB) (Peng et al., 2018). NF-κB modulates the expression of numerous chemokines, cytokines, and cell adhesion molecules that participate in inflammation. IκB, a regulatory protein, inhibits the activity of

NF-κB, although, during inflammation, IκB gets phosphorylated and consequently degraded. As a result, NF-κB can translocate from the cytoplasm to the nucleus, initiating the expression of numerous proinflammatory genes. In a previous investigation, flavonoids have been shown to control the activity of NF-κB and IκB, having a direct effect on cell activation (Chen et al., 2017).

8.6 ANTIOXIDANT ACTIVITY OF FLAVANOIDS

Tissue damage during inflammations results in the generation of free radicals like nitrogen-derived radicals (or RNS), oxygen-derived radicals (or ROS) that have deleterious effects on the function of cells (Mittal et al., 2014). These free radicals contain unpaired electrons, making them very reactive and detrimental to DNA, proteins, and lipids. Free radicals impact nucleic acids and proteins by oxidative damage and on cell membranes by lipid peroxidation. Increased production of free radicals along with low free radical scavenging activity results in oxidative stress (Nimse & Pal, 2015). The antioxidant activity of flavonoids is governed by inhibitory effects on the generation of free radicals and their scavenging function for RNS, ROS, and other reactive species. The antioxidant activity of flavonoids is due to their chemical structure, particular patterns of substitution within the structure, and the phenolic hydrogen that allows them to act as hydrogen-donating molecules (Li et al., 2016; Chen et al., 2019). Based on the previous studies, the most potent flavonoids with antioxidants are flavones and catechins (Pietta, 2000).

8.7 INHIBITION OF PROSTANOID SYNTHESIS

The inflammatory response is a highly organized series of cell activation processes, the majority of which are associated with prostanoid biosynthesis through the metabolism of arachidonic acid. During inflammation, arachidonic acid is released from phospholipids present in cellular membranes by the enzyme phospholipase A2 and is further oxidized to inflammatory mediators such as thromboxanes and prostaglandins due to the activity of LOX and COX, respectively.

Flavonoids are potent inhibitors of arachidonic acid metabolism, thereby decreasing inflammatory mediators released from this pathway. For example, a previous study by Damon et al. (1987) has shown significant inhibition of prostaglandin formation by flavonoids diosmin and hesperidin in *in-vivo* models. In addition, inhibition of phospholipase A2, COX, and

LOX by flavonoids has been reported in numerous *in-vitro* studies. Several bioflavonoids (bilobetin, amentoflavone, ginkgetin, and morelloflavone) and flavonol quercetin have been demonstrated to inhibit phospholipase A2 and COX, respectively (Lee et al., 1997; Kim et al., 1999).

8.8 INHIBITION OF HISTAMINE RELEASE

Histamine is an organic nitrogenous compound that acts as a critical chemical mediator in allergic inflammatory reactions. Previous investigations have demonstrated that flavonoids attenuate histamine release during the late phase of allergic reactions. During this phase of an allergic reaction, histamine release is modulated by leukotrienes produced by LOX catalyzed reactions. Many hydroxylated flavones, aglycones have demonstrated significant inhibition of this process, whereas methoxylated flavones have demonstrated much lower inhibition (Petkov et al., 1981).

8.9 IMPACT ON IMMUNE CELLS

The different functions and properties of flavonoids affect cell activation, maturation, signal transduction, or cytokine production in numerous immune system cells. For instance, certain flavonoids such as tea flavonoids (epigallocatechin gallate (EGCG) and apigenin) have been shown to inhibit the activation of immune cells and their effectors (cytokines and chemokines) (Cialdella-Kam et al., 2017). In addition, several studies have supported the impact of flavonoids in signal transduction via different mechanisms. One such mechanism involves the binding of flavonoids to cytokine receptors like the interleukin 17 (IL-17) RA subunit of the IL-17 receptor, resulting in attenuation of its signaling. Furthermore, the inhibitory effect of flavonoids is also seen in downstream signaling from receptors like high-affinity immunoglobulin E (IgE) receptor (FcεRI) and other receptors at the site of inflammation (Liu et al., 2017; Kim et al., 2014).

8.10 ROLE OF FLAVONOIDS IN CHRONIC INFLAMMATORY DISEASE

Chronic inflammation is associated with various progressive diseases, including cancer, metabolic disorders, CVDs, neurodegenerative diseases, asthma, and obesity. Here we will explore characteristic functions that help in combating inflammatory processes underlying these chronic conditions.

8.11 FLAVONOIDS IN CANCER

In the 18th century, Rudolf Virchow, a German pathologist, was the first to find an association between cancer development and inflammation. Since then, many epidemiological studies have confirmed that chronic inflammation is an essential component of tumor progression, including colorectal, esophageal, cystic, gastric, pulmonary hepatocellular pancreatic, ovarian, and skin cancers (Ostrand-Rosenberg & Sinha, 2009; Todoric et al., 2016). Chronic inflammation causes continuous production of detrimental ROS that damages DNA and genetic alterations, leading to the initiation of tumor growth. In addition, there is uninterrupted production of inflammatory mediators like tumor necrosis factor-α (TNF-α) or interferon-γ (IFN-γ) and proangiogenic growth factors like cytokines, promoting tumor neovascularization and bringing in the much-needed blood supply to nourish the growing tumor. Lastly, inflammation enhances the dissemination of tumors by producing extracellular matrix-degrading enzymes (Yang et al., 2008). These factors are produced either by cancer cells or by tumor-infiltrating immune cells such as lymphocytes, dendritic cells (DCs), neutrophils, natural killer cells, neutrophils, DCs, lymphocytes, and macrophages for stimulating tumor growth and survival. Consequently, utilizing therapeutic agents targeting these inflammatory factors will effectively treat and prevent cancer development.

The anti-inflammatory properties of flavonoids allow them to act as potent anti-cancer phytochemicals that exert their function by numerous mechanisms, including induction of apoptosis, carcinogen inactivation, inhibition of angiogenesis, and triggering cell cycle arrest (Akiyama et al., 1987; Constantinou et al., 1998; Fotsis et al., 1993; Markovits et al., 1989; Matsukawa et al., 1993) (Figure 8.2). Flavonoids have demonstrated the inhibition of tumor cell proliferation by inhibiting ROS formation and suppressing the activity of COX and LOX (Chahar et al., 2011). Increasing evidence has supported the role of cyclin-dependent kinases (CDKs) as potent regulators of inflammation, immune cell activation, and cell cycle progression. Further previous investigations have shown that hyper-activation of CDKs due to CDK inhibitor genes or mutation of CDK genes is associated with the development of various cancers (Schmitz & Kracht, 2016). Flavonoids have been reported to induce cell cycle arrest by inhibiting CDK in the skin and human breast cancer (Chahar et al., 2011). In addition. Flavonoids isoflavones and their metabolites induce apoptosis of cancer cells derived from human gastric cancer by different mechanisms, including decreasing ROS production, regulating the expression of heat shock proteins, and modulating signaling pathways (Matsukawa et al., 1993).

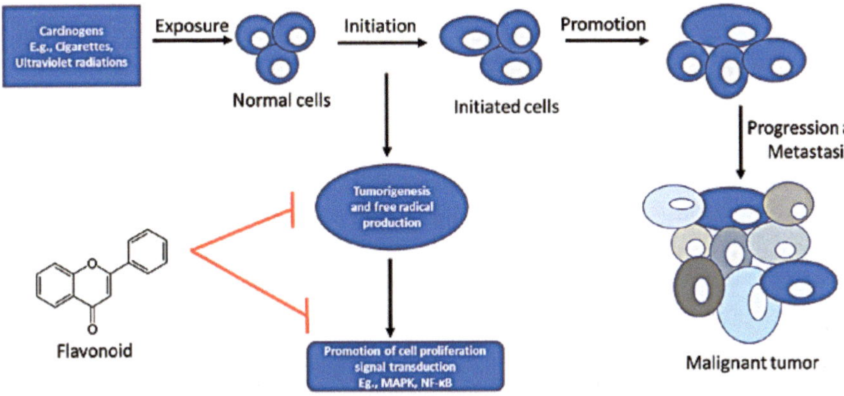

FIGURE 8.2 Role of flavonoids in cancer – the anti-inflammatory activity of flavonoids involves reduced production of free radicals such as reactive oxygen species (ROS) and decreases expression of numerous inflammatory mediators via inhibition of signal transduction pathways [MAPK: mitogen-activated protein kinase; NF-κB: nuclear factor-kappa].

8.12 FLAVONOIDS IN METABOLIC DISORDERS

Inflammation has long been associated with many metabolic disorders, including diabetes and non-alcoholic fatty liver disease (NAFLD). Inflammatory responses are considered an essential feature of metabolic dysfunction characterized by activation of signaling pathways involved in inflammation, abnormal production of proinflammatory cytokines, and increased acute-phase protein production (Hotamisligil, 2006). Type 2 diabetes is amongst the most prevalent metabolic disorders and is associated with insulin resistance and impaired insulin secretion. In type 2 diabetes, nutrient excesses like increased free fatty acids and hyperglycemia cause endoplasmic reticulum stress, oxidative stress, lipid, and amyloid deposition, glucotoxicity, and lipotoxicity stimulated by inflammation. In addition, both clinical and experimental studies have shown significant involvement of numerous inflammatory cytokines in the pathogenesis of insulin resistance (Kahn, 2003). For instance, an increased amount of inflammatory cytokine interleukin-1β (IL-1β) in type 2 diabetes has harmful effects on the activity of IL-1 receptor antagonist proteins (IL-1ra), contributing to β-cell dysfunction and insulin resistance (Malozowski & Sahlroot, 2007). Moreover, elevated levels of inflammatory cytokine interleukin-6 (IL-6) in type 2 diabetes interferes with the interaction between the suppressor of cytokine signaling (SOCS) proteins and insulin receptor, resulting in insulin resistance (Meshkani & Adeli, 2009).

Numerous investigations have reported the beneficial functions of flavonoids, promoting their use as a supplement in treating type 2 diabetes (Testa et al., 2016) (Figure 8.3). Flavonoids have been demonstrated in modulating lipid and carbohydrate metabolism, attenuation of hyperglycemia, insulin resistance, alleviation of stress-sensitive signaling pathways, oxidative stress, and inflammatory processes (Choi & Kim, 2009). In addition, flavonoids chrysin (mice), rutin (rats), and hesperidin and morin have shown their efficacy in reducing inflammatory cytokines IL-1β and IL-6 in diabetic animals, significantly improving insulin resistance, glucose tolerance, and hyperglycemia (Abuohashish et al., 2013; Niture et al., 2014; Sirovina et al., 2016; Visnagri et al., 2014). Type 2 diabetes ultimately results in secondary damage of multiple organs such as nerves, eyes, heart, and kidneys in the affected individual. Not only do flavonoids help in the restoration of glucose homeostasis to attenuate diabetes and participate in the regulation of secondary damage to the peripheral organs. For example, chrysin improves cognitive functions, preventing the progression of diabetic neuropathy. In addition, chrysin also improves renal pathology in diabetic mice (Ahad et al., 2014, Li et al., 2014).

NAFLD has emerged as the most common chronic liver disease associated with obesity, diabetes, and insulin resistance (Ahmed et al., 2015; Chao et al., 2016). The critical characteristics of NAFLD include the deposition of triglycerides and fatty acids in hepatocytes that results in further complications, including hepatocellular carcinoma, liver cirrhosis, and liver fibrosis (Chao et al., 2016; Fotbolcu & Zorlu, 2016). NAFLD comprises a wide range of disorders, such as simple steatosis that does not include inflammation and a more intense form of NAFLD called non-alcoholic steatohepatitis (NASH), defined by lobular inflammation (Bibbo et al., 2018). In addition, the accumulation of lipids in the liver results in enhanced transcription and release of C-reactive protein, TNF-α, and IL-6, causing low-grade inflammation via a decrease in the production of anti-inflammatory compound adiponectin, consequently sensitizing hepatocytes to insulin (Fotbolcu & Zorlu, 2016).

As the pathology of NAFLD is multidimensional, there is no evidence of effective treatment for this disease. In such a scenario, bioactive compounds like flavonoids that regulate multiple pathways are potential candidates for developing therapeutic measures against NAFLD (Figure 8.3). Flavonoids have demonstrated favorable effects on inflammation, oxidative stress, insulin resistance, and lipid metabolism, the critical causative agents of NAFLD (Van De Wier et al., 2017). Silymarin, a flavonoid mix derived from milk thistle, has shown its anti-inflammatory activity in animal models of NASH (Kim et al., 2012). Silibinin, the most active compound in the flavonoid mix of silymarin, reduced the levels of iNOS and ROS along with a decrease in hepatic NF-κB activation in the NASH mouse model (Salamone et al., 2018). In addition, isoflavones from soybean and their derivatives have demonstrated beneficial outcomes in treating NAFLD in animal models. For example, the isoflavones genistein, a soy phytoestrogen, has shown its anti-inflammatory function by reducing mRNA expression of proinflammatory cytokines IL-1β and TNF-α in NAFLD db/db mice (Yoo et al., 2015).

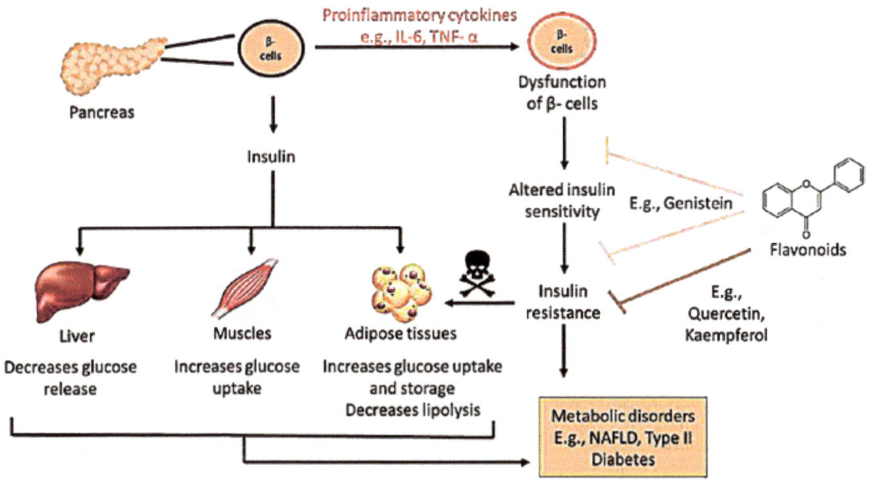

FIGURE 8.3 Role of flavonoids in mediating metabolic disorders. Insulin is produced and secreted by pancreatic β-cells and performs various functions in different tissues and organs, including reducing the glucose release in the liver, increasing glucose uptake in muscles and adipose tissues, and decreasing lipolysis in adipose tissues. Unregulated production of proinflammatory cytokines results in β-cell dysfunction affecting insulin secretion and action and subsequently promoting insulin resistance. Insulin resistance impacts the target tissues of insulin, like increased concentration of fatty acids and glucose in the liver, muscles, and adipose tissues. Flavonoids exhibit their activity via interference with proinflammatory cytokine-induced dysfunction of β-cells and cell death [IL-6: interleukin-6; TNF-α: tumor necrosis factor-α; NAFLD: non-alcoholic fatty liver disease].

8.13 FLAVONOIDS IN NEURODEGENERATIVE DISORDERS

Neurodegenerative disorders lead to progressive and irreversible loss of neurons. Several neurodegenerative disorders, such as multiple sclerosis

(MS), Parkinson's disease (PD), Huntington's disease (HD), and Alzheimer's disease (AD), are accompanied by neuroinflammation (Onasanwo et al., 2016). The complex process of inflammation requires different components of the central nervous system, such as astrocytes, mast cells, microglial cells, macrophages, and ependymal cells (Glass et al., 2010). Macrophages in the central nervous system and the brain (i.e., Microglial cells) are a primary response to a neurological injury (Dheen et al., 2007). In Alzheimer's disease, the activation of microglial cells and astrocytes is followed by activation of TFs like activator protein-1 (AP-1) and nuclear factor-kappa B (NF-κB) that trigger generation of different pro-inflammatory mediators such as TNF-α, Nitrogen oxide (NO), and interleukins as well as ROS (Glass et al., 2010; Dheen et al., 2007). This consequently leads to neurotoxicity or neuronal damage resulting in necrosis and apoptosis. In addition, proinflammatory mediators released from astrocytes and microglial cells activate each other for amplification of inflammatory signals to the brain and central nervous system.

Flavonoids have demonstrated their neuroprotective activities by inhibition of pro-inflammatory cytokine release. Flavonoids exhibit anti-inflammatory activity through interference with the production of inflammatory mediators such as TNF-α, IL-6, and IL-1β in various cell lines (Li et al., 2015). In addition, they have been shown to prevent or delay the development of neurodegenerative disorders at their effective doses in multiple *in vivo* models (Vauzour et al., 2008). Quercetin, baicalein, apigenin, luteolin curcumin, apigenin, wogonin, and many other flavonoids have been found to exert neuroprotective effects. For instance, the flavonoids wogonin exert their anti-inflammatory function by inhibiting inflammatory microglial cells by inhibiting the production of NF-κB and NO (Suk, 2007). In addition, isoflavones like genistein and daidzein have been shown to inhibit IL-1, IL-6, and TNF-α in primary astrocytes (Park et al., 2011).

8.14 FLAVONOIDS IN CARDIOVASCULAR DISEASES (CVDS)

CVDs, such as hypertension, angina, heart failure, and myocardial ischemia, are among the leading cause of death worldwide. Inflammatory processes are the most common pathology of CVDs. Multiple TFs are associated with inflammation in CVDs such as signal transducer and activator of transcription 3 (STAT3) (Kurdi et al., 2018), transcription factor Bcl11b (Daher et al., 2019), T-bet (Haybar et al., 2019), interferon regulatory factors (IRFs), activator protein 1 (AP-1) (Smale & Natoli, 2014), and transcription factor Bcl11b (Daher et al., 2019). However, a critical player in regulating the inflammatory processes is the transcription factor NF-κB (Van Der Heiden et al., 2010). Previous investigations have shown that inhibiting the NF-κB pathway has beneficial impacts on numerous CVDs, including myocardial infarction (Zhao et al., 2017), hypertension (Koeners et al., 2016), and arteriosclerosis (Wang et al., 2016). These findings suggest that NF-κB inhibition is a promising strategy for reducing complications related to CVDs.

A few cellular redox pathways participate in the progression of chronic inflammatory CVD, including NF-κB. The transcription factor NF-κB triggers the expression of inhibitor of kappa B (IκB) kinase in the cytoplasm on stimulation by inflammatory stimulus (Braiser, 2010). Subsequently, canonical and non-canonical signaling pathways result in the migration of NF-κB to the nucleus, thereby initiating the expression of inflammatory mediators in pro-inflammatory cells, T-cells, B-cells, macrophages, and monocytes. For instance, the canonical NF-κB signaling pathway responds rapidly to the inflammatory stimulus, increasing pro-inflammatory cytokines like tumor necrosis factor-alpha (TNF-α), interleukin (IL)-1β, and IL-6 that leads to cell apoptosis. Here, we discuss the anti-inflammatory activities of certain flavonoids via inhibition of the NF-κB pathway in the CVDs.

8.15 ANTI-INFLAMMATORY FUNCTIONS OF FLAVONOIDS IN CVDS THROUGH REGULATION OF NF-κB SIGNALING

8.15.1 RUTIN

The flavonol rutin is present in citrus fruits and buckwheat. In hypertensive rats, rutin administration has been demonstrated to reduce the elevation of blood pressure by enhancing the bioavailability of NO via upregulation of the transcription factor nuclear factor erythroid 2-related factor 2 (Nrf2) that is a major regulator of cellular protective genes and antioxidants and down-regulation of NFκB expression (Oyagbeni et al., 2018). Rutin was also found to decrease the activity of NFκB, increase activation of Nrf in human embryonic kidney cell lines, and promote relaxation of fetal placental arteries obtained from human chronic plates (Sthijns et al., 2017). In addition, in rats with carfilzomib-induced cardiotoxicity, rutin had a protective effect against myocardial hypertrophy by down-regulation of NFκB expression and up-regulation of IκB-α (Imam et al., 2017).

8.15.2 QUERCETIN

The flavonol quercetin is widely found in vegetables and fruits such as shallots, tomatoes, onions, apples, berries, and grapes (Li et al., 2016). In a recent clinical study on patients with chronic systemic inflammation (CSI) as a pathology of stable coronary artery disease (CAD), quercetin demonstrated anti-inflammatory functions by reducing the CSI indicators (Chekalina et al., 2018). In addition to decreasing the expression of NFκB in peripheral blood mononuclear cells (PBMCs), quercetin also decreases the levels of TNF-α and IL-1β in blood serum (Chekalina et al., 2018).

8.15.3 APIGENIN

The flavone apigenin is abundantly present in vegetables and fruits like oranges, onions, grapefruits, and celeries (Ren et al., 2018). Apigenin has been demonstrated to ameliorate fibrosis and cardiac dysfunction in diabetic cardiomyopathy. Apigenin dampened the activity of NFκB and decreased the activity of caspase 3 accompanied by a reduction in markers of oxidative stress, malondialdehyde (MAD), superoxide dismutase (SOD), and glutathione peroxidase (GSH-Px) (Huangjun et al., 2016). Furthermore, in an LPS-induced model of myocardial injury, apigenin has been shown to relieve the injury by regulating both inflammatory cytokines like IL-1β, TNF-α, macrophage inflammatory protein-2 (MIP-2), and macrophage inflammatory protein-1α (MIP-1α) and oxidative stress via NF-κB modulation (Zeng et al., 2015).

8.15.4 CHRYSIN

The flavone Chrysin is found predominantly in honey, propolis, and blue passionflower (Mantawy et al., 2017). In a monocrotaline-induced pulmonary arterial hypertension (PAH) rat model, chrysin was shown to decrease mean pulmonary artery pressure and ventricular systolic pressure. In addition, chrysin also eliminated the upregulated expression of NF-κB, collagen I, and collagen III (Li et al., 2015). Furthermore, in a recent study by Mantawy et al. (2017), chrysin has been demonstrated to prevent increased serum cardiac markers, doxorubicin (DOX)-induced cardiomyopathy, and histopathological alteration in heart of rats by downregulating mitogen-activated protein kinase (MAPK) and NFκB.

8.15.5 KAEMPFEROL

The flavonol Kaempferol is abundantly present in herbs, vegetables, and fruits, including tomatoes, tea, and grapes. In lipopolysaccharide (LPS) stimulated cardiac fibroblasts, kaempferol has been shown to decrease proinflammatory cytokines by inhibition of NFκB activation and AKT signal transduction pathway (Tang et al., 2015). In addition, Kaempferol demonstrated the prevention of cardiac damage in rats through inhibition of protein expression for NFκB, JNK, and p38 (Suchal et al., 2016).

8.15.6 FLAVONOIDS IN ASTHMA

Asthma is a chronic respiratory disorder that includes airway obstruction, inflammation, and bronchial hyperresponsiveness as its major characteristics. The pathology of asthma is complex and involves multiple immune responses as a part of the disease. For example, allergic asthma is stimulated by allergen and mediated by T-helper type 2 (Th2)-cell-associated immune response. Th2 cells generate cytokines critical for antibody IgE class switching in B-cells and recruit eosinophils and mast cells (Schatz & Rosenwasser, 2014). On this basis, flavonoids like flavonol, flavones, isoflavones, and anthocyanidins have been demonstrated to possess a favorable impact on asthma.

8.16 CONCLUSION

Unregulated inflammation has critical implications for the damage of chronic systems that can lead to inflammatory disorders like neurodegenerative disorders, CVD, asthma, or cancer. Recently, numerous studies have been conducted to gain knowledge on the influence of diet on chronic inflammatory disorders. In this relation, the function of flavonoids as an essential part of a healthy diet has become popular due to their anti-inflammatory effects. Current *in vitro* studies have provided novel insights into the numerous variables that distinct flavonoid subclasses can regulate in several stages of inflammation. In addition, *in vivo* has provided important steps towards increasing our knowledge of potential health benefits due to flavonoids. These advances have widened our understanding of the positive effects that a flavonoid-rich diet has in the treatment of chronic inflammatory diseases. However, further scientific efforts and approaches at a clinical level will provide us with a deeper understanding of the antioxidant effects and anti-inflammatory effects of flavonoids in general human health and in chronic diseases (Table 8.3).

TABLE 8.3 Effects of Flavonoid Subclasses on Asthma

Subclass	Type	Effect on Asthma	References
Flavonol	Quercetin Isoquercitrin Kaempferol	Anti-inflammatory activities in a murine model of airways allergic inflammation. Improved ocular symptoms caused by pollinosis. Decreased infiltration of eosinophils in the airways of murine models.	Rogerio et al. (2010); Hirano et al. (2009); Kawai et al. (2009); Gong et al. (2011)
Flavone	Luteolin Tetramethoxyluteolin (methlut)	Reduced histamine and prostaglandin release from human mast cells in culture. Inhibition of release of mediators involved in asthma in human cell lines.	Kimata et al. (2000); Weng et al. (2015)
Flavanol	Catechin	Reduced symptoms of Japanese cedar pollinosis.	Masuda et al. (2014)
Isoflavone	Genistein	• Stability of mast cells • Reduced airway hyperresponsiveness in murine asthmatic models or guinea pigs. • Inhibition of transcription factors GATA-3 and STAT-5. • Inhibition of pro-inflammatory cytokines in mast cells.	Duan et al. (2003); Gao et al. (2012); Kim et al. (2014)
Anthocyanidin	Cyanidin	Attenuated T-helper 17 (Th17) cells induced inflammation with a decrease in airway hyperreactivity in a mouse model of severe asthma.	Liu et al. (2017)

KEYWORDS

- **anti-inflammatory**
- **antioxidant**
- **asthma**
- **cancer**
- **cardiovascular diseases**
- **flavonoids**
- **inflammation**
- **metabolic diseases**
- **neurodegenerative diseases**

REFERENCES

Abuohashish, H. M., Al-Rejaie, S. S., Al-Hosaini, K. A., Parmar, M. Y., & Ahmed, M. M., (2013). Alleviating effects of morin against experimentally-induced diabetic osteopenia. *J. Diabetol. Metab. Syndrome, 5*(1), 1–8.

Agullo, G., Gamet-Payrastre, L., Manenti, S., Viala, C., Rémésy, C., Chap, H., & Payrastre, B., (1997). Relationship between flavonoid structure and inhibition of phosphatidylinositol 3-kinase: A comparison with tyrosine kinase and protein kinase C inhibition. *J. Biochem. Pharmacol., 53*(11), 1649–1657.

Ahad, A., Ganai, A. A., Mujeeb, M., & Siddiqui, W. A., (2014). Chrysin, an anti-inflammatory molecule, abrogates renal dysfunction in type 2 diabetic rats. *J. Toxicol. Appl. Pharmacology, 279*(1), 1–7.

Ahmed, A., Wong, R. J., & Harrison, S. A. (2015). Nonalcoholic fatty liver disease review: Diagnosis, treatment, and outcomes. *J. Clin. Gastroenterol. Hepatology, 13*(12), 2062–2070.

Akiyama, T., Ishida, J., Nakagawa, S., Ogawara, H., Watanabe, S. I., Itoh, N., & Fukami, Y., (1987). Genistein is a specific inhibitor of tyrosine-specific protein kinases. *Journal of Biological Chemistry, 262*(12), 5592–5595.

Aswad, M., Rayan, M., Abu-Lafi, S., Falah, M., Raiyn, J., Abdallah, Z., & Rayan, A., (2018). Nature is the best source of anti-inflammatory drugs: Indexing natural products for their anti-inflammatory bioactivity. *J. Inflamm. Res., 67*(1), 67–75.

Attiq, A., Jalil, J., Husain, K., & Ahmad, W., (2018). Raging the war against inflammation with natural products. *J. Front. Pharmacol., 9*, 976.

Azab, A., Nassar, A., & Azab, A. N., (2016). Anti-inflammatory activity of natural products. *Molecules, 21*(10), 1321.

Bibbò, S., Ianiro, G., Dore, M. P., Simonelli, C., Newton, E. E., & Cammarota, G., (2018). The gut microbiota as a driver of inflammation in nonalcoholic fatty liver disease. *J. Mediators of Inflamm., 2018.*

Brasier, A. R., (2010). The nuclear factor-κB–interleukin-6 signaling pathway mediating vascular inflammation. *J. Cardiovasc. Res., 86*(2), 211–218.

Calder, P. C., Albers, R., Antoine, J. M., Blum, S., Bourdet-Sicard, R., Ferns, G., & Guarner, F., (2009). Inflammatory disease processes and interactions with nutrition. *Br. J. of Nutr., 101*(S1), 1–45.

Chahar, M. K., Sharma, N., Dobhal, M. P., & Joshi, Y. C., (2011). Flavonoids: A versatile source of anticancer drugs. *J. Pharmacogn. Rev., 5*(9), 1.

Chao, C. Y., Battat, R., Al Khoury, A., Restellini, S., Sebastiani, G., & Bessissow, T., (2016). Co-existence of non-alcoholic fatty liver disease and inflammatory bowel disease: A review article. *World J. of Gastroenterol., 22*(34), 7727.

Chekalina, N., Burmak, Y., Petrov, Y., Borisova, Z., Manusha, Y., Kazakov, Y., & Kaidashev, I., (2018). Quercetin reduces the transcriptional activity of NF-kB in stable coronary artery disease. *Indian Heart J., 70*(5), 593–597.

Chen, G. L., Fan, M. X., Wu, J. L., Li, N., & Guo, M. Q., (2019). Antioxidant and anti-inflammatory properties of flavonoids from lotus plumule. *J. Food Chemistry, 277*, 706–712.

Chen, L., Teng, H., Jia, Z., Battino, M., Miron, A., & Yu, Z., (2018). Intracellular signaling pathways of inflammation modulated by dietary flavonoids: The most recent evidence. *Crit. Rev. Food Sci. Nutr., 58*(17), 2908–2924.

Choi, E. J., & Kim, G. H., (2009). 5-fluorouracil combined with apigenin enhances anticancer activity through induction of apoptosis in human breast cancer MDA-MB-453 cells. *J. Oncol. Rep., 22*(6), 1533–1537.

Cialdella-Kam, L., Ghosh, S., Meaney, M. P., Knab, A. M., Shanely, R. A., & Nieman, D. C., (2017). Quercetin and green tea extract supplementation downregulate genes related to tissue inflammatory responses to a 12-week high fat diet in mice. *J. Nutrients, 9*(7), 773.

Constantinou, A. I., Kamath, N., & Murley, J., (1998). Genistein inactivates bcl-2, delays the G2/M phase of the cell cycle, and induces apoptosis of human breast adenocarcinoma MCF-7 cells. *Eur. J. of Cancer, 34*(12), 1927–1934.

Daher, M., Mougenot, N., Rohr, O., Bausero, P., Liu, P., Li, Z., & Parlakian, A., (2019). Role of the transcription factor Bcl11b in cardiac homeostasis and remodeling. *Archives of Cardiovascular Diseases Supplements, 11*(2), 231.

Damon, M., Flandre, O., Michel, F., Perdrix, L., Labrid, C., & Crastes De, P. A., (1987). Effect of chronic treatment with a purified flavonoid fraction on inflammatory granuloma in the rat. Study of prostaglandin E2 and F2 alpha and thromboxane B2 release and histological changes. *J. Arzneimittelforschung, 37*(10), 1149–1153.

Dehmlow, C., Murawski, N., & De Groot, H., (1996). Scavenging of reactive oxygen species and inhibition of arachidonic acid metabolism by silibinin in human cells. *J. Life Sci., 58*(18), 1591–1600.

Duan, W., Kuo, I. C., Selvarajan, S., Chua, K. Y., Bay, B. H., & Wong, W. F., (2003). Anti-inflammatory effects of genistein, a tyrosine kinase inhibitor, on a guinea pig model of asthma. *Am. J. of Respir Crit. Care Med., 167*(2), 185–192.

Fotbolcu, H., & Zorlu, E., (2016). Nonalcoholic fatty liver disease is a multi-systemic disease. *World J. Gastroenterol., 22*(16), 4079.

Fotsis, T., Pepper, M., Adlercreutz, H., Fleischmann, G., Hase, T., Montesano, R., & Schweigerer, L., (1993). Genistein, a dietary-derived inhibitor of *in vitro* angiogenesis. *Proc. Natl. Acad. Sci. U S A., 90*(7), 2690–2694.

Gao, F., Wei, D., Bian, T., Xie, P., Zou, J., Mu, H., & Zhou, X., (2012). Genistein attenuated allergic airway inflammation by modulating the transcription factors T-bet, GATA-3, and STAT-6 in a murine model of asthma. *J. Pharmacology, 89*(3, 4), 229–236.

Garcıa-Lafuente, A. Guillamon, E., Villares, A., Rostagno, M. A., & Martınez, J. A., (2009). Flavonoids as anti-inflammatory agents: Implications in cancer and cardiovascular disease. *J. Inflamm. Res., 58*(9), 537.

Glass, C. K., Saijo, K., Winner, B., Marchetto, M. C., & Gage, F. H., (2010). Mechanisms underlying inflammation in neurodegeneration. *Cell, 140*(6), 918–934.

Gong, J. H., Shin, D., Han, S. Y., Kim, J. L., & Kang, Y. H., (2012). Kaempferol suppresses eosinophil infiltration and airway inflammation in airway epithelial cells and in mice with allergic asthma. *J. Nutr., 142*(1), 47–56.

Guo, Y. Q., Tang, G. H., Lou, L. L., Li, W., Zhang, B., Liu, B., & Yin, S. (2018). Prenylated flavonoids as potent phosphodiesterase-4 inhibitors from *Morus alba*: Isolation, modification, and structure-activity relationship study. *European Journal of Medicinal Chemistry, 144*, 758–766.

Haybar, H., Rezaeeyan, H., Shahjahani, M., Shirzad, R., & Saki, N., (2019). T-bet transcription factor in cardiovascular disease: Attenuation or inflammation factor? *J. Cell. Physiol., 234*(6), 7915–7922.

Hirano, T., Kawai, M., Arimitsu, J., Ogawa, M., Kuwahara, Y., Hagihara, K., & Koyanagi, M., (2009). Preventative effect of a flavonoid, enzymatically modified isoquercitrin on ocular symptoms of Japanese cedar pollinosis. *J. Allergology International, 58*(3), 373–382.

Hotamisligil, G. S., (2006). Inflammation and metabolic disorders. *Nature, 444*(7121), 860–867.

Hou, D. X., & Kumamoto, T., (2010). Flavonoids as protein kinase inhibitors for cancer chemoprevention: Direct binding and molecular modeling. *Antioxid. Redox Signal., 13*(5), 691–719.

Imam, F., Al-Harbi, N. O., Al-Harbia, M. M., Korashy, H. M., Ansari, M. A., Sayed-Ahmed, M. M., & Kazmi, I., (2017). Rutin attenuates carfilzomib-induced cardiotoxicity through inhibition of NF-κB, hypertrophic gene expression, and oxidative stress. *Cardiovasc. Toxicol., 17*(1), 58–66.

Kahn, S., (2003). The relative contributions of insulin resistance and beta-cell dysfunction to the pathophysiology of type 2 diabetes. *J. Diabetologia, 46*(1), 3–19.

Kawai, M., Hirano, T., Arimitsu, J., Higa, S., Kuwahara, Y., & Hagihara, K., (2009). Effect of enzymatically modified isoquercitrin, a flavonoid, on symptoms of Japanese cedar pollinosis: A randomized double-blind placebo-controlled trial. *Int. Arch. Allergy Immunol., 149*(4), 359–368.

Keservani, R. K., & Sharma, A. K., (2014). Flavonoids: Emerging trends and potential health benefits. *Journal of Chinese Pharmaceutical Sciences, 23*(12), 815.

Keservani, R. K., Kesharwani, R. K., Sharma, A. K., Vyas, N., & Chadoker, A., (2010b). Nutritional supplements: An overview. *International Journal of Current Pharmaceutical Review and Research, 1*(1), 59–75.

Keservani, R. K., Kesharwani, R. K., Vyas, N., Jain, S., Raghuvanshi, R., & Sharma, A. K., (2010a). Nutraceutical and functional food as future food: A review. *Der Pharmacia Lettre, 2*(1), 106–116.

Keservani, R. K., Sharma, A. K., & Kesharwani, R. K. (2020). *Nutraceuticals and Dietary Supplements: Applications in Health Improvement and Disease Management*. CRC Press. ISBN: 9781771888738.

Kim, D. H., Jung, W. S., Kim, M. E., Lee, H. W., Youn, H. Y., Seon, J. K., & Lee, J. S., (2014). Genistein inhibits pro-inflammatory cytokines in human mast cell activation through the inhibition of the ERK pathway. *Int. J. Mol. Med., 34*(6), 1669–1674.

Kim, H. K., Son, K. H., Chang, H. W., Kang, S. S., & Kim, H. P., (1999). Inhibition of rat adjuvant-induced arthritis by ginkgetin, a biflavone from Ginkgo biloba leaves. *Planta Med., 65*(5), 465–467.

Kim, M., Lim, S. J., Kang, S. W., Um, B. H., & Nho, C. W., (2014). *Aceriphyllum rossii* extracts, and its active compounds, quercetin, and kaempferol, inhibit IgE-mediated mast cell activation and passive cutaneous anaphylaxis. *J. Agric. Food Chem., 62*(17), 3750–3758.

Kim, M., Yang, S. G., Kim, J. M., Lee, J. W., Kim, Y. S., & Lee, J., (2012). Silymarin suppresses hepatic stellate cell activation in a dietary rat model of non-alcoholic steatohepatitis: Analysis of isolated hepatic stellate cells. *Int. J. Mol. Med., 30*(3), 473–479.

Kimata, M., Shichijo, M., Miura, T., Serizawa, I., Inagaki, N., & Nagai, H., (2000). Effects of luteolin, quercetin, and baicalein on immunoglobulin E-mediated mediator release from human cultured mast cells. *Journal of the British Society for Allergy and Clinical Immunology 30*(4), 501–508.

Knekt, P., Kumpulainen, J., Järvinen, R., Rissanen, H., Heliövaara, M., Reunanen, A., & Aromaa, A., (2002). Flavonoid intake and risk of chronic diseases. *Am. J. Clin. Nutr., 76*(3), 560–568.

Koeners, M. P., Wesseling, S., Sánchez, M., Braam, B., & Joles, J. A., (2016). Perinatal inhibition of NF-KappaB has long-term antihypertensive and renoprotective effects in fawn-hooded hypertensive rats. *Am. J. of Hypertens., 29*(1), 123–131.

Křížová, L., Dadáková, K., Kašparovská, J., & Kašparovský, T., (2019). Isoflavones. *Molecules, 24*(6), 1076.

Kumar, S., & Pandey, A. K., (2013). Chemistry and biological activities of flavonoids: An overview. *The Scientific World Journal, 2013.*

Kurdi, M., Zgheib, C., & Booz, G. W., (2018). Recent developments on the crosstalk between stat3 and inflammation in heart function and disease. *Front. Immunol., 9*, 3029. doi: 10.3389/fimmu.2018.03029.

Kusano, A., Nikaido, T., Kuge, T., Ohmoto, T., Delle, M. G., & Botta, B., (1991). Inhibition of adenosine 3', 5'-cyclic monophosphate phosphodiesterase by flavonoids from licorice roots and 4-aryl coumarins. *Chem. Pharm. Bull. (Tokyo), 39*(4), 930–933.

Lee, H. S., Ko, H. R., Ryu, S. Y., Oh, W. K., Kim, B. Y., Ahn, S. C., & Ahn, J. S., (1997). Inhibition of phospholipase Cγ1 by the prenylated flavonoids from *Sophora flavescens*. *Planta Med., 63*(03), 266–268.

Li, K. C., Ho, Y. L., Hsieh, W. T., Huang, S. S., Chang, Y. S., & Huang, G. J., (2015). Apigenin-7-glycoside prevents LPS-induced acute lung injury via downregulation of oxidative enzyme expression and protein activation through inhibition of MAPK phosphorylation. *Int. J. Mol. Sci., 16*(1), 1736–1754.

Li, P., Zheng, Y., & Chen, X., (2017). Drugs for autoimmune inflammatory diseases: From small molecule compounds to anti-TNF biologics. *Front. Pharmacol., 8*, 460.

Li, R., Zang, A., Zhang, L., Zhang, H., Zhao, L., Qi, Z., & Wang, H., (2014). Chrysin ameliorates diabetes-associated cognitive deficits in Wistar rats. *Neurol. Sci., 35*(10), 1527–1532.

Li, X., Jiang, Q., Wang, T., Liu, J., & Chen, D., (2016). Comparison of the antioxidant effects of quercitrin and isoquercitrin: Understanding the role of the 6 "-OH group. *Molecules, 21*(9), 1246.

Libby, P., (2007). Inflammatory mechanisms: The molecular basis of inflammation and disease. *Nutr. Rev., 65*(suppl_3), S140–S146.

Liu, C., Zhu, L., Fukuda, K., Ouyang, S., Chen, X., Wang, C., & Qin, L., (2017). The flavonoid cyanidin blocks binding of the cytokine interleukin-17A to the IL-17RA subunit to alleviate inflammation *in vivo*. *Sci. Signal., 10*(467).

Liu, H., Yang, Z., Deng, W., & Tang, Q., (2016). GW27-e0367 Apigenin attenuates the cardiac remodeling in experimental diabetic cardiomyopathy. *J. of the Am. Coll. of Cardiol., 68*(16S), C14.

Lolli, G., Cozza, G., Mazzorana, M., Tibaldi, E., Cesaro, L., Donella-Deana, A., & Sarno, S., (2012). Inhibition of protein kinase CK2 by flavonoids and tyrphostin. A structural insight. *Biochemistry, 51*(31), 6097–6107.

Malozowski, S., & Sahlroot, J. T., (2007). Interleukin-1-receptor antagonist in type 2 diabetes mellitus. *N. Engl. J. Med., 357*(3), 302–303; Author Reply 303.

Mantawy, E. M., Esmat, A., El-Bakly, W. M., ElDin, R. A. S., & El-Demerdash, E., (2017). Mechanistic clues to the protective effect of chrysin against doxorubicin-induced cardiomyopathy: Plausible roles of p53, MAPK and AKT pathways. *Scientific Reports, 7*(1), 1–13.

Markovits, J., Linassier, C., Fossé, P., Couprie, J., Pierre, J., Jacquemin-Sablon, A., & Larsen, A. K., (1989). Inhibitory effects of the tyrosine kinase inhibitor genistein on mammalian DNA topoisomerase II. *Cancer Res., 49*(18), 5111–5117.

Masuda, S., Maeda-Yamamoto, M., Usui, S., & Fujisawa, T., (2014). 'Benifuuki' green tea containing o-methylated catechin reduces symptoms of Japanese cedar pollinosis: A randomized, double-blind, placebo-controlled trial. *Allergology International, 63*(2), 211–217.

Matsukawa, Y., Marui, N., Sakai, T., Satomi, Y., Yoshida, M., Matsumoto, K., & Aoike, A., (1993). Genistein arrests cell cycle progression at G2-M. *Cancer Res., 53*(6), 1328–1331.

Medzhitov, R., (2008). Origin and physiological roles of inflammation. *Nature, 454*(7203), 428–435.

Mennen, L. I., Sapinho, D., De Bree, A., Arnault, N., Bertrais, S., Galan, P., & Hercberg, S., (2004). Consumption of foods rich in flavonoids is related to a decreased cardiovascular risk in apparently healthy French women. *J. of Nutr., 134*(4), 923–926.

Meshkani, R., & Adeli, K., (2009). Hepatic insulin resistance, metabolic syndrome, and cardiovascular disease. *Clin. Biochem., 42*(13, 14), 1331–1346.

Mittal, M., Siddiqui, M. R., Tran, K., Reddy, S. P., & Malik, A. B., (2014). Reactive oxygen species in inflammation and tissue injury. *Antioxid. Redox Signal., 20*(7), 1126–1167.

Nimse, S. B., & Pal, D., (2015). Free radicals, natural antioxidants, and their reaction mechanisms. *RSC Advances, 5*(35), 27986–28006.

Niture, N. T., Ansari, A. A., & Naik, S. R., (2014). *Anti-hyperglycemic Activity of Rutin in Streptozotocin-Induced Diabetic Rats: An Effect Mediated Through Cytokines, Antioxidants and Lipid Biomarkers.*

Onasanwo, S. A., Velagapudi, R., El-Bakoush, A., & Olajide, O. A., (2016). Inhibition of neuroinflammation in BV2 microglia by the biflavonoid kolaviron is dependent on the Nrf2/ARE antioxidant protective mechanism. *Mol. Cell Biochem., 414*(1, 2), 23–36.

Ostrand-Rosenberg, S., & Sinha, P., (2009). Myeloid-derived suppressor cells: Linking inflammation and cancer. *J. of Immunol., 182*(8), 4499–4506.

Oyagbemi, A. A., Omobowale, T. O., Ola-Davies, O. E., Asenuga, E. R., Ajibade, T. O., Adejumobi, O. A., & Ayodeji, F., (2018). Ameliorative effect of rutin on sodium

fluoride-induced hypertension through modulation of Kim-1/NF-κB/Nrf2 signaling pathway in rats. *Environ. Toxicol., 33*(12), 1284–1297.
Pahwa, R.; Goyal, A.; Bansal, P.; Jialal, I. Chronic Inflammation; StatPearls Publishing: Tampa, FL, USA, 2019. Available online: https://www.ncbi.nlm.nih.gov/books/NBK493173.
Pan, M. H., Lai, C. S., Dushenkov, S., & Ho, C. T., (2009). Modulation of inflammatory genes by natural dietary bioactive compounds. *J. Agric. Food Chem., 57*(11), 4467–4477.
Panche, A., Diwan, A., & Chandra, S. (2016). Flavonoids: An overview. *J. Nutr. Sci., 5*.
Peng, H. L., Huang, W. C., Cheng, S. C., & Liou, C. J. (2018). Fisetin inhibits the generation of inflammatory mediators in interleukin-1β–induced human lung epithelial cells by suppressing the NF-κB and ERK1/2 pathways. *Int. Immunopharmacol., 60*, 202–210.
Petkov, E., Nikolov, N., & Uzunov, P., (1981). Inhibitory effect of some flavonoids and flavonoid mixtures on cyclic AMP phosphodiesterase activity of rat heart. *Planta Med., 43*(10), 183–186.
Pietta, P. G., (2000). Flavonoids as antioxidants. *J. Nat. Prod., 63*(7), 1035–1042.
Ren, K., Jiang, T., Zhou, H. F., Liang, Y., & Zhao, G. J., (2018). Apigenin retards atherogenesis by promoting ABCA1-mediated cholesterol efflux and suppressing inflammation. *Cell Physiol. Biochem., 47*(5), 2170–2184.
Rogerio, A. P., Dora, C. L., Andrade, E. L., Chaves, J. S., Silva, L. F., Lemos-Senna, E., & Calixto, J. B., (2010). Anti-inflammatory effect of quercetin-loaded microemulsion in the airways allergic inflammatory model in mice. *Pharmacol. Res., 61*(4), 288–297.
Salamone, F., Galvano, F., Cappello, F., Mangiameli, A., Barbagallo, I., & Volti, G. L., (2012). Silibinin modulates lipid homeostasis and inhibits nuclear factor kappa B activation in experimental nonalcoholic steatohepatitis. *Transl. Res., 159*(6), 477–486.
Saponara, R., & Bosisio, E., (1998). Inhibition of cAMP-phosphodiesterase by biflavones of *Ginkgo biloba* in rat adipose tissue. *J. Nat. Prod., 61*(11), 1386, 1387.
Schatz, M., & Rosenwasser, L., (2014). The allergic asthma phenotype. *J. Allergy Clin. Immunol. Pract., 2*(6), 645–648.
Schmitz, M. L., & Kracht, M., (2016). Cyclin-dependent kinases as coregulators of inflammatory gene expression. *Trends Pharmacol. Sci., 37*(2), 101–113.
Sirovina, D., Oršolić, N., Končić, M., Kovačević, G., Benković, V., & Gregorović, G., (2013). Quercetin vs chrysin: Effect on liver histopathology in diabetic mice. *Hum. Exp. Toxicol., 32*(10), 1058–1066.
Smale, S. T., & Natoli, G., (2014). Transcriptional control of inflammatory responses. *Cold Spring Harb. Perspect. Biol., 6*(11), a016261.
Sthijns, M. M., Schiffers, P. M., Janssen, G. M., Lemmens, K. J., Ides, B., Vangrieken, P., & Arnér, E. S., (2017). Rutin protects against H_2O_2-triggered impaired relaxation of placental arterioles and induces Nrf2-mediated adaptation in human umbilical vein endothelial cells exposed to oxidative stress. *Biochimica et Biophysica Acta (BBA) – General Subjects, 1861*(5), 1177–1189.
Suchal, K., Malik, S., Gamad, N., Malhotra, R. K., Goyal, S. N., & Chaudhary, U. (2016). Kaempferol attenuates myocardial ischemic injury via inhibition of the MAPK signaling pathway in an experimental model of myocardial ischemia-reperfusion injury. *Oxid. Med. Cell. Longev., 2016*.
Tang, X. L., Liu, J. X., Dong, W., Li, P., Li, L., Hou, J. C., & Ren, J. G., (2015). Protective effect of kaempferol on LPS plus ATP-induced inflammatory response in cardiac fibroblasts. *Inflammation, 38*(1), 94–101.
Testa, R., Bonfigli, A. R., Genovese, S., De Nigris, V., & Ceriello, A., (2016). The possible role of flavonoids in the prevention of diabetic complications. *Nutrients, 8*(5), 310.

Thameem, D. S., Kaur, C., & Ling, E. A., (2007). Microglial activation and its implications in brain diseases. *Curr. Med. Chem., 14*(11), 1189–1197.

Todoric, J., Antonucci, L., & Karin, M., (2016). Targeting inflammation in cancer prevention and therapy. *Cancer Prev. Res. (Phila), 9*(12), 895–905.

Van De, W. B., Koek, G. H., Bast, A., & Haenen, G. R., (2017). The potential of flavonoids in the treatment of non-alcoholic fatty liver disease. *Crit. Rev. Food Sci. Nutr., 57*(4), 834–855.

Van, D. H. K., Cuhlmann, S., Luong, L. A., Zakkar, M., & Evans, P. C., (2010). Role of nuclear factor κB in cardiovascular health and disease. *Clin. Sci. (Lond)., 118*(10), 593–605.

Vauzour, D., Vafeiadou, K., Rodriguez-Mateos, A., Rendeiro, C., & Spencer, J. P., (2008). The neuroprotective potential of flavonoids: A multiplicity of effects. *Genes Nutr., 3*(3, 4), 115–126.

Visnagri, A., Kandhare, A. D., Chakravarty, S., Ghosh, P., & Bodhankar, S. L., (2014). Hesperidin, a flavanoglycone, attenuates experimental diabetic neuropathy via modulation of cellular and biochemical markers to improve nerve functions. *Pharm. Biol., 52*(7), 814–828.

Wahlang, B., McClain, C., Barve, S., & Gobejishvili, L. (2018). Role of cAMP and phosphodiesterase signaling in liver health and disease. *Cell Signal., 49*, 105–115.

Weng, Z., Patel, A. B., Panagiotidou, S., & Theoharides, T. C. (2015). The novel flavone tetramethoxyluteolin is a potent inhibitor of human mast cells. *J. Allergy Clin. Immunol., 135*(4), 1044–1052. e1045.

Yang, L., Huang, J., Ren, X., Gorska, A. E., Chytil, A., Aakre, M., & Lin, P. C., (2008). Abrogation of TGFβ signaling in mammary carcinomas recruits Gr-1+ CD11b+ myeloid cells that promote metastasis. *Cancer Cell, 13*(1), 23–35.

Yao, J., Han, C., Yang, J., Chaudhry, M. T., Wang, S., et al., (2016). Quercetin, inflammation and immunity. *Nutrients, 8*(3), 167.

Yokoyama, T., Kosaka, Y., & Mizuguchi, M., (2015). Structural insight into the interactions between death-associated protein kinase 1 and natural flavonoids. *J. Med. Chem., 58*(18), 7400–7408.

Yoo, N. Y., Jeon, S., Nam, Y., Park, Y. J., Won, S. B., & Kwon, Y. H. (2015). Dietary supplementation of genistein alleviates liver inflammation and fibrosis mediated by a methionine-choline-deficient diet in db/db mice. *J. Agric. Food Chem., 63*(17), 4305–4311.

Zeng, P., Liu, B., Wang, Q., Fan, Q., Diao, J. X., & Tang, J., (2015). Apigenin attenuates atherogenesis through inducing macrophage apoptosis via inhibition of AKT Ser473 phosphorylation and downregulation of plasminogen activator inhibitor-2. *Oxidative Medicine and Cellular Longevity, 2015.*

Zhao, H., Shen, R., Dong, X., & Shen, Y. (2017). Murine double minute-2 inhibition attenuates cardiac dysfunction and fibrosis by modulating the NF-κB pathway after experimental myocardial infarction. *Inflammation, 40*(1), 232–239.

Zuo, H. J., Wang, W. H., Deng, L. Q., & Su, J. L., (2018). Control of cardiovascular disease risk factors among patients with type II diabetes in a primary-care setting in Beijing. *Journal of the American Society of Hypertension, 12*(2), 128–134.

CHAPTER 9

CURRENT TRENDS IN THE HEALTH BENEFITS OF FLAVONOIDS

HARSH MOHAN,[1] MONIKA CHAUHAN,[1] AJAY KUMAR,[2] PRAGATI SAINI,[2] and DIWAKAR CHAUHAN[3]

[1]*Department of Forensic Science, School of Basic and Applied Sciences, Galgotias University, Greater Noida, Uttar Pradesh, India*

[2]*Department of Life Science, School of Basic and Applied Sciences, Galgotias University, Greater Noida, Uttar Pradesh, India*

[3]*Department of Chemistry, School of Basic and Applied Sciences, Galgotias University, Greater Noida, Uttar Pradesh, India*

ABSTRACT

Fruits, tea, stems, bark, roots, grains, flowers, vegetables, and wine all contain flavonoids, a collection of usual compounds with varying phenolic structures. Flavonoid chemicals are plant-derived molecules that can be found in various sections of the plant in nature. Vegetables utilize flavonoids to help them develop and protect themselves against plaque. Plants, animals, and microbes all use flavonoids for a range of biological functions. Flavonoids have long been known to be synthesized in specific locations in plants, and they are dependable for the aroma and color of flowers, as well as the color and aroma of fruits, which draw pollinators and, as a result, fruit dispersion, which aids in seed and spore germination, as well as the growth and growth of seedlings. The health benefits of flavonoids derived from dietary sources have been the subject of current study. These ordinary compounds are well-known for their human health benefits, and attempts are currently being conducted to segregate the components. Flavonoids have become an

Flavonoids as Nutraceuticals. Rajesh K. Kesharwani, Deepika Saini, Raj K. Keservani, and Anil Kumar Sharma (Eds.)

essential factor in a wide range of nutraceutical, pharmacological, cosmetic, therapeutic, and uses. However, due to the intricacy of flavonoids' existence in diverse food sources, the diversity of dietary cultures, and the incidence of a vast quantity of flavonoids in nature, precisely quantifying daily flavonoid consumption remains a challenge.

9.1 INTRODUCTION

Flavonoids are a broad collection of plant natural products produced from phenylpropanoid and acetate precursors that play essential roles in plant growth and development, as well as defense against microbes and pests (Rice-Evans et al., 1997). Chalcones, flavanones, flavanols, flavonols, flavones, anthocyanidins, isoflavones, and their glycosides, condensed tannins, and so on are all subfamilies of flavonoids (Williams & Graver, 2004; Keservani & Sharma, 2014) They are phenolic chemicals with a low molecular weight that are found throughout the plant world. They are one of the most distinctive groups of chemicals found in higher plants (Havsteen, 2002). Flavonoids are bioactive polyphenols with a low molecular weight that play a significant role in photosynthesizing cells (Cushnie & Lamb, 2000). Flavonoids are a class of plant phenolics that includes roughly 10,000 distinct chemicals with a chemical structure that consists of two aromatic rings linked by a three-carbon chain, producing a heterocyclic ring. Flavonoids are an essential component in a number of nutraceuticals (Keservani et al., 2010a, b, 2020), pharmacological, medical, and cosmetic applications (Metodiewa et al., 1997) and are linked to a wide range of health-promoting benefits. Flavonoids are the most common polyphenols in human diets, accounting for more than half of the 8,000 naturally occurring phenolic chemicals found in blackberries, black currants, blues, grapes, strawberries, cherries, plums, cranberries, and pomegranates (Balasundram et al., 2006). Since of their multifaceted health impacts on animal and human health, as well as their ubiquity in the plant kingdom, flavonoids have sparked a lot of attention in the last decade. Because of their possible involvement in improving health and avoiding chronic degenerative illnesses, they've been dubbed "functional components" and "health-promoting biomolecules" in new fiction (Niiveldt et al., 2001). Flavonoids' functional hydroxyl groups act as antioxidants by scavenging free radicals or chelating metal ions. This aids in the prevention of radical production, which damages biomolecules and causes oxidative stress, as well as a variety of diseases. Flavonoids provide protection against diseases like cancer, cardiovascular and respiratory ailments, arthritis, and premature aging. They help the human body's antioxidant defense system and also activate human defensive enzyme systems (Kumar et al., 2013).

9.2 CLASSIFICATION OF FLAVONOIDS

Flavonoids are a type of in nature occurring polyphenolic chemical found across the plant kingdom (Brodowska, 2017). Flavonoids are classified into subgroups depending on the place of the B ring in relation to the C ring, as well as the degree of corrosion and saturation of the heterocyclic ring (Teles et al., 2018). Isoflavone is a unique subgroup of flavonoids that has the B ring attached at position 2 on the C ring, whereas flavonol, flavones (Keservani et al., 2010a), flavanone, flavanonol, anthocyanidin, and flavanol have the B ring attached at position 2 on the C ring (Rana & Gulliva, 2019). Natural flavonoids, on the other hand, are frequently changed enzymatically throughout process such as anthocyanidin, hydroxylation glycosylation, prenylation, methylation, sulfation, and acetylation, resulting in a plethora of aglycone derivatives with distinct biochemical properties (Rauter et al., 2007). The occurrence of a double bond among the carbons at positions 2 and 3 on the C ring, as well as a ketone group at position 4 on the C ring, is a common flavone feature. There are about 400 different forms of aglycone flavones, including approximately 500 O-glycosyl and 300 C-glycosyl flavones (Zhang et al., 2013). Flavonols have a similar structure to flavones, with the exception of an additional hydroxyl group at C-3. C-glycosides are rare in flavonols, although aglycones (450 types) and O-glycosides (900 types) are found in most plants (Jiang et al., 2016). Flavanonols are flavanones' 3-hydroxy derivative. Dihydroflavanonols are another name for flavanonols. Taxifolin, a flavanonol, has a long list of therapeutic pharmacological qualities, including enhancing capillary microcirculation, reducing damage to diabetic vascular systems, and increasing blood flow in the retinal region of the eye (Pietta et al., 2003). Flavanonols, like flavones, flavanones, and flavonols are common in citrus fruits. The fundamental structure of anthocyanidins differs somewhat from that of the other flavonoid subgroups. Anthocyanidins are a kind of flavylium ion with a positive charge on the first oxygen atom on the C ring. Anthocyanidins, dissimilar flavanones, and flavonols lack a ketone group at position 4 on the C ring (Panche et al., 2016). Anthocyanidins are rarely found in fresh plants because they are unstable. Anthocyanidins are most typically found as anthocyanins in their glycosylated form. Flavanols lack the ketone group at position 4 of the C ring and the double bond between C-2 and C-3. Because they have a hydroxyl group linked to position 3 on the C ring, they are also recognized as flavan-3-ols.

9.3 SOURCE OF FLAVONOIDS

Flavonoids are the generally widespread and extensively dispersed category of plant phenolic chemicals, with flavonoids (Keservani & Sharma, 2014) found in almost all plant components, especially photosynthesizing plant cells. They have an important role in both animal and human nutrition (Harborne & Turner, 1984). Flavonoids are plant phytochemicals that cannot be produced by animals or humans. Animal flavonoids are thought to come from the plants that the animals eat rather than being biosynthesized in the wild (Clifford & Cuppett, 2000). Around 5,000 distinct flavonoids produced from plants have been identified (Cook & Samman, 1996). The three most frequent flavonols are myricetin, kaempferol, and quercetin, which are the most prevalent flavonoids in foods. Citrus fruits contain flavanones, whereas celery has flavones. Black and green teas, as well as red wine, are high in catechins, whereas anthocyanins are originated in strawberries and other berries. Soy meals are virtually entirely made up of isoflavones (Ho et al., 1994) flavonoids, which are found in all plant diets, are a significant coloring component of flowering plants (Peterson & Dwyer, 1998). Flavonoids in food are liable for color, flavor, fat oxidation avoidance, and vitamin and enzyme protection. Flavonoid distribution in plants is influenced by a number of variables, including variety and light exposure. Light accelerates the production of higher oxidized flavonoids. Flavanones are originated mostly in citrus fruits and iso-flavonoids in legumes (Huang et al., 1994) and flavones primarily in herbs as coloring agents, whereas catechins and anthocyanins are found in vegetables, fruits, and teas. More food flavonoids from diverse plants are likely to be recognized in the future, without a doubt.

9.4 BIOLOGICAL AND CHEMICAL ACTIVITIES OF FLAVONOIDS

Flavonoids' chemical properties are determined by their degree of hydroxylation, structural class, other conjugations and replacements, and polymerization degree. They are different in arrangement around the heterocyclic oxygen ring, but they all contain the same carbon skeleton (C6-C3-C6) (Vessal et al.,

2003; Ghasemzadeh, 1986). Flavonoids are phytonutrients that belong to the polyphenol family. Polyphenols have been utilized in Ayurvedic and Chinese medicine for centuries. They are linked to cognitive function, blood pressure management, blood sugar, and skin protection, as well as antioxidant and anti-inflammatory action, according to the Worldwide Healing Center. Oranges were used to isolate the chemical. Vitamin P was given to it since it was thought to be associated with a novel class of vitamins at the time. It was later discovered that this molecule was a flavonoid (rutin), and there are now over 4,000+ kinds of flavonoids (Middleton, 1998). Aside from their basic chemical property, flavonoids have a broad range of biological activities that contribute to human health. These actions include anti-inflammatory, anti-ulcer, antiviral, anti-cancer, and anti-diabetic properties, among others (Clifford & Cuppett, 2000; Cook & Samman, 1996). Flavonoids and their metabolites' metabolic activity are determined by their chemical structure and the relative direction of different moieties within the molecule.

9.5 BIOLOGICAL EFFECTS OF FLAVONOIDS ON HUMAN HEALTH

olive and soybean oils, green vegetables, chocolate, red wine, fruits, and teas all contain flavonoids, which contribute to their antioxidant qualities. A range of life actions have been described for flavonoids, including antiallergic, anti-inflammatory, antiviral, antiproliferative, and anticarcinogenic properties, as well as impacts on mammalian metabolism (Ren et al., 2003). Flavonoids have gotten a lot of press because of their antioxidant properties in the prevention of human illnesses, including cancer and cardiovascular disease, as well as some pathological problems like duodenal and gastric ulcers, vascular fragility, allergies, and viral and bacterial infections (Zand et al., 2002). Overall, flavonoids have been reported to have antidiabetic, antiallergic, antiviral, antioxidative, gastroprotective, anti-inflammatory, and antineoplastic effects (Duthie et al., 2000; Lman et al., 1997).

9.5.1 ANTIOXIDANT

Flavonoids, through inhibiting oxidative damage, may protect against cancer and anticarcinogenesis. *In vitro* and in human models, flavonoids are demonstrated to exhibit both pro-oxidant and antioxidant properties. Flavonoids have been classified as "high-level" ordinary antioxidants due to their ability to neutralize active oxygen species and free radicals (Klahorst, 2002; Unno et al., 2000). Flavonoids' antioxidant mechanisms include free radical chain breaking, metal chelation, and oxygen quenching, as well as enzymatic inhibition. When oxidized by peroxidase/hydrogen peroxide, flavonoids, including naringenin, hesperetin, and apigenin, formed pro-oxidant metabolites which damaged glutathione and NADH (Duthie et al., 1999). Flavonoids are shown to chelate copper and iron, which might explain some of their antioxidant properties. Flavonoids' antioxidant processes include free radical chain breaking, metal chelation, and oxygen quenching, as well as enzymatic inhibition (Birt et al., 2001). When oxidized by peroxidase/hydrogen peroxide, flavonoids, including naringenin, hesperetin, and apigenin, formed pro-oxidant metabolites that oxidized glutathione and NADH. Flavonoids chelate copper and iron, which might explain some of their antioxidant properties (Bors et al., 1996).

9.5.2 ANTI-INFLAMMATORY

Inflammation is the body's complicated biological reaction to adverse stimuli, including tissue injury, chemical irritation, pathogen infection, and damaged cells (Nijveldt et al., 2001). Immune cells, blood arteries, and chemical mediators all play a role in this protective response. The discharge of chemical mediators at the spot of tissue injury triggers the immigration of immune cells from blood vessels. Inflammatory illnesses like leukemia, sepsis, asthma, psoriasis, ileitis/colitis, sclerosis, atherosclerosis, rheumatoid arthritis, allergic rhinitis, and others have all been linked to flavonoids (Hein et al., 2002). This is followed by the need for inflammatory cells, the production of reactive oxygen species (ROS), reactive nitrogen species (RNS), and proinflammatory cytokines in order to remove invading invaders and heal damaged tissues (Cos et al., 1998). Inflammation is normally self-limiting and quick, but abnormal resolution and persistent inflammation can lead to a variety of chronic diseases (Pan et al., 2010). Anti-inflammatory properties are known to exist in Hesperidin, Luteolin, and Quercetin. They primarily impact enzyme systems involved in inflammatory process generation (Mishra et al., 2013). Phosphodiesterases involved in cell activation are similarly inhibited by flavonoids. It transfers hydrogen and peroxynitrite radicals, peroxyl, and an electron to hydroxyl, stabilizing them and giving birth to a moderately stable flavonoid radical (Cao et al., 1997). This makes it a determinant of ROS and RNS scavenging. Flavonoids also protect lipid-peroxidized cell membranes. Thus, flavonoids have an important role as antioxidants in the protection of oxidative stress-related illnesses (Ramchoun et al., 2009).

9.5.3 ANTI-BACTERIAL

Flavonoids are extremely efficient antibacterial compounds against a wide variety of pathogens because plants may manufacture them in response to microbial infection. Plants that have a high number of flavonoids have antibacterial action, according to several studies (Mishra et al., 2011). Antimicrobial action has been demonstrated in apigenin, glycosides, flavone flavanones, isoflavones, flavonol galanin, and chalcones (Pandey et al., 2010). The capacity of antimicrobials to enzymes inactivate microbial adhesins, cell envelope conveys proteins, and other proteins may be connected to their mechanism of action. Lipophilic flavonoids have also been linked to bacterial membrane disruption (Cushnie et al., 2005; Cowan, 1999; Mishra et al., 2009).

9.5.4 ANTIVIRAL

Flavonoids have powerful antiviral properties. They aid in the reserve of several enzymes involved in the viral biological cycle (Prasad et al., 2010). It has been discovered that flavonoids and their enzyme-inhibitory action have a structural and functional connection. Flavon-3-ol was shown to be more efficient than flavonones and flavones in inhibiting HIV infections (HIV1, HIV2), as well as other immunodeficiency viruses (Kreft et al., 1999; Yao et al., 2004).

9.5.5 ANTI-CANCER

Cancer is a multi-step illness involving metabolic, chemical, physical, environmental, and genetic variables, all of which have a part in cancer's initiation and progression. Many polyphenolic chemicals, together with phenolic acids, anthocyanidins, flavonoids, and tannins, have a wide range of medical properties, as well as anticancer properties (Namiki, 1990). By modulating diverse receptors and enzymes in the indication transduction pathway linked to apoptosis, differentiation, cellular proliferation, inflammation, metastasis, angiogenesis, and hitch of multidrug opposition, they have been reported to hinder the initiation, progression, and promotion of cancer. Flavonoids have been shown to have potential uses in anti-cancer treatments because of their various molecular modes of exploit (Dixon et al., 1983). Flavonoids have a significant impact on the immunological processes that occur during the genesis and progression of cancer. They have the ability to influence a variety of biological processes in cancer, including cell proliferation, cell differentiation, vascularization, and apoptosis (Walker et al., 2000). Flavonoid-induced kinase regulation has a high connection with apoptosis, cell proliferation, and tumor cell invasive behavior *in vitro* (Kuhnau, 1976). Flavonoids primarily promote the carcinogenicity start and support phases, as well as impacts on development and hormonal activity (Rice-Evanas, 1976).

9.6 CONCLUSION

Over the last 10 years, flavonoids have garnered a lot of attention in the scientific literature, and a range of possible positive benefits have been discovered. Flavonoids are usually harmless and have a wide range of biologically useful properties. Flavonoids may be found in abundance in the human diet, such as tea, red wine, fruits, and vegetables. Because of its health advantages, this category of chemicals is being studied extensively. Dietary flavonoids' function in cancer prevention is hotly debated. Flavonoids are mostly found in fruits and vegetables, with tea and wine serving as secondary sources. Flavonoids' functions are attributed to scavenging (chelating) capacities, antioxidative activities, and interactions with enzyme systems, while a variety of medical possessions have been investigated or are being tested (e.g., anti-HIV activities and anticancer, avoidance of blood vessel disorders, and coronary heart disease), with some promising results. The bioavailability of flavonoids, as well as the purpose assessment of oxidative compensation *in vivo*, must be the emphasis of flavonoid research. More study is essential to establish an accurate and compelling system or model for evaluating human flavonoid consumption and metabolism, as well as their purported health effects. Future studies will focus on the exchanges of flavonoids with receptor molecules in the conduct of chronic and acute illnesses.

KEYWORDS

- **anti-inflammatory activity**
- **dietary sources**
- **flavonoids**
- **human health**
- **metabolic disorders**
- **microbes**
- **phosphodiesterases**
- **vegetables**
- **wine**

REFERENCES

Balasundram, N., Sundram, K., & Samman, S., (2006). Phenolic compounds in plants and agri-industrial by-products: Antioxidant activity, occurrence, and potential uses. *Food Chemistry, 99*, 191–203.

Birt, D. F., Hendrich, S., & Wang, W., (2001). Dietary agents in cancer prevention: Flavonoids and isoflavonoids. *Pharmacology and Therapeutics, 90,* 157–177.

Bors, W., Heller, W., Michel, C., & Stettmaier, K., (1996). Flavonoids and polyphenols: Chemistry and biology. In: Cadenas, E., & Packer, L., (eds.), *Handbook of Antioxidants* (pp. 409–468). New York: Marcel Dekker Inc.

Brodowska, K. M., (2017). Natural flavonoids: Classification, potential role, and application of flavonoid analogs. *European Journal of Biological Research, 7,* 108–123.

Cao, G., Sofic, E., & Prior, R. L., (1997). Antioxidant and prooxidant behavior of flavonoids: Structure-activity relationships. *Free Radical Biology and Medicine, 22*(5), 749–760.

Clifford, A. H., & Cuppett, S. L., (2000). Review: Anthocyanins—Nature, occurrence and dietary burden. *Journal of Science and Food Agriculture, 80,* 1063–1072.

Cook, N. C., & Samman, S., (1996). Review: Flavonoids—Chemistry, metabolism, cardioprotective effects, and dietary sources. *Journal of Nutritional Biochemistry, 7,* 66–76.

Cos, P., Ying, L., Calomme, M., Hu, J. P., Cimanga, K., & Van, P. B., (1998). Structure-activity relationship and classification of flavonoids as inhibitors of xanthine oxidase and superoxide scavengers. *Journal of Natural Products, 61,* 71–76.

Cowan, M. M., (1999). Plant products as antimicrobial agents. *Clinical Microbiology Reviews, 12*(4), 564–582.

Cushnie, T. P. T., & Lamb, A. J., (2005). Antimicrobial activity of flavonoids. *International Journal of Antimicrobial Agents, 26*(5), 343–356.

Dixon, R. A., Dey, P. M., & Lamb, C. J., (1983). Phytoalexins: Enzymology and molecular biology. *Advances in Enzymology and Related Areas of Molecular Biology, 55,* 1–136.

Duthie, G. G., Duthie, S. J., & Kyle, J. A. M., (2000). Plant polyphenols in cancer and heart disease: Implications as nutritional antioxidants. *Nutritional Research Review, 13,* 79–106.

Duthie, S. J., & Dobson, V. L., (1999). Dietary flavonoids protect human colonocyte DNA from oxidative attack *in vitro*. *European Journal of Nutrition, 38,* 28–34.

Ghasemzadeh, A., & Jaafar, H. Z. E., (1986). Anticancer and antioxidant activities of young Malaysian ginger (*Zingiber officinale Roscoe*) varieties grown under different CO_2 concentrations. *Journal of Medicinal Plant Research, 5,* 3247–3255.

Harborne, J. B., & Turner, B. L., (1984). *Plant Chemosystematics*. London: Academic Press.

Havsteen, B., (2002). The biochemistry and medical significance of the flavonoids. *Pharmacology Therapy, 96,* 67–202.

Heim, K. E., Tagliaferro, A. R., & Bobilya, D. J., (2002). Flavonoid antioxidants: Chemistry, metabolism and structure-activity relationships. *Journal of Nutritional and Biochemistry, 13,* 572–584.

Ho, C. T., Osawa, T., Huang, M. T., & Rosen, R. T., (1994). *Food Phytochemicals for Cancer Prevention. II: Teas, Spices, and Herbs* (p. 370). Austin, TX: American Chemical Society, Washington, DC.

Huang, M. T., Osawa, T., Ho, C. T., & Rosen, R. T., (1994). *Food Phytochemicals for Cancer Prevention; I: Fruits and Vegetables* (p. 370). Austin, TX: American Chemical Society, Washington, DC.

Jiang, N., Doseff, A. I., & Grotewold, E. (2016). Flavones: From biosynthesis to health benefits. *Plants, 5,* 27–35.

Keservani, R. K., & Sharma, A. K., (2014). Flavonoids: Emerging trends and potential health benefits. *Journal of Chinese Pharmaceutical Sciences, 23*(12), 815.

Keservani, R. K., Kesharwani, R. K., Sharma, A. K., Vyas, N., & Chadoker, A., (2010b). Nutritional supplements: An overview. *International Journal of Current Pharmaceutical Review and Research, 1*(1), 59–75.

Keservani, R. K., Kesharwani, R. K., Vyas, N., Jain, S., Raghuvanshi, R., & Sharma, A. K., (2010a). Nutraceutical and functional food as future food: A review. *Der Pharmacia Lettre, 2*(1), 106–116.

Keservani, R. K., Sharma, A. K., & Kesharwani, R. K. (2020). *Nutraceuticals and Dietary Supplements: Applications in Health Improvement and Disease Management*. CRC Press. ISBN: 9781771888738.

Klahorst, S., (2002). Exploring antioxidants. *World Food Ingredients*, 54–59.

Kreft, S., Knapp, M., & Kreft, I., (1999). Extraction of rutin from buckwheat (*Fagopyrum esculentum* Moench) seeds and determination by capillary electrophoresis. *Journal of Agricultural and Food Chemistry, 47*(11), 4649–4652.

Kuhnau, J., (1976). The flavonoids. A class of semi-essential food components: Their role in human nutrition. *World Review of Nutrition and Dietic, 24,* 117–191.

Kumar, S., & Pandey, A. K., (2013). Antioxidant mediated protective effect of *Parthenium hysterophorus*; against oxidative damage using in-vitro models. *BMC Complementary and Alternative Medicine, 13,* 120–125.

Lman, P. C. H., Van, T. J. M. P., Buysman, M. N. C. P., Gaag, M. S., Mengelers, M. J. B., De Vries, J. H. M., & Katan, M. B., (1997). Relative bioavailability of the anti-oxidant flavonoid quercetin from various foods in man. *FEBS Letters, 418,* 152–156.

Metodiewa, D., Kochman, A., & Karolczak, S., (1997). Evidence for antiradical and antioxidant properties of four biologically active N, N, diethylaminoethyl ethers of flavanone oximes: A comparison with natural polyphenolic flavonoid (rutin) action. *Biochemistry and Molecular Biology, 41,* 1067–1075.

Middleton, E. J., (1998). Effect of plant flavonoids on immune and inflammatory cell function. *Advances in Experimental Medicine and Biology, 439,* 175–182.

Mishra, A. K., Mishra, A., Kehri, H. K., Sharma, B., & Pandey, A. K., (2009). Inhibitory activity of Indian spice plant *Cinnamomum zeylanicum* extracts against *Alternaria solani* and *Curvularia lunata,* the pathogenic dematiaceous molds. *Annals of Clinical Microbiology and Antimicrobials, 8,* 9–15.

Mishra, A., Kumar, S., & Pandey, A. K., (2013). Scientific validation of the medicinal efficacy of *Tinospora cordifolia. The Scientific World Journal, 8,* 29–34.

Mishra, A., Kumar, S., Bhargava, A., Sharma, B., & Pandey, A. K., (2011). Studies on in-vitro antioxidant and anti-staphylococcal activities of some important medicinal plants. *Cellular and Molecular Biology, 57*(1), 16–25.

Namiki, M., (1990). Antioxidants/antimutagens in food. *Critical Reviews in Food Science and Nutrition, 29*(4), 273–300.

Nijveldt, R. J., Van, N. E., Van, H. D. E. C., Boelens, P. G., Van, N. K., & Van, L. P. A. M., (2001). Flavonoids: A review of probable mechanisms of action and potential applications. *American Journal of Clinical Nutrition, 74,* 418–425.

Pan, M. H., Lai, C. S., & Ho, C. T., (2010). Anti-inflammatory activity of natural dietary flavonoids. *Food and Function, 1*(1), 15–31.

Panche, A. N., Diwan, A. D., & Chandra, S. R. (2016). Flavonoids: An overview. *Journal of Nutrition and Science, 5,* 47–55.

Pandey, A. K., Mishra, A. K., Mishra, A., Kumar, S., & Chandra, A., (2010). Therapeutic potential of *C. zeylanicum* extracts: An antifungal and antioxidant perspective. *International Journal of Biological and Medical Research, 1,* 228–233.

Peterson, J., & Dwyer, J., (1998). Flavonoids: Dietary occurrence and biochemical activity. *Nutritional Research, 18,* 1995–2018.

Pietta, P., Minoggio, M., & Bramati, L., (2003). Plant polyphenols: Structure, occurrence, and bioactivity. *Stud. Natural Product Chemistry, 28,* 257–312.

Prasad, N. K., Xie, H., & Hao, J., (2010). Antioxidant and anticancer activities of 8-hydroxypsoralen isolated from wampee [*Clausena lansium* (Lour.) Skeels] peel. *Food Chemistry, 118*(1), 62–66.

Ramchoun, M., Harnafi, H., Alem, C., Benlys, M., Elrhaffari, L., & Amrani, S., (2009). Study on antioxidant and hypolipidemic effects of polyphenol-rich extract from *Thymus vulgaris* and *Lavendula multifida. Pharmacognosy Research, 1,* 106–112.

Rana, A. C., & Gulliya, B., (2019). Chemistry and pharmacology of flavonoids: A review. *Indian Journal of Pharmacy & Education Research, 53,* 8–20.

Rauter, A. P., Lopes, R. G., & Martins, A., (2007). C-glycosyl flavonoids: Identification, bioactivity and synthesis. *Natural Product and Communities, 2,* 1175–1196.

Ren, W., Qian, Z., Wang, H., Zhu, L., & Zhang, L., (2003). Flavonoids: Promising anticancer agents. *Medicinal Res. Rev., 23*(4), 519–534.

Rice-Evans, C. A., Miller, N. J., & Paganga, G., (1976). Structure antioxidant activity relationships of flavonoids and phenolic acids. *Free Radical Biology and Medicine, 20*(7), 933–956.

Rice-Evans, C. A., Miller, N. J., & Paganga, G., (1997). Antioxidant properties of phenolic compounds. *Trends in Plant Science, 2,* 152–159.

Teles, Y. C. F., Souza, M. S. R., & Souza, M. D. F. V. D., (2018). Sulfated flavonoids: Biosynthesis, structures, and biological activities. *Molecules*, *23*, 480.

Unno, T., Sugimoto, A., & Kakuda, T., (2000). Scavenging effect of tea catechins and their epimers on superoxide anion radicals generated by a hypoxanthine and xanthine oxidase system. *Journal of the Science of Food Agriculture, 80,* 601–606.

Vessal, M., Hemmati, M., & Vasei, M., (2003). Antidiabetic effects of quercetin in streptozocin-induced diabetic rats. *Comparative Biochemistry and Physiology, 135,* 357–364.

Walker, E., Pacold, M., & Perisic, O., (2000). Structural determinations of phosphoinositide 3-kinase inhibition by wortmannin, LY294002, quercetin, myricetin, and staurosporine. *Molecular Cell, 6,* 909–919.

Williams, C. A., & Grayer, R. J., (2004). Anthocyanins and other flavonoids. *Natural Products Reports, 21,* 539–573.

Yao, L. H., Jiang, Y. M., & Shi, J., (2004). Flavonoids in food and their health benefits. *Plant Foods for Human Nutrition, 59*(3), 113–122.

Zand, R. S. R., Jenkins, D. J. A., & Diamandis, E. P., (2002). Flavonoids and steroid hormone-dependent cancers. *Journal of Chromatography, 777,* 219–232.

Zhang, Q., Zhao, X., & Qiu, H., (2013). Flavones and flavonols: Phytochemistry and biochemistry. In: *Natural Products: Phytochemistry, Botany and Metabolism of Alkaloids, Phenolics and Terpenes* (pp. 1821–1847). Springer: Heidelberg, Germany.

CHAPTER 10

ANALYSIS OF COLOR FASTNESS PROPERTIES OF NATURAL DYE EXTRACTED FROM *RHUS PARVIFLORA* (TUNG) ON WOOL FIBERS USING A COMBINATION OF NATURAL AND SYNTHETIC MORDANTS

SHYAM VIR SINGH

Department of Chemistry, Shri Guru Ram Rai (PG) College, Pathribagh, Dehradun, Uttarakhand, India

ABSTRACT

The color fastness properties of colorant on wool fibers dyed by natural dye extracted (the active dye constituent in this species is quercetin which is a yellow plant flavonol from the flavonoid group) from the fruit of *Rhus parviflora* have been studied using a combination of mordants such as white vinegar + copper sulfate, white vinegar + potassium dichromate, white vinegar + ferrous sulfate and white vinegar + stannous chloride in the ratio of 3:1, 1:2, and 1:3 separately. Dyeing along with mordanting techniques, which included pre-mordanting, simultaneous mordanting, and post-mordanting, has been carried out. A study on fastness tests of dyed clothes is also undertaken. A large range of shades is obtained because of varying mordant ratios and combinations. The wash, rub, light, and perspiration fastness of the dyed samples have also been evaluated, giving fair to excellent fastness grades, and this evaluation is also useful for textile industries.

Flavonoids as Nutraceuticals. Rajesh K. Kesharwani, Deepika Saini, Raj K. Keservani, and Anil Kumar Sharma (Eds.)

10.1 INTRODUCTION

The concept of natural dyes is by no means new. The primitive people had revealed the tinctorial properties of the juices of leaves, fruits, crushed flowers, roots, bark, etc., from the stains left on their hands while collecting food. Natural dyes have been used in most of the ancient civilizations of the world, like India, China, Mesopotamia, Egypt, Greece, the Aztecs, and others. The discovery of red ochre in very ancient burial sites indicates that the use of natural dyes for esthetic and other purposes is at least 15,000 years old. The art of dyeing cloth is believed to have been known since 3000 BC in China and 2500 BC in India. At relatively the same period (2000 BC), dyeing of cloth in yellow, red, blue, and green was also practiced in Egypt. Indigo is perhaps the oldest natural dye used by man. It has been known in India for about 4,000 years. In northern Europe, another blue dye known as the woad (*Isatis tinctoria*) has been in use since the Bronze Age (2500–800 BC). Another ancient dye, the Tyrian purple, derived from the Mediterranean shellfish of the genera Purpura and Murex, was probably the most expensive dyestuff in history. The Phoenician towns of Tyre and Sidon were the centers of this dye industry in about 800 BC, and the Greek dye factories that produced purple existed all along the Mediterranean coasts. The Tyrian purple was so precious that an extract of it was often dyed over the purple made from a Lichen genus *Roccella*. Gradually, the use of Murex died out until the Lichen (*Roccella*) alone provided the purple dye. Natural dyes produce an extraordinary diversity of rich colors that complement each other (Gaur, 2008). Natural dyes from plants may also have dozens of compounds, and their properties vary with soil type and the weather. In India, Rajasthan and Kutch still possess a rich tradition in the use of natural dyes for textile dyeing. In many places in India, traditional wool and woolen products are dyed with natural dyes. Certain problems with the use of natural dyes in textile dyeing are color yield, compressibility of the dyeing process, reproducibility results, limited shades, blending problems, and inadequate fastness properties (Dayal & Dobhal, 2001). India has a rich biodiversity, and it is not only one of the world's 12 mega-diversity countries but also one of the eight major centers of origin and diversification of domesticated taxa.

Mordants are metal salts that produce an affinity between the fabric and the dye (Nishida & Kabayashi, 1992), and alum, chrome, stannous chloride, copper sulfate, and ferrous sulfate are the commonly used mordants. Natural dyes have the ability to produce a wide range of tints and shades with the same dye material (Vinod et al., 2010). A generally active dye constituent in this species is quercetin (Figure 10.1), which is a yellow natural colorant isolated from the *Rhus parviflora* plant, and it's a flavonol from the flavonoid group of polyphenols. *Rhus parviflora* species is very active in giving natural dyes properties, and these species give us good fastness grades with respect to the grayscale. Literature review shows that isolation of natural dyes from this species has not been done till now (Kumaresan et al., 2011; Gulrajani & Gupta, 1992). The present study has been undertaken to revive the age-old dyeing with natural dyes (Anderson, 1971). In this work, the fruit extract of *Rhus parviflora* is used to dye wool at optimized dyeing conditions, using a combination of mordants and dyed samples are evaluated for the color fastness of the dyed samples to wash, rub, perspiration, and light (Bains et al., 2005; Anitha & Prasad, 2007). Colorfastness is the resistance of a material to change any of its color characteristics or the extent of transfer of its colorants to adjacent white materials in touch. Anciently, the purpose of coloring textiles was initiated using colors from natural sources until synthetic colors/ dyes were invented and commercialized. Almost all the synthetic colorants being synthesized from petrochemical sources through hazardous chemical processes pose a threat to their eco-friendliness. Up to the end of the 19th century, natural dyes were the main colorants for textiles. Recently, interest in the use of natural dyes has been growing rapidly due to the result of stringent environmental standards imposed by many countries in response to toxic and allergic reactions associated with synthetic dyes.

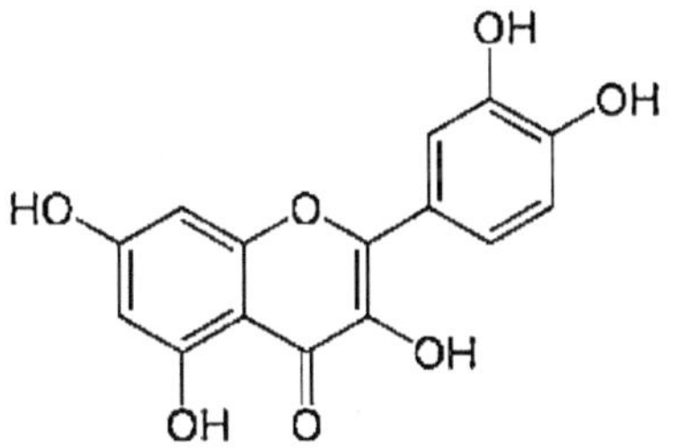

FIGURE 10.1 Quercetin (a yellow natural dye constituent).

In 1856, William Perkins accidently synthesized a basic dye; with the advent of synthetic dyes, the use of natural dyes declined tremendously because the existing natural dyes failed to full fill the demand of the market. The widely and commercially used synthetic dyes impart strong colors but cause carcinogenicity and inhibition of benthic photosynthesis (Adeel et al., 2009). The District Chamoli, Uttarkashi, and Pithoragarh of Uttarakhand are traditional wool and woolen products, and they were still used for dyeing by natural dyes. Certain problems with the use of natural dyes in textile dyeing are color yield, compressibility of the dyeing process, reproducibility results, limited shades, blending problems, and inadequate fastness properties (Sachan & Kapoor, 2007; Siva, 2007).

Mordants are metal salts which produce an affinity between the fabric and the dye (Vankar et al., 2009; Samanta & Agarwal, 2009). Alum, chrome, stannous chloride, copper sulfate, ferrous sulfate, etc., are the commonly used mordants (Mahangade et al., 2009). *Rhus parviflora* species is very rare to give natural dyes properties, and these species give us good fastness grades with respect to the grayscale. Isolation of natural dyes from these species could not be done till now because recent data of the literature did not show these results of natural dyes properties in the past. The present study has been undertaken so as to revive the age-old area of dyeing with natural dyes. Colorfastness is the resistance of a material to change any of its color characteristics or the extent of transfer of its colorants to adjacent white materials in touch. The natural dyes present in plants and animals are pigmentary molecules (Bains et al., 2002) which impart color to the materials. There are several plants that provide natural dyes which are used in the textile industry. However, the common drawbacks of natural dyes are their nonreproducible and non-uniform shades, poor to moderate color fastness, and lack of scientific information on the chemistry of dyeing and standardized dyeing methods (Gulrajani et al., 2003). Many reports are available on the application of natural dyes on wool fabrics (Anderson, 1971; Kumaresan & Palanisamy, 2010; Kumaresan et al., 2011). In the present scenario, the environmental consciousness of people about natural products, the renewable nature of materials, less environmental damage, and sustainability of natural products have further revived the use of natural dyes in the dyeing of textile materials. Natural dyes have some inherent advantages:

- No health hazard;
- Easy extraction and purification;
- No effluent generation;
- Very high sustainability;
- Mild dyeing condition;
- Renewable sources.

There are some technical issues and disadvantages related to the application of natural dyes which reduced its applications that are:

- Mostly applicable to natural fibers (cotton, linen, wool, and silk);
- Poor color fastness properties;

- Poor reproducibility of shades;
- No standard color recipes and methods available;
- Use of metallic mordants, some of which are not eco-friendly.

Hill (1997) gave his views that research work with natural dyes is inadequate, and there is a need for significant research work to explore the potential of natural dyes before their important application to textile substrates. In India, initially Alps Industries Ghaziabad (Uttar Pradesh, India) and later Ama Herbals, Lucknow, and Bio-Dye, Goa, done extensive work for industrial research and production of natural dyes and natural dyed textiles. Textile-based handicraft industries in many countries engaged local people to dye textile yarn with natural dyes and weave them to produce specialty fabrics. Printing of textile fabrics with natural dyes in India is specially done in Rajasthan and Madhya Pradesh.

10.1.1 THE CLASSIFICATION OF NATURAL DYES BASED ON ORIGIN/SOURCE

- Vegetable origin;
- Animal origin;
- Mineral origin.

For vegetable origin of natural dyes, the best sources of natural dyes are the different parts of plants and trees. Most natural dyes are extracted from different parts of plants and trees. Natural dyes and pigments are taken from the following parts of plants/trees:

- Seed;
- Root;
- Stem;
- Barks;
- Leaves;
- Flowers.

Natural dyes have wide applications in the coloration of most natural fibers, e.g., cotton, linen, wool, and silk fiber, and to some extent, for nylon and polyester synthetic fiber. However, the major issues for natural dyed textiles are reproducibility of shade, non-availability of well-defined standard procedures for application, and poor lasting performance of shade under water and light exposure. To achieve good color fastness to washing and light are also a challenge to the dyer. Several researchers have proposed different dyeing methods and process parameters, but still, this information is inadequate, so this calls for the need for research to develop some standard dye extraction techniques and standardization of the whole process of natural dyeing on textiles.

10.1.2 IMPORTANCE OF NATURAL DYES

First, the colors produced by natural dyes and pigments are **vibrant**. Next, they are not only biodegradable but nontoxic and nonallergic too. This means that they are much better for the environment and for use around humans. It is easy to extract the natural color from plants, fruits, or flowers. Many natural dyes also have antimicrobial properties, making them safer for kids in particular. Additionally, natural dyes neither contain harmful chemicals nor carcinogenic components, common to artificial or synthetic dyes. By using natural dyes over these other choices, you are helping preserve the environment and lowering human dependence on harmful products. When toxic runoff and residuals from the textile manufacturing and dyeing process often end up in our delicate oceans, we should do all we can to ensure we are using the nontoxic alternative, natural dyes. Furthermore, the products used in producing natural dyes, particularly plants, produce no waste, unlike the products used in the synthetic dyeing process. This is because plants bypass the entire production process it takes to create synthetic dyes. This is yet another reason why natural dyes are infinitely better for the environment. By using natural dyes rather than synthetic dyes, you are able to be closely connected to nature and recognize the importance it plays in all of our lives. Another interesting advantage of natural dyes is that they provide **higher UV absorption** in the fabrics they are used on. By wearing clothes dyed naturally, you are able to more fully protect your skin from the sun's harmful rays.

10.1.3 LIMITATIONS OF NATURAL DYES

Tedious extraction of coloring components from the raw material, low color value, and long dyeing time push the cost of dyeing with natural dyes considerably higher than with synthetic dyes. In the case of sappan wood, prolonged exposure to air converts the colorant baseline to brasilein, causing a color change from red to brown. To overcome this drawback, we used a sonicator and found that the dye extraction was much faster. Some of the natural dyes are fugitive and need a mordant for the enhancement of their fastness properties. Some of the metallic mordants are hazardous.

10.1.4 FIXATION BETWEEN FABRIC AND NATURAL DYE CONSTITUENT

Natural dyes work best with natural fibers such as cotton, linen, wool, silk, jute, ramie, and sisal. Among these, wool takes up dyes most easily, followed by cotton, linen, silk, and then the coarse fibers such as sisal and jute. Nearly all of them require some sort of a mordant. The trick is to choose the right dye from the right source that gives not only beautiful tones but color-fast shades as well. The chemistry of bonding (Vankar, 2000) of dyes to fibers is complex. It involves direct bonding, H-bonds, and hydrophobic interactions. Mordants help binding of dyes to fabric by forming a chemical bridge from dye to fiber, thus improving the staining ability of a dye along with increasing its fastness properties. Mordants form insoluble compounds of the dye within the fiber. The presence of certain functional groups in suitable positions in the dye molecule causes its coordination with the metal ion. Generally, two hydroxy groups or a hydroxy group with a carbonyl, nitroso, or azo group in adjacent positions are responsible for coordination. The mordant dyes produce a wide range of hues of remarkable resistance to wet treatments, but the shades lack brilliance.

10.1.5 MORDANTS AND MORDANTING

The natural dyes, having limited substantively for the fiber (Gupta, 2019), require the use of the mordant, which enhances the fixation of the natural colorant on the fiber by the formation of the complex with the dye. Some of the important mordants used are alum, potassium dichromate, ferrous sulfate, copper sulfate, zinc sulfate, tannin, and tannic acid. Although these metal mordants contribute to developing a wide gamut of hues after complexing with the natural coloring compounds, most of these metals are toxic in nature, and only in trace quantity is their presence found to be safe for the wearer. The word mordant comes from the Latin word "mordere," meaning "to bite." A mordant is a chemical which can itself be fixed on the fiber and also forms a chemical bond with the natural colorants. It helps in absorption and fixation of natural dyes and also prevents bleeding and fading of colors, i.e., improves the fastness properties of the dyed fabrics.

This complex may be formed by first applying the mordant and then dyeing (pre-mordanting process) or by simultaneous application of the dye and the mordant (meta-mordanting process), or by after-treatment of the dyed material with the mordant (post-mordanting process). There are three types of mordants, namely Metal salts or Metallic mordants, tannic acid (Tannins), and Oil mordants.

10.1.5.1 METAL SALTS OR METALLIC MORDANTS

Metal salts of aluminum, chromium, iron, copper, and tin are used. Some of the common mordants used are Alum, Copper sulfate, Ferrous sulfate, Potassium dichromate, Stannous Chloride and Stannic Chloride. Based on the final color produced with the natural dyes, these metallic mordants are further divided into two types, i.e., Brightening Mordants and Dulling Mordants. Alum, Potassium dichromate, and Tin (Stannous chloride) falls under the category of brightening mordants, and Copper sulfate and Ferrous sulfates are dulling mordants.

10.1.5.2 PLANT-BASED MORDANTS

Certain plant materials contain high concentrations of tannic acid, or tannin, which works well as a mordant to bond color to plant-based fiber. Tannin as a mordant, especially in combination with alum, can provide a greater color range with more successful results on most vegetable fibers. Certain tannin-bearing plant materials work especially well as mordants, such as horse chestnuts, pine bark, certain roots, some leaves, acorns, oak galls, pomegranate rind, and some fruits. Among the plant-based mordants, oak galls contain the highest amount of tannic acid. Some tannin substances will bind to the fiber and stay clear, allowing the true color of the dye source to saturate the fiber. But some tannins can alter the color by making it dull, especially if the dyes are yellow, pink, or brown tones.

10.1.5.3 OIL MORDANTS

Oil mordants are used mainly in dyeing of Turkey Red color from madder. The main function of the oil mordants is to form a complex with alum used as the main mordant. Since alum is soluble in water and not it has an affinity for cotton, it is easily washed out from the treated fabric. The naturally occurring oil contains fatty acids such as palmitic, stearic, oleic, etc., and their glycerides.

10.1.6 STATUS OF NATURAL DYES AND DYE-YIELDING PLANTS IN INDIA

Indians have been considered as forerunners in the art of natural dyeing (Siva, 2007). Natural dyes find use in the coloring of textiles, drugs, cosmetics, etc. Owing to their non-toxic effects, they are also used for coloring various food products. In India, there are more than 450 plants that can yield dyes. In addition to their dye-yielding characteristics, some of these plants also possess medicinal value. Though there is a large plant resource base, little has been exploited so far. Due to the lack of availability of precise technical knowledge on the extracting and dyeing technique, it has not commercially succeeded like the synthetic dyes. Although indigenous knowledge system has been practiced over the years in the past, the use of natural dyes has diminished over generations due to lack of documentation. Also, there is not much information available on databases of their dye-yielding plants or their products. Recently, an international workshop on "quality standards and certification of natural dye" was organized in Hyderabad. Visva Bharati University conducted a seminar and workshop on the application of vegetable dyes on textiles. Several national and international workshops were held on natural dyes as a part of the UNDP program of technical cooperation among developing countries. During these workshops, it was concluded that there is a great potential for the revival of the use of natural dyes in Asia, particularly in India.

Therefore, efforts should be made to promote the use of natural dyes, extend the range of their application, and encourage their commercial use rather than restricting it to cottage scale and a need to carry out R&D on natural dyes, develop extraction techniques, standardize applications on synthetic as well as natural fibers, leather, and also to evaluate them for their toxicity.

10.1.6.1 RECENT TRENDS OF NATURAL DYEING

Between January and September 2010, exports of natural dyes grew to an impressive annual rate of 181.0%, mainly boosted by the higher price of carmine cochineal and set off by the growing international demand. This report presents the latest information on the performance of the production of inputs used in the production of natural dyes (Vankar & Shukla, 2019), such as paprika, marigold, annatto, and turmeric. It provides information on the average yield of these crops, farm-gate prices, and global market analysis of dyes and the development of Peruvian exports and imports of natural colors. In these years, the demand for natural dyes and the interest in these followed much of the fashion trend, with ups and downs recurrent. Currently, we are in one phase of increase. The fields of industry that today are more interested in introducing natural dyes are intimate dress, children's clothes, and the interior, fields where naturalness is more important and where the problems of allergies are greater and for which it's needed to use eco-friendly natural dyes for dyeing fabrics.

10.1.7 IMPROVING THE QUALITY OF NATURAL DYES

Poor light fastness of some of the natural dyes is attributable to photo-oxidation of the chromophore. We have tried to prevent and minimize such photo-oxidation by forming a complex of the dye with a transition metal. We have improved the washing fastness of natural dyes by treatment with eco-friendly mordants such as alum, stannic chloride, stannous chloride, and ferrous sulfate. We have also used tannins with mordants. Treatment with metal salts alters the light absorption characteristics of tannins in addition to making them insoluble in water, with the fabric acquiring washing fastness.

10.1.7.1 GENERAL EXTRACTION METHODS OF DYE

The extraction method of vegetable dyes basically depends on the method in which the dye is extracted. There are mainly four methods used in the extraction of natural dyes:

1. **Aqueous Method:** Boil the dyestuff in soft water at 100C. Filter the dye solution and record the optical density.
2. **Alkaline Method:** Prepare 1% alkaline solution with the addition of sodium carbonate or sodium hydroxide in water. Enter the dye material in it and boil the same at 100°C. Filter the dye solution and record the optical density.

3. **Acidic Method:** Prepare 1% of acidic solution by adding HCl in soft water. Enter the dye material and boil it at 100°C. Filter the dye solution and record the optical density.
4. **Alcoholic Method:** Alcoholic solution is made by adding an equal amount of alcohol and water. Enter the dye material and boil it at 100°C. Filter the dye solution.

The present investigation deals with the aqueous extraction of natural dyes from the fruit of *Rhus parviflora* grow in almost all cold and dense parts of Garhwal Himalaya in Uttarakhand, India. The aim of the present work has been carried out to prepare eco-friendly natural dyes of the fruit of *Rhus parviflora* and then apply them to wool fabrics. In the present work, an attempt has been made to study the effect of mordanting and dyeing properties of wool fabrics such as washing, rubbing, light fastness, and perspiration 13 and also to visualize the effect of metallic mordants have been undertaken.

10.2 MATERIALS AND METHODS

Rhus parviflora is an evergreen tree or shrub found in the Himalaya region of India. The fruit (light greenish-yellow) was used for the extraction of dye. Bleached plain weave wool fabric purchased from the market of Gopesh war, Uttarakhand, was used for the study. Analytical reagents (AR) grade ferrous sulfate, copper sulfate, potassium dichromate, stannous chloride, commercial grade acetic acid, common salt and sodium carbonate were used as such. A natural mordant 'white vinegar' was used for the study. Depending upon the mordant used, the color obtained on textiles from the fruit of *Rhus parviflora* extract may give different shades. The white vinegar mixed with a known volume of water and heated at 80°C for 30 min was also used for mordanting. The resulting solution was cooled and filtered. The filtrate was used for mordanting. A known quantity of fruit was dried, powdered, and soaked in warm water overnight. The extract was obtained by boiling it in the same water and allowed to cool, finally filtered, and used for dyeing. The dyeing was carried out at optimized dyeing conditions, such as a dye extraction time of 60 min, material-to-liquor ratio of 1:20, and dyeing time of 50 min.

The mordant combinations viz. white vinegar: copper sulfate, white vinegar: potassium dichromate, white vinegar: ferrous sulfate, white vinegar: stannous chloride was used in the ratio of 3:1, 1:2, and 1:3. The total amount of two mordants used in each combination was 5% on the weight of the fabric, i.e., 5 g of the mordant/100 g of the fabric. Each of the four mordant combinations in three different ratios mentioned above was used for all three mordanting methods, namely pre-mordanting, simultaneous mordanting, and post-mordanting for dyeing (Gulrajani & Gupta, 1992; Adeel et al., 2009). After dyeing, the solution was allowed to cool, removed from the dye bath, rinsed under running water to remove excess dye particles, and shade dried.

10.2.1 EXTRACTION OF COLOR COMPONENT

For optimizing the extraction method, the aqueous extraction of dye liquor was carried out under varying conditions, such as time of extraction, temperature of extraction bath, and material-to-liquor ratio. In each case, the optical density or absorbance value at a particular maximum absorbance wavelength (λ_{420} nm) for the ethanol extract of plant parts was estimated by using a Hitachi-U-2000 UV-VIS absorbance spectrometer.

10.2.2 DYEING OF WOOL FABRICS WITH THE EXTRACT OF FRUIT OF RHUS PARVIFLORA AND MORDANTING

The wetted-out wool samples were entered into dye baths containing the required amount of dye extract and water. After 15 minutes, the required amount of sodium carbonate and sodium chloride were added. The dyeing was carried out for one hour at 60°C. The dyed samples were dried in air without washing to make them ready for pre, simultaneous, and post-mordanting using metallic salts.

10.2.2.1 PRE-MORDANTING OF WOOL FABRICS WITH METALLIC SALTS

Soaked wool fabric with or without pre-mordanting was further mordanted prior to dyeing using 1–3% of any one of the chemical mordants, such as aluminum sulfate, potassium dichromate, stannous chloride, and ferrous sulfate, at 60°C for 30 minutes with a material-to-liquor ratio of 1:20. The samples treated with metal salts were dyed with the dye extract.

10.2.2.2 SIMULTANEOUS MORDANTING OF WOOL FABRICS WITH METALLIC SALTS

Soaked wool fabrics were treated with both dye extract and metal salts simultaneously, using 1–3% of any one of the chemical mordants, such as aluminum sulfate, potassium dichromate, stannous chloride, and ferrous sulfate, at 60°C for 30 minutes with a material-to-liquor ratio of 1:20.

10.2.2.3 POST-MORDANTING OF WOOL FABRICS WITH METALLIC SALTS

Soaked wool fabrics were dyed with dye extract. The wetted-out wool samples were entered into different dye baths containing the required amount of dye extract and water. After 15 min. required amount of sodium sulfate was added. After 30 minutes required amount of sodium chloride was added. The dyeing was carried out for one hour at 60°C. The dyed samples were taken out, squeezed, and used for treatment with metal salts process without washing. The dyed wool fabrics samples were treated with different metal salts using 1–3% of anyone of the chemical mordants, such as aluminum sulfate, potassium dichromate, stannous chloride, and ferrous sulfate, at 60°C for 30 minutes with material-to-liquor ratio of 1:20. In all the above three methods, after the dyeing is over, the dyed samples were repeatedly washed with water and then dried in air. Finally, the dyed samples were subjected to soaping with 2 gpL soap solution at 50°C for 10 minutes, followed by repeated water washing and drying under the sun.

10.2.3 MEASUREMENTS

10.2.3.1 DETERMINATION OF SURFACE COLOR STRENGTH (K/S VALUE)

The K/S value of the undyed and dyed wool fabrics was determined by measuring surface reflectance of the samples using a computer-aided Macbeth 2020 plus reflectance spectrophotometer, using the following Kubelka-Munk equation with the help of relevant software:

$$K/S = (1 - R_{\lambda max})^2/2R_{\lambda max}$$

where; K is the coefficient of absorption; S is the coefficient of scattering; and $R_{\lambda max}$ is the surface reflectance value of the sample at a particular wavelength, where maximum absorption occurs for a particular dye/color component.

10.2.3.2 EVALUATION OF COLOR FASTNESS

Color fastness to washing of the dyed fabric samples was determined as per IS: 764-1984 method using a Sasmira launder-*O*-meter following Is-3 wash fastness method. The wash fastness rating was assessed using the grayscale as per ISO-05-A02 (loss of shade depth) and ISO-105-AO3 (extent of staining), and the same was cross-checked by measuring the loss of depth of color and staining using Macbeth 2020 plus computer-aided color measurement system attached with relevant software. Colorfastness to rubbing (dry and wet) was assessed as per IS: 766-1984 method using a manually operated crock meter and grayscale as per ISO-105-AO3 (extent of staining), and the test samples were graded for change in color and staining using gray scales.

10.3 RESULTS AND DISCUSSION

10.3.1 MORDANT COMBINATION – WHITE VINEGAR: STANNOUS CHLORIDE

The color fastness to light, washing, rubbing, and perspiration on dyed wool samples treated with white vinegar: stannous chloride combination in aqueous medium is presented in Table 10.1. All the treated samples subjected to light show good (4) light fastness for all ratios of mordant combinations. The treated samples for pre-mordanting show fair (2–3) washing fastness grades, but they ranged between excellent and good (4–5 to 4) for all the treated samples for simultaneous and post-mordanting. There is no color staining. The color change to dry and wet rubbing for all the treated samples is found to be excellent (5). There is no color staining ranging between no staining and negligible staining (5 to 4–5) in dry rubbing. The perspiration fastness grades range between 4 and 5 and 4, except for the 3:1 mordant proportion in the pre-mordanting method, where it is fair (3) for all samples in both acidic and alkaline media. There is no color staining (5) for all the treated samples in both acidic and alkaline media (Table 10.1).

TABLE 10.1 Fastness Grades of Wool Fibers Dyed with *Rhus parviflora* Dye at Optimum Dyeing Conditions Using WV:SC Mordant Combination

Mordanting Method	Mordant Proportions	Light Fastness	Wash Fastness		Rub Fastness				Perspiration Fastness		
			CC	CS	Dry		Wet		Acidic		Alkaline
					CC	CS	CC	CS	CC	CS	CC
Pre-mordanting	3:1	4	2–3	5	4–5	5	4	5	3	5	3
	1:2	4	2–3	5	5	4–5	5	5	4–5	5	4
	1:3	4	2–3	5	5	4–5	5	5	4–5	5	4
Simultaneous mordanting	3:1	4	4–5	5	5	5	5	5	4	5	4
	1:2	4	4	5	5	5	5	5	4	5	4
	1:3	4	4	5	5	5	5	5	4	5	4
Post-mordanting	3:1	4	4	5	5	4–5	5	5	4	5	4–5
	1:2	4	4–5	5	5	4–5	5	5	4–5	5	4–5
	1:3	4	4–5	5	5	4–5	5	5	4–5	5	4–5

Note: WV:SC: White vinegar:Stannous chloride; CC: Color change; CS: Color staining.

10.3.2 MORDANT COMBINATION – WHITE VINEGAR: COPPER SULFATE

Table 10.2 provides the assessment of the dyed wool samples treated with the white vinegar: copper sulphate combination in an aqueous solution for color fastness to light, washing, rubbing, and perspiration.

Nearly all of the treated samples that were exposed to light exhibit (4) fairly strong light fastness for all ratio mordant combinations. While wash fastness grades ranged from outstanding to good (4-5) for all treated samples for simultaneous and post-mordanting, treated samples for pre-mordanting showed medium (4–5) wash fastness grades. No color stains are present. For all of the treated samples, the color change upon dry and wet rubbing is found to be excellent (5). There is no color staining, ranging from none to barely any staining (5 to 4-5), when dry rubbing is used. For all samples in both acidic and alkaline media, the sweat fastness grades varied from 4 to 5, with the exception of the 1:3 mordant proportion used in the pre-mordanting procedure, where it is only fair (4). There is no color staining (5) for all the treated samples in both acidic and alkaline media, as mentioned in Table 10.2.

TABLE 10.2 Fastness Grades of Wool Fibers Dyed with *Rhus parviflora* Dye at Optimum Dyeing Conditions Using WV: CS Mordant Combination

Mordanting Method	Mordant Proportions	Light Fastness	Wash Fastness		Rub Fastness				Perspiration Fastness		
			CC	CS	Dry		Wet		Acidic		Alkaline
					CC	CS	CC	CS	CC	CS	CC
Pre-mordanting	3:1	4	3–4	5	5	4–5	5	5	4–5	5	4
	1:2	3–4	4	4–5	5	5	5	5	3	5	4
	1:3	4	3–4	5	5	5	5	5	3	5	4
Simultaneous mordanting	3:1	4	4	5	5	5	5	5	3	5	4
	1:2	4	4	5	5	4–5	5	5	3	5	4
	1:3	4	4	4–5	5	4–5	5	5	4	5	4
Post-mordanting	3:1	4	4	4–5	5	4–5	5	5	4	5	4–5
	1:2	4	4	4–5	5	4–5	5	5	4	5	4–5
	1:3	3–4	4	4–5	5	4–5	5	5	4	5	4–5

Note: WV: CS: White vinegar: copper sulfate; CC: Color change; CS: Color staining.

10.3.3 MORDANT COMBINATION – WHITE VINEGAR: POTASSIUM DICHROMATE

The evaluation of color fastness to light, washing, rubbing, and perspiration of dyed wool samples treated with white vinegar: potassium dichromate combination in aqueous medium is presented in Table 10.3. Almost all the treated samples subjected to light show fairly good (4) light fastness for all ratio mordant combinations. The wash fastness grades show fairly good (3–4) for all the treated samples except for the 1:3 mordant proportion in the pre-mordanting method, where it is fair (2–3). The color change to dry and wet rubbing for all the treated samples is found to be excellent (5). The color staining ranges between no staining and negligible staining (4–5) in dry and wet rubbing except for pre-mordanting method where it shows fair (5). Most of the treated samples showed excellent fastness grade to color change, except for 1:3 mordant proportion in pre-mordanting methods, where it is good (4–5). There is no color staining (5) for all treated samples in both acidic and alkaline media. There are no significant results from simultaneous and post-mordanting methods with respect to excellency in the fastness properties (Table 10.3).

TABLE 10.3 Fastness Grades of Wool Fibers Dyed with *Rhus parviflora* Dye at Optimum Dyeing Conditions Using WV: PD Mordant Combination

Mordanting Method	Mordant Proportions	Light Fastness	Wash Fastness		Rub Fastness				Perspiration Fastness		
					Dry		Wet		Acidic		Alkaline
			CC	CS	CC	CS	CC	CS	CC	CS	CC
Pre-mordanting	3:1	4	3	5	5	4–5	3–4	5	4–5	5	3
	1:2	3–4	2–3	5	5	4–5	5	5	4	5	4
	1:3	4	2–3	5	5	4–5	5	5	4–5	5	4
Simultaneous mordanting	3:1	4	3	5	5	5	5	5	4–5	5	4
	1:2	4	4	5	5	5	5	5	4–5	5	4
	1:3	4	4	5	5	5	5	5	4	5	4–5
Post-mordanting	3:1	3–4	3	5	5	5	5	5	4	5	4–5
	1:2	3–4	4–5	5	4–5	5	5	5	4–5	5	4
	1:3	3–4	4–5	5	4–5	5	5	5	4–5	4–5	4–5

Note: WV: PD: White vinegar: Potassium dichromate; CC: Color change; CS: Color staining.

10.3.4 MORDANT COMBINATION – WHITE VINEGAR: FERROUS SULFATE

The evaluation of color fastness to light, washing, rubbing, and perspiration of dyed wool samples treated with white vinegar: ferrous sulfate combination in aqueous medium is presented in Table 10.4. The treated samples subjected to light show fairly good (4) light fastness for all ratio mordant combinations. The wash fastness grades ranged between excellent and good (3 to 4–5) for all the treated samples. The color change to dry and wet rubbing for all the treated samples is found to be excellent (5). The color staining in dry rubbing is almost fair (5). Most of the treated samples show excellent fastness grade to color change, except for 1:3 mordant proportion in a simultaneous mordanting method, where it is good (4–5) for all samples in both acidic and alkaline media. There is no color staining (5) for all the treated samples in both acidic and alkaline media (Table 10.4). Extracted natural dye from the fruit of *Rhus parviflora* shows yellow color. There are many shades of color obtained after dyeing by applying different mordanting methods (Mahangade et al., 2009; Samanta & Agarwal, 2009; Vankar et al., 2009; Madison, 1973). Different shades of colors are obtained by using different mordants viz. $K_2Cr_2O_7$, $CuSO_4$, $SnCl_2$, and $FeSO_4$. Generally, as synthetic or chemical mordants, $K_2Cr_2O_7$ gives pale yellow color, $CuSO_4$ shows light green color, $FeSO_4$ gives brown color, and $SnCl_2$ shows light yellow or cream color with dyes on wool fibers. A number of shades are obtained by mordanting the wool with varying concentrations of mordants. It is found that different metallic mordants give different shades of colors as compared to all natural dyed samples (Table 10.5).

TABLE 10.4 Fastness Grades of Wool Fibers Dyed with *Rhus parviflora* Dye at Optimum Dyeing Conditions Using WV: FS Mordant Combination

Mordanting Method	Mordant Proportions	Light Fastness	Wash Fastness		Rub Fastness			Perspiration Fastness			
					Dry		Wet	Acidic		Alkaline	
			CC	CS	CC	CS	CC	CS	CC	CS	CC
Pre-mordanting	3:1	4	3–4	4–5	5	5	5	5	5	5	4
	1:2	4	4	4–5	5	5	5	4–5	5	5	4–5
	1:3	4	3–4	4–5	5	5	5	5	5	5	5
Simultaneous mordanting	3:1	4	3–4	4–5	5	4–5	5	5	5	5	5
	1:2	4	3–4	4–5	5	4–5	5	5	5	5	5
	1:3	4	3–4	4–5	5	5	5	5	5	5	5
Post-mordanting	3:1	4	3–4	4–5	5	5	5	5	4	5	5
	1:2	4	4	4–5	5	5	5	5	4	5	4–5
	1:3	4	4	4–5	5	5	5	5	4	5	4–5

Note: WV: FS: White vinegar: Ferrous sulfate; CC: Color change; CS: Color staining.

TABLE 10.5 Dyed Samples with and Without Mordanting Using Combination of Synthetic and Natural Mordants

Mordanting Methods	Without Chemical Mordant	Combination of Synthetic and Natural Mordants (White Vinegar)
Pre-mordanting		
Simultaneous mordanting		

TABLE 10.5 *(Continued)*

Mordanting Methods	Without Chemical Mordant	Combination of Synthetic and Natural Mordants (White Vinegar)
Post-mordanting		

10.4 CONCLUSION

It is found that extracted dye from the fruit of *Rhus parviflora* can be successfully used for dyeing wool to obtain a wide range of soft and light colors by using a combination of natural and chemical mordants. With regards to color fastness, test samples exhibit excellent fastness to washing (except for pre-mordanting using white vinegar: potassium dichromate combination and white vinegar: stannous chloride combination), excellent fastness to rubbing, and good to excellent fastness to perspiration in both acidic and alkaline media and fairly good fastness to light and these data also helpful for textile industries.

KEYWORDS

- **color fastness**
- **dye constituent**
- **mordants**
- **natural dye**
- ***Rhus parviflora***
- **wool fabrics**

REFERENCES

Adeel, S., Ali, S., Bhatti, I. A., & Zsila, F., (2009). Dyeing of cotton fabric using pomegranate (*Punica granatum*) aqueous extract. *Asian J. Chem., 21*(5), 3493–3499.

Anderson, B., (1971). *Creative Spinning, Weaving and Plant Dyeing* (pp. 24–28). Angus and Robinson, Singapore.

Anderson, B., (1971). *Creative Spinning, Weaving and Plant Dyeing* (pp. 24–28). Angus and Robinson, Singapore.

Anitha, K., & Prasad, S. N., (2007). Developing multiple natural dyes from flower parts of Gulmohur. *Current Science*, *92*(12), 1681, 1682.

Bain, S., Singh, O. P., & Kang, M., (2002). Dyeing of cotton with Arjun (*Arjuna terminelia*) dye. *Man-made Textiles in India*, *45*(8), 315.

Bains, S., Kang, S., & Kaur, K., (2005). Dyeing of wool with *Prunus persica* dye using a combination of mordants. *Journal of the Textile Association*, 127–131.

Dayal, R., & Dobhal, P. C., (2001). Natural dye from some Indian plants. *Colorage, 48*, 33–38.

Gaur, R. D., (2008). Traditional dye-yielding plants of Uttarakhand, India. *Nat. Prod. Rad., 7*(2), 154–165.

Gulrajani, M. L., & Gupta, D., (1992). *Introduction to Natural Dyes* (pp. 81–96). Indian Institute of Technology, Delhi.

Gulrajani, M. L., & Gupta, D., (1992). *Natural Dye and Their Application to Textiles* (p. 25). Department of Textile Technology, IIT, Delhi.

Gulrajani, M. L., Gupta, D., & Gupta, P., (2003). Application of natural dyes on bleached coir yarn. *Indian Journal of Fiber and Textile Research*, *28*(4), 466–470.

Gupta, V. K. (2019). *Fundamentals of Natural Dyes and its Application Textile Substances (A Book Chapter)*. doi: 10.5772/intechopen.89964.

Hill, D. J., (1997). Is there a future for natural dyes? *Review of Progress in Coloration and Related Topics, 27*, 18.

Kumaresan, et al., (2011). Application of eco-friendly natural dye on silk using a combination of mordants. *Int. J. Chem. Res., 2*(1), 11–14.

Kumaresan, M., & Palanisamy, P. N., (2010). Dyeing of silk with the stem of Achras sapota using a combination of mordants. *Int. J. Appl. Eng. Res., 5*(12), 2031–2037.

Kumaresan, M., Palanisamy, P. N., & Kumar, P. E., (2011). Application of eco-friendly natural dye on silk using a combination of mordants. *Int. J. Chem. Res., 2*(1), 11–14.

Madison, J. H., (1973). *Natural Plant Dyeing: A Handbook, Brooklyn Botanic Garden* (Vol. 29, No. 2). Brooklyn, N.Y. Brooklyn Botanic Garden, 1000. Washington Ave., Brooklyn, N.Y. 11225.

Mahangade, R. R., Varadarajan, P. V., Verma, J. K., & Bosco, H., (2009). New dyeing technique for enhancing color strength and fastness properties of cotton fabric dyed with natural dyes. *Ind. J. Fib. Tex. Res., 34*, 279–282.

Mahangade, R. R., Varadarajan, P. V., Verma, J. K., & Bosco, H., (2009). New dyeing techniques for enhancing color strength and fastness properties of cotton fabric dyed with natural dyes. *IJFTR, 34*, 279–282.

Nishida, K., & Kabayashi, K., (1992). Dyeing properties of natural dyes from vegetable sources. *Am. Dyestuffs Rep., 81*(9), 26.

Sachan, K., & Kapoor, V. P., (2007). Optimization of extraction and dyeing conditions for traditional turmeric dye. *Ind. J. Tradit. Knowl., 6*(2), 270–278.

Samanta, A. K., & Agarwal, P., (2009). Application of natural dyes on textiles. *Indian J. Fib. Text Res., 34*, 384–399.

Samanta, A. K., & Agarwal, P., (2009). Application of natural dyes on textiles. *IJFTR, 34*, 384–399.

Siva, R., (2007). Status of natural dyes and dye-yielding plants in India. *Current Science, 92*(7), 916–919.

Vankar, P. S., & Shukla, D., (2019). *A Handbook New Trends in Natural Dyes for Textiles* (I[st] edn.). Elsevier-Woodhead Publishing.

Vankar, P. S., (2000). Chemistry of natural dyes. *Resonance, 5*(10), 73–80.

Vankar, P. S., Shankar, R., & Wijayapala, S., (2009). Dyeing cotton, silk, and wool yarn with extract of *Garcinia mangostana* pericarp. *Journal of Textile and Apparel, Technology and Management*, *6*(1), 1–11.

Vinod, K. N., Puttaswamy, Gowda, K. N. N., & Sudhakar, R., (2010). Kinetic & adsorption studies of Indian siris natural dye on silk. *IJFTR, 35*, 159–163.

CHAPTER 11

FLAVONOIDS IN TREATING PREGNANCY-INDUCED DISORDERS

NIHARIKA DEWANGAN[1] and ALKA MISHRA[2]

[1]*Kalinga University, Naya Raipur, Chhattisgarh, India*

[2]*Government VYTPG Autonomous College, Durg, Chhattisgarh, India*

ABSTRACT

In nature, there are many such natural elements present that are essential and responsible for health care one of them are Flavonoids. Flavonoids are a group of plant metabolites that provide health benefits through cell signaling and pathway and their antioxidant effects. It is present in a variety of fruits, vegetables, food products, and beverages. Flavonoids help to regulate cellular activity, and fight free radical, which is the cause of oxidative stress in the body and toxin. Many nutritive elements are required during pregnancy, which is responsible for the growth of the fetus. Some medical conditions occur during pregnancy, that are hypertension, asthma, hypertensive disorders, gestational hypertension, preeclampsia, and chronic hypertension, which can affect the growth of the fetus. In this condition, Flavonoids are helpful in decreasing nausea, vomiting, and other gastrointestinal problems and also prepare labor. It also acts as conventional drugs. In this chapter, we give the details about the effects of flavonoids on Pregnancy disorders.

11.1 INTRODUCTION

Many nutritive elements which are responsible for the growth of the fetus are required during pregnancy. Because some medical conditions occur during pregnancy, that as hypertension, asthma, hypertensive disorders, gestational

Flavonoids as Nutraceuticals. Rajesh K. Kesharwani, Deepika Saini, Raj K. Keservani, and Anil Kumar Sharma (Eds.)

hypertension, preeclampsia, and chronic hypertension, and they can affect the growth of the fetus. There are many elements present in nature, such as natural bioactive compounds from plant sources, which are responsible for health care and also regulate cellular activity during pregnancy. Plants are rich sources of natural bioactive compounds such as secondary metabolites and antioxidants. Medicinal components produced are stored in different plant parts. One of the compounds is Flavonoid. Flavonoids are the secondary metabolites present in plant parts.

11.2 FLAVONOIDS

Flavonoids are secondary metabolites that are widely distributed in the plant kingdom and found in many plant parts. They are bioactive compounds having low molecular weight, belonging to a class of low molecular weight phenolic compounds (Kim et al., 2003). Flavonoids (Keservani & Sharma, 2014; Keservani et al., 2020) are also found in foods and beverages of plant origin, such as fruit, vegetables, tea, cocoa, and wine. They have several subgroups, having unique major sources that include chalcones, flavones, flavonols, and isoflavones. For example, onions and tea are important dietary sources of flavonols and flavones. They are involved in the production of flower pigments, such as the blue color in the petals, due to the presence of anthocyanin. They are also able to act as natural UV filters (Takahashi et al., 2004). This ability in flavonoids comes from their absorption in the 280–315 nm region.

11.3 ROLE OF FLAVONOIDS IN HUMAN HEALTH

Flavonoids are effective in the promotion of healing and are used as an important component in a variety of nutraceuticals (Keservani et al., 2010a, b), pharmaceutical, medicinal, and also in cosmetic products (Halliwell et al., 1989). Flavonoids are present as antioxidant, anti-inflammatory, antimutagenic, and anticarcinogenic properties; it enhances the cellular enzyme functions along with it. They also have potential inhibiting properties for some enzymes, such as xanthine oxidase (XO), cyclo-oxygenase (COX), lipoxygenase, and phosphoinositide 3-kinase. Flavonoids show little antioxidant activity in the body; thus, the increase in the antioxidant capacity of the blood can be observed when one consumes food rich in flavonoids. Flavonoids do not get properly absorbed in the human body. They quickly metabolized into smaller fragments. The properties of these fragments are not well known. Flavonoids are quickly eliminated from the body (Figure 11.1).

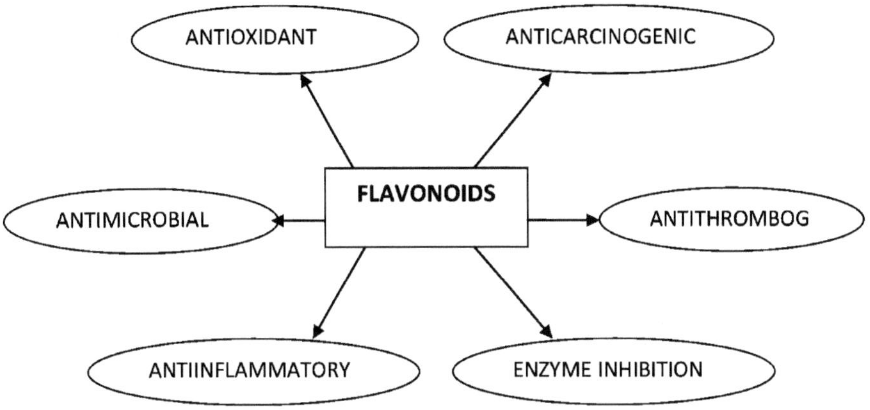

FIGURE 11.1 Role of flavonoids.

11.3.1 ANTI-INFLAMMATORY ACTIVITY

Cyclooxygenase is an endogenous enzyme responsible for formation of prostaglandins and thromboxanes conversion of arachidonic acid (Smith et al., 2000). Inhibiting cyclooxygenase reduces pain and inflammation. Therefore, a class of compounds with good anti-inflammatory activity that inhibits cyclooxygenase activity needs to be developed. The commercially available flavonoids were tested for cyclooxygenase inhibitory activity (O'Leary et al., 2004). These flavonoids include silbinin, galangin, hesperidin, scopoletin, genistein, daidzein, taxifolin, esculatin, naringenin, and celecoxib (Madeswaran et al., 2012).

11.3.2 ANTIOXIDANT ACTIVITY

In various studies about the antioxidant properties of Flavonoid, it is observed that it is used as a drug in preventing oxidative stress (Kitagawa et al., 1992; Ishikawa, 1997). Flavonoids have been shown to be effective in preventing lipid peroxidation since the peroxidation of lipids leads to diseases such as diabetes, atherosclerosis, hepatotoxicity, aging, and inflammation (Halliwell, 1991). Flavonoid quercetin, a plant pigment found in onion, apple, and berries, helps to reduce lipid peroxidation (Letan, 1966). Some other flavonoids, such as rutin, myricetin, and quercitrin, also help to check the production of superoxide radicals (Grace, 1994).

11.4 DISORDERS IN HUMAN HEALTH

11.4.1 DURING PREGNANCY

Hypertensive disorders during pregnancy are common in most of the women. These disorders include chronic hypertension, gestational hypertension, preeclampsia, and chronic hypertension with superimposed preeclampsia. It results in about 10% of pregnancy complications in the USA (Lai et al., 2017). Asthma is another common preexisting condition that affects about 9% of pregnant women (Kwon et al., 2006). Epilepsy is also a disease that occurs in pregnant women. It has been reported that the use of antiepileptic drugs by pregnant women may result in adverse outcomes such as miscarriage, antepartum, and post-partum hemorrhages. The foods that contributed the most to the intake of total flavonoids in the pregnant women's diet were beans, oranges, chocolate powder, and orange juice.

11.4.2 CANCER

Cancer is one of the diseases in which the body cells grow uncontrollably and is causing across the globe. Not only do very expensive treatments were there to cure cancer also cause several side effects, which result in co-morbidities, i.e., simultaneously, two or more diseases or medical conditions are seen among the cancer survivors. Flavones, a sub-class of flavonoids, have been reported to act as anticancer drugs. Flavones modulate signal transduction pathways in carcinogenesis. Flavones have been found to control cell cycle progression, oxidative stress, angiogenesis, and metastasis, along with some molecular signaling pathways that ultimately prevent disease progression. Flavones have been shown to be cancer preventatives.

Foods rich in flavonoids, such as apples, can inhibit *in vitro* tumorigenesis and human breast cancer cell growth (Schiavano et al., 2015). Pelingo-type apple juice induced cell accumulation in the G2/M phase of the cell cycle. It leads to autophagy, inhibition of the activity of extracellular signal-regulated kinases 1/2 (ERK1/2), and causes an increase in lipidated microtubule-associated protein 1 light chain 3 (LC3B). Thus, it can be used as a source of bioactive compounds with potential chemopreventive activity (Wallace et al., 2016).

11.4.3 HORMONAL ACTIVITY OF FLAVONOIDS

Some of the flavonoids are reported to have hormone-like activities that are similar to steroid hormones, particularly with estrogen. This class of flavonoids is found in fruits and vegetables, tea, red wine, and grains (Shrivastava et al., 2015). These hormones, especially estrogen, are known to protect against various chronic diseases, which have neuroprotective effects on the brain. Various flavonoids such as genistein, daidzein, and equol have been reported for their estrogenic activity in clinical studies. They are effective in treating various chronic diseases such as cancer, cardiovascular disease, and osteoporosis (Wiseman, 2000). Further research concludes that the flavonoid genistein has the greatest potential for preventing postmenopausal bone loss in women. A number of flavonoids used in foods for their dietary importance have beneficial effects on atherosclerosis, including lipoprotein oxidation, platelet aggregation, and cardiovascular reactivity (Tham, 1998).

11.5 CONCLUSION

Flavonoids play a significant role as an antioxidant in plants. As a result of oxidative stress, cellular damage occurs, which affects health and causes diabetes, neurodegenerative disorders, aging, etc. Thus, during pregnancy, it is very much important to take care of these disorders for the mother. This review provides supportive information on the role of flavonoids and consuming flavonoid foods during pregnancy.

KEYWORDS

- **anti-inflammatory activity**
- **antioxidant**
- **disorders**
- **flavones**
- **flavonoids**
- **hormonal activity**
- **pregnancy**

REFERENCES

Grace, P. A., (1994). Ischaemia–reperfusion injury. *Br. J. Surg., 81*, 637–647.

Halliwell, B., (1991). Drug antioxidant effects. A basis for drug selection? *Drugs, 42*, 569–605.

Halliwell, B., (1991). Reactive oxygen species in living systems: Source, biochemistry, and role in human disease. *Am. J. Med., 91*, 14S–22S.

Halliwell, B., (1995). How to characterize an antioxidant: An update. *Biochem. Soc. Symp., 61*, 73–101.

Ishikawa, T., Suzukawa, M., Ito, T., et al., (1997). Effect of tea flavonoid supplementation on the susceptibility of low-density lipoprotein to oxidative modification. *Am. J. Clin. Nutr., 66*, 261–266.

Keservani, R. K., & Sharma, A. K., (2014). Flavonoids: Emerging trends and potential health benefits. *Journal of Chinese Pharmaceutical Sciences, 23*(12), 815.

Keservani, R. K., Kesharwani, R. K., Sharma, A. K., Vyas, N., & Chadoker, A., (2010b). Nutritional supplements: An overview. *International Journal of Current Pharmaceutical Review and Research, 1*(1), 59–75.

Keservani, R. K., Kesharwani, R. K., Vyas, N., Jain, S., Raghuvanshi, R., & Sharma, A. K., (2010a). Nutraceutical and functional food as future food: A review. *Der Pharmacia Lettre, 2*(1), 106–116.

Keservani, R. K., Sharma, A. K., & Kesharwani, R. K. (2020). *Nutraceuticals and Dietary Supplements: Applications in Health Improvement and Disease Management*. CRC Press. ISBN: 9781771888738.

Kim, D., Jeond, S., & Lee, C., (2003). Antioxidant capacity of phenolic phytochemicals from various cultivars of plums. *Food Chem., 81*, 321–326.

Kitagawa, S., Fujisawa, H., & Sakurai, H., (1992). Scavenging effects of dihydric and polyhydric phenols on superoxide anion radicals, studied by electron spin resonance spectrometry. *Chem. Pharm. Bull., 40*, 304–307.

Kwon, H. L., Triche, E. W., Belanger, K., & Bracken, M. B., (2006). The epidemiology of asthma during pregnancy: Prevalence, diagnosis, and symptoms. *Immunol. Allergy Clin. North Am. 26*(1), 29–62.

Lai, C., Coulter, S. A., & Woodruff, A. (2017). Hypertension and pregnancy. *Tex. Heart Inst. J., 44*(5), 350, 351. doi: 10.14503/THIJ-17-6359.

Letan, A., (1966). The relation of structure to antioxidant activity of quercetin and some of its derivatives. *J. Food Sci., 31*, 395–399.

Madeswaran, A., Umamaheswari, M., Asokkumar, K., et al., (2012). *In-silico* docking studies of cyclooxygenase inhibitory activity of commercially available flavonoids. *Asian J. Pharm. Life Sci., 2*, 174–181.

O'Leary, K. A., De Pascual-Teresa, S., De Pascual-Tereasa, S., Needs, P. W., Bao, Y. P., O'Brien, N. M., & Williamson, G., (2004). Effect of flavonoids and vitamin E on cyclooxygenase-2 (COX-2) transcription. *Mutation Research, 551*(1, 2), 245–254.

Schiavano, G. F., De Santi, M., Brandi, G., et al., (2015). Inhibition of breast cancer cell proliferation and *in vitro* tumorigenesis by a new red apple cultivar. *Plos One, 10*, e0135840.

Smith, R., DeWitt, D., & Garavito, R., (2000). Cyclooxygenases: Structural, cellular, and molecular biology. *Ann. Rev. Biochem., 69*, 145–182.

Srivastava, N., & Bezwada, R., (2015). *Flavonoids: The Health Boosters.* White Paper. Hillsborough, NJ: Indofine Chemical Company.

Takahashi, A., & Ohnishi T. (2004). The significance of the study about the biological effects of solar ultraviolet radiation using the exposed facility on the international space station. *Biol. Sci. Space, 18*, 255–260.

Tham, D., Gardner, C., & Haskell, W., (1998). Clinical review 97: Potential health benefits of dietary phytoestrogens: A review of the clinical, epidemiological and mechanistic evidence. *J. Clin. Endocrinol. Metab., 83*, 2223–2235.

Wallace, T. C., Slavin, M., & Frankenfel, C. L., (2016). A systematic review of anthocyanins and markers of cardiovascular disease. *Nutrients, 8*, 32.

Wiseman, H., (2000). The therapeutic potential of phytoestrogens. *Exp. Opin. Investig. Drugs, 9*, 1829–1840.

CHAPTER 12

THE CLASSES AND BIOSYNTHESIS OF FLAVONOIDS

MADHURI PATIL and CHANDRASHEKHAR MURUMKAR

Post-Graduate Research Center, Department of Botany, Tuljaram Chaturchand College of Arts, Science, and Commerce, Baramati (Autonomous), Maharashtra, India

ABSTRACT

Plants, in addition to primary metabolites, produce some novel compounds by a series of chemical reactions under enzymatic control, and these secondary metabolites are activated more during particular stages of growth as well as in response to biotic and abiotic stress. Flavonoids comprise the largest group of naturally occurring secondary metabolites as phenolic compounds responsible for much of the flavor and color of flowers. Large numbers of flavonoids are generally yellow in color. They play an important role in signaling molecules, UV protection, growth, and development, defense against herbivores and pathogens, pollinators, and seed dispersers. This group of natural products is divided into three classes: flavonoids, isoflavonoids, and neoflavonoids, based on the position of the linkage of the aromatic ring to the benzopyrano moiety. These are derived from subunits supplied by the acetate and Shikimate pathways. The present chapter deals with flavonoids, their function and classification, along with their biosynthesis.

12.1 INTRODUCTION

Plant secondary metabolites and their distribution is of prime importance in defense mechanism often related to survival restricted to taxonomically

Flavonoids as Nutraceuticals. Rajesh K. Kesharwani, Deepika Saini, Raj K. Keservani, and Anil Kumar Sharma (Eds.)

related groups (Patil et al., 2021). These signal compounds play an important role in particular life strategies embedded in a given phylogenetic framework (Wink et al., 2003). Flavonoids are secondary metabolites involved in several physiological responses to the environment, such as defense against herbivores, pollination attractors and symbionts, UV radiation, and pathogens (Agati & Tattini, 2010; Schulz et al., 2015; Wink, 1998; Harborne, 1993; Dixon & Pasineti, 2010).

More and more information has become increasingly available about the distribution of various types of flavonoids in plants, but a great deal about their biosynthetic interrelationships has been still unanswered. An examination of flavonoid structure, function, evolution, classes, biosynthesis, and their importance to human health is dealt with in this chapter. A review of the flavonoid's skeleton, its appearance, and function along with evolution, different classes of flavonoids based on their degree of oxidation, and biosynthesis of these flavonoids through different biochemical pathways are worth studying the major aspect of research to know its potentiality.

12.2 DEFINITION AND STRUCTURE OF FLAVONOIDS

Flavonoids, the term is derived from the Latin word "*Flavus*," meaning "yellow," as a large number of flavonoids are yellow in color. More than 6,000 different flavonoids have been identified still so far, and still this number is increasing (Ferrer et al., 2008). The basic skeleton of flavonoids is the flavan nucleus containing 15 carbon arranged in two aromatic rings connected by a three-carbon bridge (C_6-C_3-C_6) (Figure 12.1).

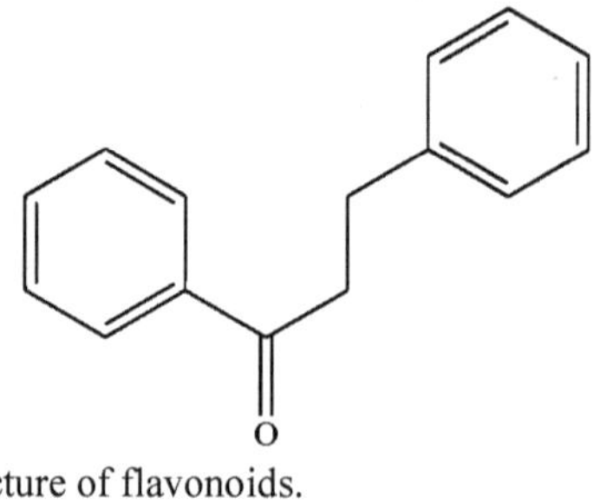

FIGURE 12.1 Basic structure of flavonoids.

They are also known as plant pigments or co-pigments, responsible for various red, blue, and purple color pigmentation found in plants and are being studied for their association with the health benefits of wine, chocolate, as well as a diet rich in fruits and vegetables. Those are rich in bark, leaves, flowers, fruit, and seeds through their pigmentation in plants. These colors attract the insects, serving as pollinating agents. They occur in the cell sap of young tissues as they are sap soluble, occurring in the free state as well as glycosides. Flavonoids, which occur as aglycones, are sparingly soluble in water but soluble in organic solvents.

The flavonoids have always been of interest to botanists and plant taxonomists since they occur in all land plants, unlike, for example, the alkaloids, which are less widespread, are widely used as taxonomic markers. These are common in higher plants belonging to families Solanaceae, Leguminoseae, Rutaceae, Primulaceae, Polygonaceae, Salicaceae, Pinaceae, Rosaceae, Asteraceae, Lamiaceae, Bignoniaceae, Moraceae, Betulaceae, Rubiaceae, Myrtaceae, Aristolochiaceae, etc. (Bhat, 2010). Different classes of flavonoids and their conjugates have numerous functions during the interactions of plants with the environment, both in biotic and abiotic stress conditions (Linda, 1999; Close & Beadle, 2003; Patil et al., 2021).

In fact, it has been amply demonstrated that the flavonoids of the plants provide a clear indication of its revolutionary status from primitive to advanced. In many instances, hybridization of two plants containing two different flavonoids produces a new strain of plant which is capable of synthesizing the flavonoids typical of both parents. They play different roles in plants, like changes in the color of petals, specifically anthocyanins helpful to plant breeders through hybridization, as phytoalexins which are produced by the plants in response to microbial invasion, e.g., pistachin in pea plants, phaseolin in *Phaseolus vulgaris*, Phlorizin root bark component of apple tree confer disease resistance on the apple plant, Glyceolins in Soyabean and isoflavone daidzein in alfalfa (Goyal, 2012), Narigenin from grapefruit peel, Hesperidin from the orange peel are mediators of plant-insect interactions (repellants), condensed tannins produced from the polymerization of flavon diols often protective effect play a similar part in repelling herbivores (Mann, 2005).

Flavonoids (Keservani et al., 2010a, b, 2020; Keservani & Sharma, 2014) also exhibit various important biological activities in mammals, as antioxidant, antimicrobial, mitochondrial adhesion inhibition, antiulcer, antiarthritic, estrogenic receptor binding, antiangiogenic, anticancer, protein kinase inhibition, prostaglandin synthesis inhibition, DNA synthesis cell cycle arrest, topoisomerase inhibition and many more (Oyvind & Kenneth, 2006).

Plants have developed flavonoids to protect themselves against pathogenic fungal parasites, herbivores, pathogens, and ultraviolet (UV) radiation. These are recognized by insects, birds, and animals for seed dispersal. The initial evolution of the flavonoid pathway thus probably occurred during colonization of land plants. In contrast to flavonols, isoflavonoids occur only sporadically throughout the land plants (Mann, 2005). Isoflavonoids have been reported in the moss *Bryum capillare* (Anhut et al., 1984). The evolution of the flavonoid pathway provides an excellent example of how a biochemically complex trait may be built up in stages, with each addition being adaptive. Intelligent design through complex biochemical adaptations, which are irreducibly complex, cannot be resulted by a gradual process of natural selection. The complex biochemical adaptation is flower color produced by anthocyanin pigments is through complexity lying in the fact that anthocyanin production requires at least six sequential biochemical reactions enabled by six different enzymes. Looking at this character by itself, there is no question that removal of one enzyme limits the production of anthocyanins. It is only because the intermediate products that were formed along the way the diverse flavonoids in land plants retained their important ecological and physiological role and so have not been superseded by other secondary metabolites. Thus, we are able to recognize that the gradual irreducible complexity of floral pigment production evolved. The anthocyanin glycosides of anthocyanidins are primarily responsible for the blue and violet color of fruits. They are believed to derive from dihydroflavonols, but the final stage of the biosynthetic origin remains to be elucidated.

The anthocyanins play a vital role as mediators of these interactions. The variety of hues associated with anthocyanins has increased in the evolutionary process. Flower color evolution appears to involve *loss-of-function* mutations in the anthocyanin pathway; it is also well-known that this pathway has contributed significantly to the evolution of novel characters. Arguably, the most important process yielding a new function is gene duplication, followed by the evolution of a novel functional biomolecule in one of the duplicate copies (neofunctionalization). An evolutionary analysis of flavonoid gene families suggests that this process has repeatedly given rise to several novel classes of secondary compounds in plants (Raushur, 2006).

Flower color, in large part, is determined by the floral branches that are most active in a species. Evolutionary transitions in flower color frequently are accompanied by which changes in floral morphology that are believed to enhance the efficiency of interactions with new pollinators. "pollinator syndromes" have been recognized by plant evolutionary biologists for decades (Faegri & van der Pijl, 1966). For example, bee-pollinated flowers are typically blue-purple, have relatively short, broad tubes, broad limbs that serve as landing platforms, small amounts of concentrated nectar, and inserted anthers and stigmas. By contrast, hummingbird-pollinated flowers usually have reddish flowers, long narrow tubes, small limbs, copious dilute nectar, and exerted anthers and stigmas. Moth and bat-pollinated flowers tend to be white, fragrant, and open at night. Many evolutionary changes in flower color thus seem to be the adaptations associated with pollinator attraction (Dewick, 1985). Polyphenols and flavonoids are double bond cyclic aromatic groups representing primitive traits like in pteridopsida and gymnopsida. It also prevails the formation of resins that are saturated hydrocarbons which can be further broken down and deposited in the older parts of the plants. If they are soluble, they are categorized as tannins. Anthocyanins represent the group of organic compounds which are responsible for absorbing the UV-radiations. The high amount of flavonoids and tannins in *T. cordifolia*, along with the large amount of anthocyanins in *A. bracteolata*, reflect the advancement of the group unisexual, which now is considered under a separate series Daphnales (Patil et al., 2021).

The development of metabolic pathways that resulted in flavonoids would thus have been benefited to plants that emerged from the primeval oceans. It is remarkable that the marine plants do not produce flavonoids. A shorter lifetime of the anthocyanins is hardly surprising in view of the shorter life span of flowers when compared to the life cycle of the plant. Winkel-Shirley (1999) reported that it is an extensive class of low molecular weight characterized by the flavan nucleus with the presence of characteristic blue, purple, and red anthocyanin pigments of plant tissues.

Medicinally, flavonoids are acting as antiproliferative, antioxidant, antitumor, anti-cancer, anti-pro apoptotic activities, and anti-inflammatory compounds (Kim et al., 2008). Flavonoids from *Vaccinium* species exhibited anti-cancer activity (Katsube et al., 2003), protecting the human body contrary to highly reactive oxygen species (ROS) and endogenous scavenging compounds (Jadhav et al., 2008; Kerry & Abbey, 1997; Willam et al., 2004). Flavonoids such as epicatechin are inhibitors of nitrous acid-dependent nitration and *in vitro* DNA deamination (Oldreive et al., 1998).

12.3 CLASSIFICATION OF FLAVONOIDS

Flavonoids are typical phenolic compounds that act as potent antioxidants and metal chelators. Overall, several of these flavonoids are effective anticancer promoters and cancer chemopreventive agents. The flavonoids, ubiquitous in plants, with a common structure of diphenyl propane (C_6-C_3-C_6), consisting of two aromatic rings linked through three carbons.

Flavonoids comprise a large group of secondary metabolites which are derived from subunits supplied by the acetate and shikimate pathways. They occur exclusively in higher plants and are responsible for much of the flavor of food and the color of flowers. A basic C_{15} unit is invariably present. It is established from tracer experiments that the ArC_3 subunit is derived from Shikimate and that the aromatic Ring A is of polyketide origin (Bhat, 2010; Keservani & Sharma, 2014).

A flurry of research began in an attempt to isolate the various individual flavonoids and to study the mechanism by which flavonoids act originated in 1930 when a new substance isolated from oranges, believed to be a member of a new class of vitamins, as vitamin P. Afterwards it is clear that this substance was a flavonoid (rutin) (Rice-Evans et al., 1997). Flavonoids are classified into different groups primarily on the basis of degree of oxidation of three carbon bridges like anthocyanins, flavones, flavonols, isoflavones, and flavonones. A structural variation in each group is partly due to the degree and pattern of hydroxylation, methoxylation, or glycosylation.

Flavonoids contain conjugated double bonds and groups (hydroxyl or other substituents) that can donate electrons through resonance to stabilize the free radicals, which originate in the electronic spectra of flavonoids (Gupta et al., 2016). Flavonoids are classified into four main groups depending on the position of the linkage of the aromatic ring to the benzopyrano moiety: (i) flavonoids (2-phenylbenzopyrans) **1**; (ii) Isoflavonoids (3-benzopyrans) **2**; (iii) Neoflavonoids (4-benzopyrans) **3** (Figure 12.2). These groups usually share a common chalcone precursor and thus genetically and structurally closely related. Chalcones and Aurones are with C_6-C_3-C_6 Backbone named as minor flavonoids. Anthocyanins possess flavylium salt structures and are the glycosides of anthocyanidins. Flavones and isoflavones also occur as hydroxyl derivatives and as glycosides.

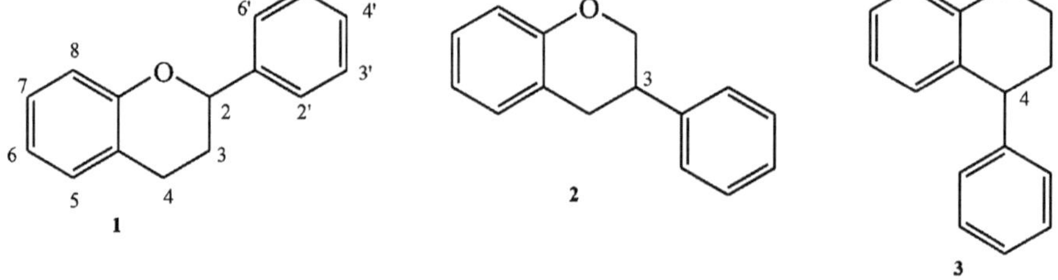

FIGURE 12.2 Major groups of flavonoids.

12.3.1 FLAVONOIDS

Flavonoids with C_6-C_3-C_6 backbone are known as *2-Phenylbenzopyrans.* Based on the degree of oxidation and saturation present in the heterocyclic C-ring, the flavonoids may be divided into flavan, Flavanone, Flavone, Flavonol, Flavanonols (Dihydroflavonol), Flavan-3-ol, Flavan-4-ol and flavan 3,4-diol (Leucoanthocyanidins) (Figure 12.3).

Flavanones are key intermediates in flavonoid chemistry as they can be converted into flavones, isoflavone, and Flavanols (Dihydroflavonol), flavan 3,4-diol (Leucoanthocyanidins) and flavan-4-ols. Flavanones contain only one functional group, kenotic carbonyl at 4-position, and two aromatic rings. Flavanones are well-known components of *Citrus* fruits, and they are present in solid wastes and residues obtained during their industrial processing. Butrin is a yellow-orange flavanone obtained from *Butea monosperma* flowers used in dying silk and cotton. Myricetin, a yellow-colored dihydro flavonol obtained from *Myrica rubra* used in tanning, Santal a red-colored isoflavone obtained from *Pterocarpus santalinus* wood, useful for dying cotton, wool, leather, and wood. Daidzein, a yellow-colored isoflavone from *Glycin max* berries used as a food supplement.

Flavones (Keservani et al., 2010a) contain two basic functional groups carbon-carbon double bond in conjugation with the carbonyl functional group and the ketonic carbonyl group. Flavone and flavonols occurring as glycosides on hydrolysis yield glucose as rhamnose and a sugar-free aglycon as anthoxanthidin. The luteolin isolated from Weld (*Reseda luteola*) leaves and seed is the oldest yellow dyestuff (flavones) used in Europe for dying silk, wood, and textiles. Some common flavones found in flowers, leaves, and seeds of various *Primulas*, Chrysin in buds of Poplar, Apigenin in yellow *Dahlias* (Mann, 2005; Bhat, 2010).

Flavonols are very widely distributed in nature, both in plants and insects. The yellow dye in the wings of the butterfly (*Melanargea gelatea*), Galangin in *Galanga* root, Kampfero in blue *Delphinium* flowers, Quercetin in the bark of American oak (*Overcus tinctoria*), Myricetin in myricaceae family plants, Rutin in *Sophora japonica* flower. Flavone and flavonol with the basic unit as γ pyrone present as benzo γ pyrone, known as pyrone pigments. Catechins are colorless crystalline compounds obtained from Catechu (Harborne, 1982).

Flavonols and flavones are not colored. They do absorb strongly in the UV and, although invisible to the human eye, can be seen by insects. They often occur at the center of flowers and probably acts as "honey guides" able to attract insect in their nectars. In return, the insect is accessory in the process of pollination, carrying away from the flower, not the nectar but also pollen particles to transfer to other plants (Mann, 2005). They also emitted as signal substances in order to induce *Rhizobia* from leguminous roots for expression of genes required for nodulation (Liu, 2016).

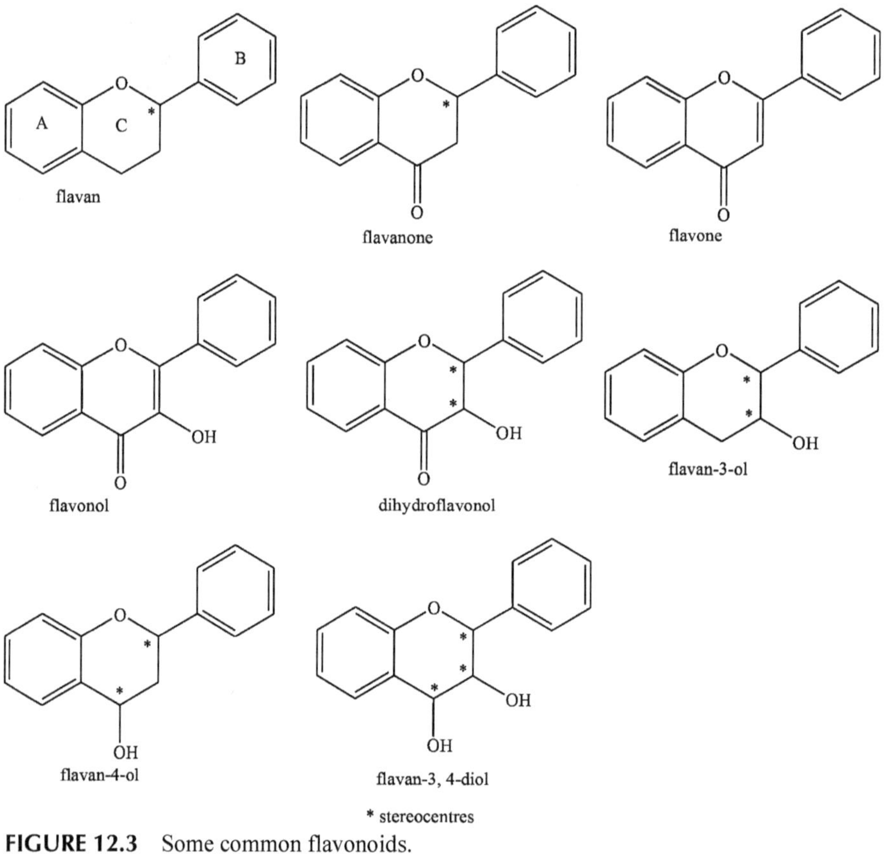

FIGURE 12.3 Some common flavonoids.

12.3.2 ISOFLAVONOIDS

The isoflavonoids are a distinctive subclass of the flavonoids. These compounds possess a 3-phenyl chroman skeleton biogenetically derived by 1,2-aryl migration in a 2-phenyl chroman precursor. Despite their limited distribution in the plant kingdom, isoflavonoids are remarkably diverse as far as structural variations are concerned. Isoflavonoids are subdivided into the following groups: isoflavones, isoflavone, isoflavone, isoflav-3-ene, isoflavonol, rotenoid, coumestans, 3-arylcoumarin, coumaranochromene, coumaranochromone, pterocarpon (Figure 12.4).

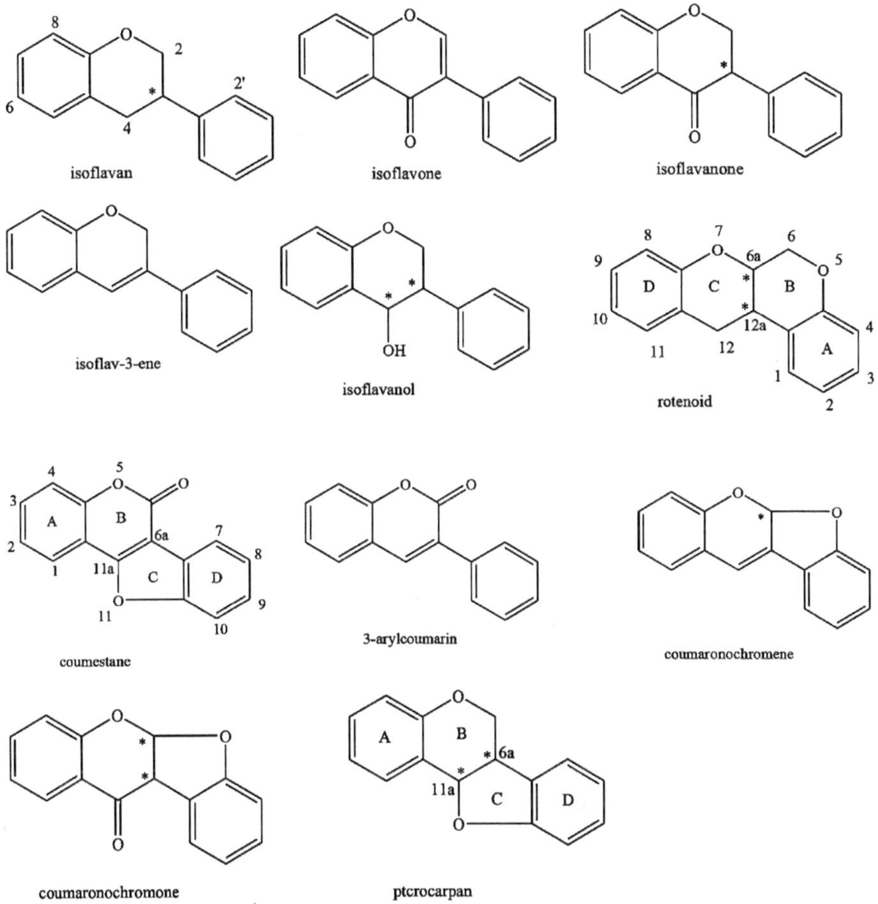

FIGURE 12.4 Some common isoflavonoids.

Isoflavones are also found in either free or a glycoside. Replacement of hydrogen atom from C_3 in benzo-γ pyrone ring by phenyl group isoflavone, the first member of the class isoflavone, is formed. Other isoflavones are hydroxyl and methyl or methoxy derivatives are isoflavone. The glycosides on hydrolysis yield sugar-free isoflavones. Isoflavone, Daidzein, Genistein, and Irigenin are commonly occurring isoflavones, of which Daidzein and Genistein are found in soybean, Medicarpin (Isoflavone) from alfalfa (*Lucerne*) is a phytoalexin. Isoflavones found in certain forage plants like legumes cause infertility in sheep grazing on it has a similar effect on estrogens in animals and are thus named phytoestrogens. Genistein (isoflavone) has a strong estrogen effect.

12.3.3 NEOFLAVONOIDS

The neoflavonoids are structurally and biogenetically closely related to the flavonoids and isoflavonoids, comprise the 4-arylcoumarins (4-aryl-2*H*-1-benzopyran-2-ones), 3,4-dihydro-4-arylcoumarins, and neoflavones (Figure 12.5).

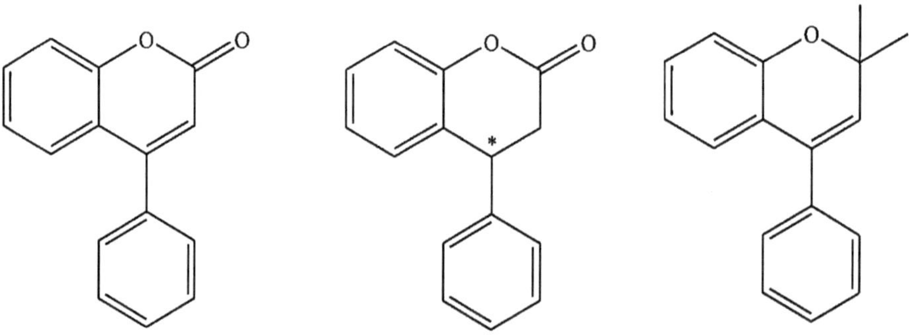

4-arylcoumarin 3,4-dihydro-4-arylcoumarin neoflavene

*stereocenters

FIGURE 12.5 Neoflavonoids.

12.3.4 MINOR FLAVONOIDS

Natural products such as chalcones and aurones containing C_6-C_3-C_6 backbone are considered to be minor flavonoids. These groups of compounds include the 2′-hydroxychalcones, 2′-OH-dihydrochalcones, 2′-OH-*retro*-chalcone, aurones (2-benzylidenecoumaranone), and aerosols (Figure 12.6).

The name aurone has been derived from the Latin name '*Aureus*' means golden, and accordingly, they are golden yellow in color. They possess the skeleton of 2 benzylidene coumaranone containing five-membered with exocyclic carbon-carbon double bond, widely present in fruits and flowers where they play a significant role in the pigmentation of the part of the plant. Sulfuretin is a known example of aurone.

Chalcones are related to flavonoids in the sense that the pyrin ring is cleaved in chalcone. The extended conjugation in chalcones is responsible

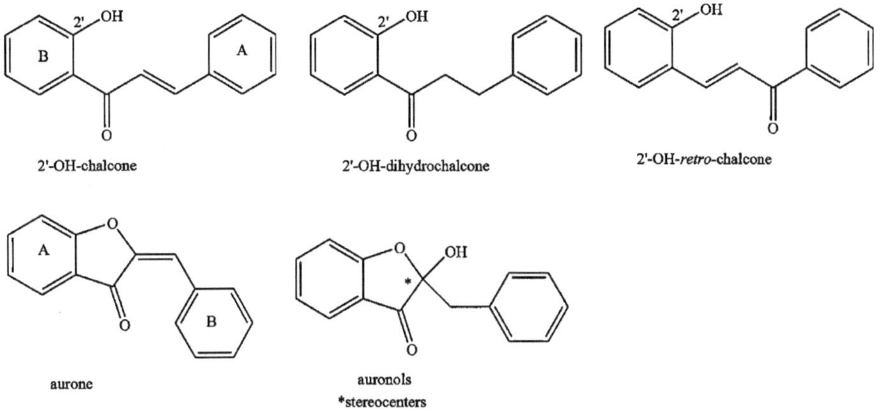

FIGURE 12.6 Minor flavonoids chalcones and aurones.

for their yellow color. As they do not contain a γ pyrone ring and hence are open-chain flavonoids in which two aromatic rings are joined by three carbon α,β-unsaturated carbonyl system. The corresponding dihydro derivative is called dihydrochalcone. They generally co-occur with another class of orange-yellow colored aurones in petals of Asteraceae family flowers. *Carthamus tinctorius* (safflower) is a rare example of isomerization of chalcone to flavanones where a yellow pigment in petals (chalcone) and a red pigment carthamin (flavanone) in flower ages.

12.3.5 ANTHOCYANINS

Anthocyanins are glycosides of aglycones. Anthocyanidin occurs in nature as a special class of flavonoids. Anthocyanins without their sugars are known as anthocyanidins (Taiz & Zeiger, 2006). Different color of flowers, fruits, stems, and leaves are due to the presence of anthocyanins and other co-pigments such as flavones and flavonols responsible for various shades of blue, purple, mauve, maroon, magenta, and red. Their acidic salts exhibit red color, metallic (basic) salts show blue color, and neutral display violet color. There are six major types of anthocyanins with the basic structure of 2 Phenylbenzopyrillum or flavylium widely spread in nature and differ in their degree of hydroxylation. On hydrolysis, they yield aglycons like pelargonidin, Cyanidin, peonidin, delphinidin, petunidin, malvidin, apigenidin, luteonidin, and Cynidin (Figure 12.7).

The variegated color of petals almost certainly acts as stimuli for pollinating agents. In *Ipomoea*, blue, and purple-flowered species tend to produce almost exclusively cyanidin-based anthocyanins. Pathway flux in these species is almost entirely down the second branch; mutations that knock out the enzyme F3′H redirect flux down the pelargonidin branch, resulting in red flowers. Red-flowered *Ipomoea* species also almost always produce pelargonidin-based rather than cyanidin-based anthocyanins (Zufall & Rausher, 2003). In *Penstemon* (Scrophulariaceae), blue/purple, bee-pollinated flowers

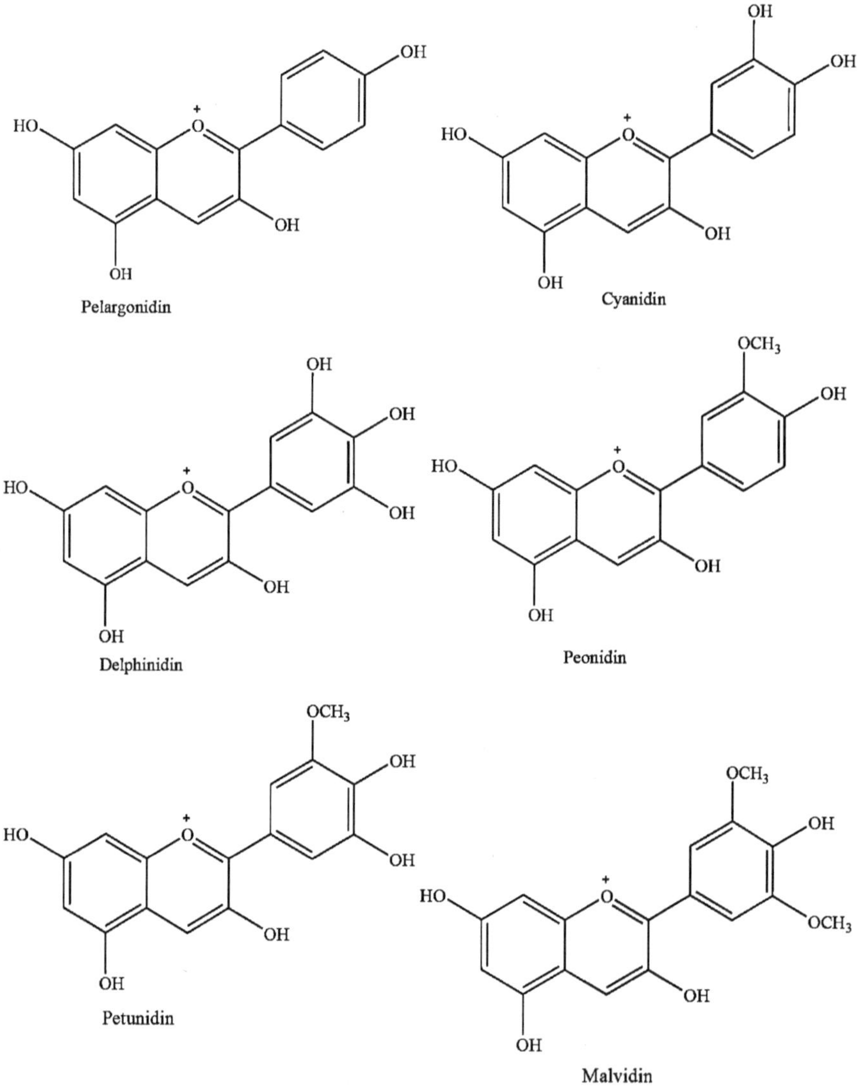

FIGURE 12.7 Anthocyanins.

tend to produce delphinidin-derived anthocyanins, while red, pollinated by hummingbird flowers tend to produce pelargonidin (Scogin & Freeman, 1987). Adjusting the relative amounts of flux down the different pathway branches thus seems to be a common way of altering flower color. Thus natural selection seems to have led to the production of scarlet hues, typical of pelargonidin or blue of delphinidin, through modification of primitive pigment cyaniding by loss or gain of hydroxyl group. Anthocyanins appear at specific developmental stages and may be induced by a number of environmental factors, including visible and UV-B radiation, cold temperature, and water stress (Linda, 1999; Nikam, 2007).

12.4 BIOSYNTHESIS OF FLAVONOIDS

From a biogenetic point of view, the C_{15} carbon framework of flavonoids can be divided into two parts, one part consisting of six carbon atoms, which forms ring A, while the other part consists of nine carbon atoms known as phenylpropanoid moiety (C_6-C_3). These phenyl propanoid moieties also serve as the precursor of a number of amino acids and phenolic compounds occurring in nature. It is well established that the biogenesis of ring A proceeds via the acetate or polyketide route, while that ring B proceeds via the Shikimate pathway (Bhat et al., 2010).

The majority of flavonoids are synthesized by a multifarious metabolic enzyme situated on the cytoplasmic shell of the endoplasmic reticulum of plant cells (Burbulis, 1999). Conversely, some flavonoids (flavonols and flavanols) and a small number of flavonoid biosynthesis enzymes furthermore originate in the nuclei of plant cells (Wang, 2005). This suggests that flavonoids are synthesized in diverse cell compartments in an array to maintain particular physiological functions.

The function of Shikimic acid is not restricted to the generation of amino acids for protein biosynthesis. It also provides a precursor for a large amount of variety of other substances formed by plants in huge quantities, particularly phenylpropanoids like flavonoids and lignins. The shikimic acid pathway converts simple carbohydrate precursors (Erythrose 4 Phosphate) and phosphoenol pyruvate (PEP) to the aromatic ring containing amino acid phenylalanine, tyrosine, and tryptophan. The most abundant class of phenolic compounds in plants is derived from deamination of phenylamine to cinnamic acid by phenylalanine lyase (PAL).

The condensation of PEP and D-erythrose-4-Phosphate was catalyzed by the 3 deoxy-D-arabino-heptulosonic acid-7-phosphate (DHAP) and inorganic phosphate. The ring closure of DHAP catalyzed by 3 hydroxyquinonate synthase takes place to form an alicyclic intermediate. The enzyme 3-dehydroduinate dehydratase catalyzes the dehydration to introduce a double bond between C_4 and C_5 to afford 3-dehydroshikimate dehydrogenase occurs to give the Shikimate. Dehydrogenase enzyme is NADP specific, and the reaction involves transfer of hydride from the nicotinamide ring in NADPH.

The shikimate is converted to Shikimate-3-monophosphate through action of Shikimate kinase. The Shikimate-3-monophosphate on condensation with PEP catalyzed by the PEP: 3-phosphoshikimate-5-O (I carboxyl vinyl) transferase (5EPSP, synthase) produces 5-enolpyruvyl-3-phospho-shikimate. The final and the seventh step in this shikimate pathway consist of dehydration to introduce a second double bond in the six-membered ring to form chorismate. The dehydration is catalyzed by chorismate synthase. Chrosimate represents a branch point for two biosynthetic pathways (Figure 12.8).

Tryptophan is formed via four reactions, prephenate is formed by rearrangement where the side chain is transferred to the 1′-position of the ring, and arogenate is formed after transamination of the keto group. Removal of water results in the formation of the third double bond and phenylalanine is formed by decarboxylation. The enzyme chorismate mutase catalyzes the conversion of chorismate to prephenate by a pericyclic reaction known as the unimolecular intermolecular rearrangement. Prephenate is further transformed into phenylpyruvate by the catalytic action of prephenate dehydratase. Transamination of phenylpyruvate catalyzed by the enzyme aromatic amino acid aminotransferase produces phenylalanine. In this *trans*amination, glutamate serves as the donor of the amino group. The enzymatic deamination of phenylalanine catalyzed by L-phenylalanine Ammonia Lyase (PAL) takes place stereospecifically with the loss of NH_2 and pro-S-hydrogen atom. The L-amino acid affords trans-cinnamate, which is phenyl propanoid moiety (C_6-C_3), the precursor of ring B in flavonoids. Hydroxylation at C_2 and C_4 of trans-cinnamic acid yields p-coumaric acid and 2,4-hydroxycinnamic acid. *Trans* cinnamic acid is converted to p-coumarate by hydroxylation, and the condensation of p-coumaroyl coenzyme-A with 3 molecules of malonyl Co-A (Acetate units) catalyzed by chalcone synthase (CHS) results in the formation of adduct which is converted to chalcone naringenin. Conversion of naringenin on decarboxylation to stilbene is so called the Malonate pathway.

Chalcones are converted into aurones by aureusidin synthase (AS) and into flavanone by Chalcone isomerase (CHI). The ring structure is formed by the addition of phenolic hydroxyl group to the double bond of the

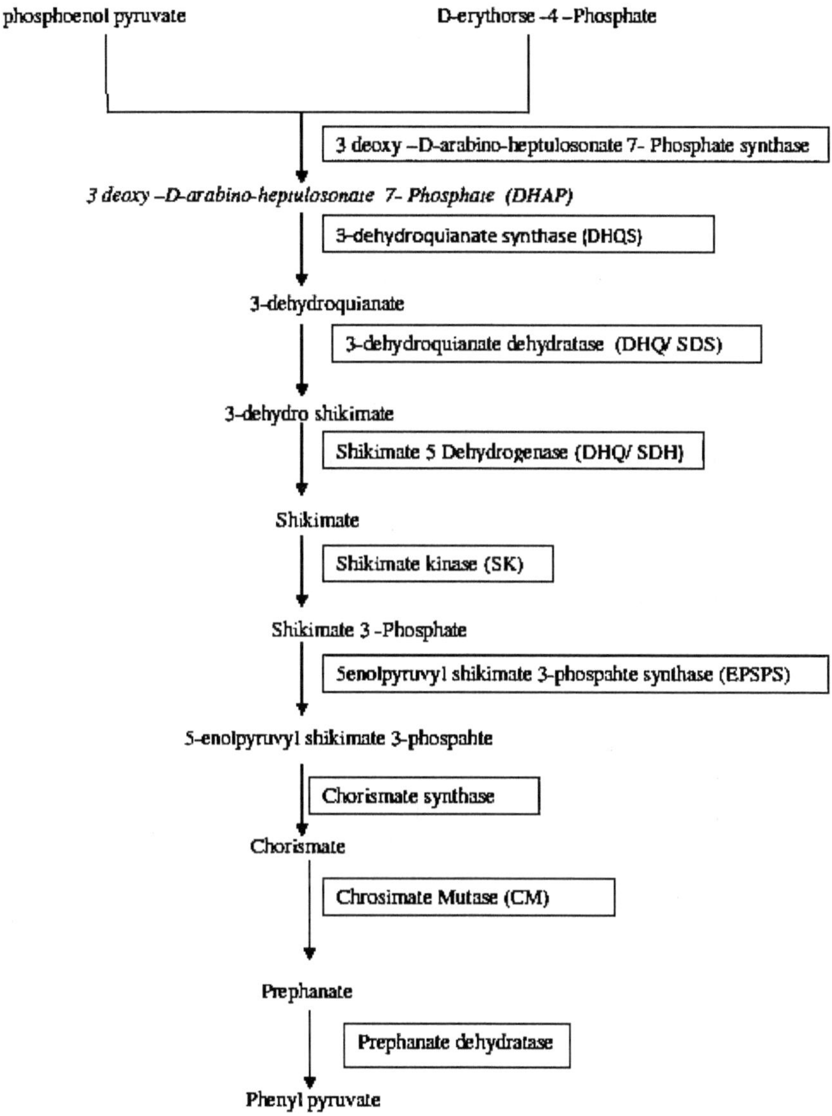

FIGURE 12.8 Biosynthesis of flavonoids.

carbon chain connecting the two phenolic rings. Flavanone is a precursor for a variety of flavonoids. As a key enzyme of flavonoid biosynthesis, the synthesis of the enzyme protein of CHI is under strict control. It is induced, like PAL and CHS, by elicitors. Dihydroflavonol is formed by the action of the enzyme CHI to yield flavanone and flavanone-3-hydroxylase (F-3-H); further hydroxylation in ring C yields flavonols. The oxidation of Flavanone yields flavone. Dihydroflavonol also serves as the precursor of various anthocyanins. The degree of blue coloration of the flowers is due to the different hydroxylation patterns of anthocyanidins. They are controlled by microsomal cytochrome p-450 enzymes such as flavonoids-3-hydroxylases and flavonoids 3,5-hydroxylases. Recently by genetic engineering manipulation, the colors of the anthocyanidins in *Petunia* flowers have been altered in an excellent way.

12.4.1 BIOGENESIS OF ISOFLAVONOIDS

It has been found that chalcone serves as the intermediate for the biogenesis of isoflavonoids. In the first step, chalcone is converted into chalcone oxides which rearrange as per the established mechanism put forward to explain the acid-catalyzed rearrangement of α, β oxy ketones to give aldehyde the final cyclization gives isoflavone.

12.5 CONCLUSION

Flavonoid is an important group of secondary metabolites influencing the various metabolic activities of primary and secondary compounds. These ultimately are responsible for plant growth and development through reproductive biology. The evolution of flavonoids in plant history is highly significant as it is the molecule that controls and regulates the correlation and co-existence of plants and the animal world. Because of their photosensitive property and unique metabolism of synthesis and role, these play a vital role in physiology, biochemistry, and taxonomy in cladistic analysis. Primary and secondary metabolic systems, thus, are intertwined with flavonoid metabolism. The complexity of cellular metabolism is nowadays becoming easy to understand due to sophisticated tools available for determining biochemical and structural characteristics and opening new eras of research for flavonoid metabolism. Flavonoids are the most promising group of chemical markers to look for in plant identification. The employment of flavonoid chemistry as an integral part of taxonomic revision programs is worth studying as this will help to use the practical implications of agronomic and nutritional traits in plants along with many more fields of applications in human healthcare.

KEYWORDS

- **anthocyanins**
- **biogenesis**
- **flavonoids**
- **flavonoids isoflavonoids**
- **isoflavonoids**
- **minor flavonoids**
- **neoflavonoids**

REFERENCES

Agati, G., & Tattini, M., (2010). Multiple functional roles of flavonoids in photoprotection. *New Phytol., 186*, 786–793. [PubMed] [Google Scholar].

Anhut, S., Zinsmeister, H. D., Mues, R., Barz, W., Mackenbrock, K., Koster, J., & Markham, K. R., (1984). The first identification of isoflavones from a bryophyte. *Phytochem., 23*, 1073–1075.

Bhat, S. V., Nagasampagi, B. A., & Sivakumar, M., (2010). *Chemistry of Natural Products* (6th edn.). Narosa publishing house private Ltd., New Delhi.

Close, D. C., & Beadle, C. L., (2003). The ecophysiology of foliar anthocyanin. *Botanical Review, 69*, 149–161.

Dewick, P. M., (1985). The biosynthesis of shikimate metabolites. *Natural Product Reports, 2*(6), 495–611.

Dixon, R. A., & Giulio, M. P., (2010). Flavonoids and isoflavonoids: From plant biology to agriculture and neuroscience. *Plant Physiology, 154*(2), 453–457.

Faegri, D. T., & Van, D. P. L., (1966). *The Principles of Pollination Ecology*. New York: Pergamon Press.

Ferrer, J., Austin, M., Stewart, C. J., & Noel, J., (2008). Structure and function of enzymes involved in the biosynthesis of phenylpropanoids. *Plant Physiol. Biochem., 46*, 356–370.

Goyal, S., Lambert, C., Cluzet, S., & Jean-Michel, M., & Ramawat, K., (2012). *Secondary Metabolites and Plant Defense.* 10.1007/978-94-007-1933–0_5.

Gupta, J., Gupta, A., & Gupta, A. K., (2016). Flavonoids: Its working mechanism and various protective roles. *International Journal of Chemical Studies, 4*, 190–198.

Harborne, J. B., & Baxter, H., (1993). *Phytochemical Dictionary: A Handbook of Bioactive Compounds from Plants*. Taylor & Francis, London.

Harborne, J. B., Mabry, T. J., & Mabry, H., (1982). *The Flavonoids* (2nd edn.). Chapman and Hall, London.

Jadhav, G. B., Upasani, C. D., & Patil, R. A., (2008). Overview of Flavonoids. P*harmaceutical Reviews, 6*(6). https://journaldatabase.info/articles/overview_flavonoids.html (accessed on 24 July 2023).

Katsube, N., Iwashita, K., Tsushida, T., Yamaki, K., & Kobori, M., (2003). Induction of apoptosis in cancer cells by bilberry (*Vaccinium myrtillus*) anthocyanins. *J. Agri. Food. Chem., 50*, 519–525.

Kerry, N. L., & Abbey, M., (1997). Red wine and fractionated phenolic compounds prepared from red wine inhibit low-density lipoprotein oxidation *in vitro*. *Atherosclerosis, 135*(1), 93–102.

Keservani, R. K., & Sharma, A. K., (2014). Flavonoids: Emerging trends and potential health benefits. *Journal of Chinese Pharmaceutical Sciences, 23*(12), 815.

Keservani, R. K., Kesharwani, R. K., Sharma, A. K., Vyas, N., & Chadoker, A., (2010b). Nutritional supplements: An overview. *International Journal of Current Pharmaceutical Review and Research, 1*(1), 59–75.

Keservani, R. K., Kesharwani, R. K., Vyas, N., Jain, S., Raghuvanshi, R., & Sharma, A. K., (2010a). Nutraceutical and functional food as future food: A review. *Der Pharmacia Lettre, 2*(1), 106–116.

Keservani, R. K., Sharma, A. K., & Kesharwani, R. K. (2020). *Nutraceuticals and Dietary Supplements: Applications in Health Improvement and Disease Management*. CRC Press. ISBN: 9781771888738.

Kim, J. Y., Park, S. J., Yun, K. J., Cho, Y. W., Park, H. J., & Lee, K. T., (2008). Isoliquiritigenin isolated from the roots of *Glycyrrhiza uralensis* inhibits LPS-induced NOS and COX-2 expression via the attenuation of NF kappa B in RAW 264.7 macrophages. *Eur. J. Pharmacol, 584*, 175–184.

Linda, C. S., (1999). Environmental significance of anthocyanins in plant stress responses.

Liu, C. W., & Murray, J. D., (2016). The role of flavonoids in nodulation host-range specificity: An update. *Plants (Basel), 5*(3), 33.

Mann, J., (2005). *Secondary Metabolism* (2nd edn., p. 275). Oxford university press Inc. New York. ISBN 0-19-855229-6.

Nikam, V. K., (2007). *Physiological Studies in a Medicinal Plant Chlorophytum borivilianum* Sant. and Fernad. Thesis submitted to Shivaji University, Kolhapur.

Oldreive, C., Zhao, K., Paganga, G., Halliwell, B., & Rice-Evans, C., (1998). Inhibition of nitrous acid-dependent tyrosine nitration and DNA base deamination by flavonoids and other phenolic compounds. *Chem. Res. Toxicol., 11*, 1574–1579.

Oyvind, M. A., & Kenneth, R. M., (2006). *Flavonoids: Chemistry, Biochemistry, and Applications.* CRC group Taylor & Francis Group, LLC, 6000 Broken Sound Parkway NW, Suite 300, Boca Raton, FL 33487-2742.

Patil, M., Poonam, K., & Chandrashekhar, M. (2021). Adoptive mechanism in some plants of tropical arid region of Maharashtra. *Indian Journal of Applied and Pure Biology, 36*(1), 187–190. ISSN:0970-2091.

Rausher, M. D., (2008). Evolutionary transitions in floral color. *International Journal of Plant Sciences, 169*(1), 7–21.

Rice-Evans, C. A., Miller, M. J., & Paganga, G., (1997). Antioxidant properties of phenolic compounds. *Trends in Plant Sci., 2*, 153–159.

Schulz, E., Tohge, T., Zuther, E., Fernie, A. R., & Hincha, D. K., (2015). Natural variation in flavonol and anthocyanin metabolism during cold acclimation in *Arabidopsis thaliana* accessions. *Plant Cell Environ., 38*, 1658–1672. [PubMed] [Google Scholar].

Scogin, R., & Freeman, C. E., (1987). Floral anthocyanins of the genus *Penstemon*: Correlations with taxonomy and pollination. *Biochem. Syst. Ecol., 15*, 355–360.

Taiz, L., & Zeiger, E., (2006). *Plant Physiology* (IVth edn.). (Publ.) Sinauer Associates, Inc., Publishers, Massachusetts.

Tzin, V., & Galili, G., (2010). New insights into the shikimate and aromatic amino acids biosynthesis pathways in plants. *Molecular Plant, 3*, 956–972. 10.1093/mp/ssq048.

Wang, H. L., Wang, W., Zhang, P., Pan, Q. H., Zhan, J. C., & Huang, W. D., (2005). Gene transcript accumulation, tissue, and subcellular localization of anthocyanidin synthase (ANS) in developing grape berries. *Plant Science, 179*, 103–113.

William, R. J., Spencer, J. P. E., & Rice, E. C., (2004). Flavonoids: Antioxidant or signaling molecules? *Free Rad. Biol. Med., 36*, 838–849.

Wink, M., (2003). Evolution of secondary metabolites from an ecological and molecular phylogenetic perspective. *Phytochemistry, 64*, 3–19.

Wink, M., Schmeller, T., & Latz-Bru¨ning, B., (1998). Modes of action of allelochemical alkaloids: Interaction with neuroreceptors, DNA and other molecular targets. *J. Chemical Ecology, 24*, 1881–1937.

Winkel-Shirley, & Burbulis, I. E. B., (1999). Interactions among enzymes of the Arabidopsis flavonoid biosynthetic pathway. *Proceedings of the National Academy of Sciences of the United States of America, 96*, 12929–12934.

Zufall, R. A., & Rausher, M. D., (2003). The genetic basis of a flower color polymorphism in the common morning glory (*Ipomoea purpurea*). *The Journal of Heredity, 94*(6), 442–448.

CHAPTER 13

PLANT-BASED FLAVONOIDS AS PROMISING TOOLS TO COMBAT THE COVID-19 INFECTION

ARUN DEV SHARMA and INDERJEET KAUR

PG Department of Biotechnology, Lyallpur Khalsa College, Jalandhar, Punjab, India

ABSTRACT

SARS CoV-2 (COVID-19) is a positive sense ssRNA virus that belongs to the family coronavirus and is disseminating its appendages throughout the world because of the nonexistence of drugs at present-day. Due to its association with respiratory distress, fever, and cough, the mortality rate is more than 15% worldwide. Since 2020, an inspiring number of scientists, biologists, pharmacologists, virologists, immunologists, and molecular biologists, are doing work on the development of biotechnological tools, specifically monoclonal antibodies and vaccines, along with the rational design of pharmaceutical drugs for remedial approaches. Although some vaccines were synthesized, unfortunately, no acceptable remedial approach or anti-COVID-19 preventive has thus far been developed and fully accepted. On the other hand, among all possible ways to combat COVID-19 inhibition or alleviation, there is another way, that is, plant bioactive, which has been given miniature consideration to date. Indeed, in the plants (edible parts) providing our foodstuff, there is an impartial quantity of secondary metabolites. Among all secondary metabolites, flavonoids are the largest class of phytochemicals that are frequently available in aromatic plants. Flavonoids are a large class of secondary metabolites endowed with antiviral properties

Flavonoids as Nutraceuticals. Rajesh K. Kesharwani, Deepika Saini, Raj K. Keservani, and Anil Kumar Sharma (Eds.)

via their capability to constrain viral pathogenesis at an initial phase of the life cycle of the virus, as cited in previous reports on the use of flavonoids on different RNA viruses, like Ebola, influenza, middle east respiratory syndrome (MERS), severe acute respiratory syndrome (SARS) and human immunodeficiency virus (HIV). These antiviral activities potentiate the use of Flavonoids as a treatment agent against the SARS-CoV-2. In addition, flavonoids are recognized to have pharmacological features and, via their well-being advantageous activities, for instance, immune-stimulating or else anti-inflammatory actions may perhaps play an important role in subsidizing to a certain extent to preclude or aggravate the virus infection and/or neutralize the progress of SARS prompted by the new coronavirus (CoV). To that end, an inclusive chapter on the role of flavonoids and their antiviral prospective against COVID-19 has been undertaken.

13.1 INTRODUCTION

A novel coronavirus (2019-n-CoV) caused a pulmonary disease pandemic (COVID-19) in Wuhan and has since spread worldwide (Huang et al., 2020). The virus has been so-called SARS-CoV-2, as the RNA genome of virus is 82% similar to the SARS coronavirus (SARS-CoV). CoVs belong to the Coronaviridae family and are positive RNA genome-enveloped viruses that are alienated in four (α, β, γ, and δ) genera. The SARS-CoV-2 particularly belongs to the β-genus. Genome SARS-CoV-2 is 88% identical to SARS-like CoVs, and 50% similar to MERS-CoV. Structurally SARS-CoV-2 proteins are 90%–100% homologous to SARS-CoV. SARS-CoV-2 comprises four structural proteins: Spike protein (S), envelope protein (E), Membrane protein (M), and nucleocapsid protein (N). These are the major host interacting proteins that interact with targets of the host cell (for instance, CD26, ACE2, cyclophilins, ezrin, and cell adhesion factors) significant for virus entry into the host cell, cell adhesion, and virulence (Lu 2020; Millet et al., 2012). The SARS-CoV-2 N and S-protein has a conserved receptor binding domain (RBD) which identifies receptors of host cells like CD26, ACE2, cyclophilins, ezrin, and additional cell adhesion factors and contributes to the cell receptor binding, tissue tropism, and pathogenesis (Millet et al., 2012). Consequently, because of its important role, SARS-CoV-2 spike protein (S), envelope protein (E), membrane protein (M), and nucleocapsid proteins (N) are considered to be an appropriate target for the development of viral inhibitors. Inhibiting the activity of SARS-CoV-2 S-protein activity would obstruct the replication of the virus. Although some vaccines have been synthesized, unfortunately, no acceptable remedial approach or anti-COVID-19 preventive has thus far been developed and fully accepted. Presently, no particular therapies are available for COVID-19, and research concerning the management of COVID-19 is limited (Lipinski, 2004). Certain initial studies have scrutinized possible combinations that consist of anti-HIV vaccines and antimalarial drugs (hydroxychloroquine) that could be used for the treatment of COVID-19 infections. In addition to the use of antiviral drugs, clinicians are using MERS-CoV and SARS-CoV-2 neutralizing antibodies affecting the S1 domain of the SARS-CoV-2 glycoprotein of the spike, which has been proposed as a key therapeutic target for drug development against COVID-19 treatment (Liu & Wang, 2020). Unfortunately, no specific vaccine is available to date, so a dire crucial prerequisite to treat COVID-19 has led researchers to Spike proteins as potential drug targets to combat this disease.

In synergy with other strategies comprising therapeutic treatments and vaccines, among all possible ways to combat COVID-19 inhibition or alleviation, we propose that there is one another way, which is a diet containing plant bioactive, which has been paid less attention to date. Indeed, in the plants (edible parts) providing our foodstuff, there is an impartial quantity of secondary metabolites that have the potential to inhibit or mitigate the signs of illness. As a substitute and supplementary preventive or therapeutic approach, different *in vitro* and *in-silico* studies have shown that naturally occurring bioactive molecules can inhibit SARS-CoV-2 spike proteins are measured as an additional approach to combat COVID-19 (Bhardwaj et al., 2020). Recently, Sharma et al. (2020a–c), using *in-silico* approaches, have demonstrated that various polyphenolics-like compounds can prevent the replication of the COVID-19 virus by blocking N and S proteins. From the prehistoric era, several medicinal herbs and plants are advantageous in drug therapeutics because they are harmless substitutes being used by human beings. Formerly, various novel drug formulations are obtained from naturally available products. In Traditional Ayurvedic and Chinese medicines, natural products of plants have been utilized immensely as antiviral management. In addition, naturally occurring compounds are too a primary source for modern drugs. Chloroquine and hydroxychloroquine, imperative natural compounds, are derived from the Cinchona tree's secondary metabolites that are under scientific trial and have displayed probable properties against SARS-CoV-2 (Wu et al., 2020). Among all natural metabolites, flavonoids are phenolic phytochemicals (Solnier & Fladerer, 2020) large class of phytochemicals that are not only found in medicinal plants but also found in fruits, vegetables, nuts, honey, beverages, such as red wine and tea, etc., with various pharmacological activities including antiviral potential (Ahmad et al., 2015; Yahia, 2019). The major classes of flavonoids are illustrated in Figure 13.1. Flavonoids pose various potent biological activities like antioxidant, anti-inflammatory, and antiviral activities (Figure 13.2) (Krych & Gebicka, 2013; Ragab et al., 2014; Zhang et al., 2020). It was demonstrated that these bioactive molecules have the potential to inhibit viral pathogenesis targeting essential stages of the viral life cycle (Chikhale et al., 2020). Catechins Quercetin, baicalein, and kaempferol are the key examples in the flavonoid family exhibiting antiviral properties (Ngwa et al., 2020; Ahmadian et al., 2020). Therefore, the intent of this chapter was to provide focused, valuable, and comprehensive insights on the role of flavonoids and their antiviral prospective against COVID-19 or SARS-CoV-2-related viruses. Particularly, the focus will be on the type of flavonoids which are described to be able to considerably mitigate entry of CoV or infection and hence may perhaps too play a crucial role in fortification against the COVID-19.

FIGURE 13.1 Biological activities of flavonoids.

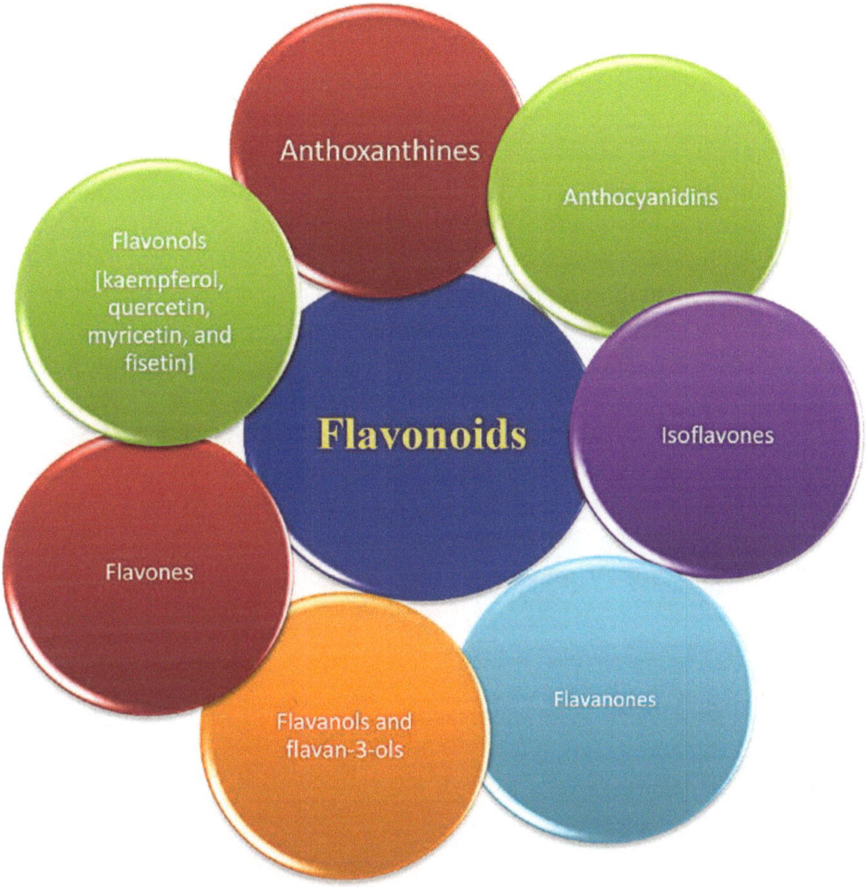

FIGURE 13.2 The major types of flavonoids.

13.2 FLAVONOIDS-MEDIATED MITIGATION MECHANISMS OF COVID-19: *IN SILICO* STUDY

Several *in-silico* studies revealed the role of various flavonoids depicting interaction with SARS-CoV-2 proteins. The results are summarized in Table 13.2. It was studied that M^{pro} and S proteins are key proteins involved in replication of COVID-19 virus (Mahmoud et al., 2020). Earlier genome-wise studies have shown that there is a high resemblance between the genome of SARS-CoV and the SARS-CoV-2 virus. Hence these proteins can be suitable therapeutic drug targets. A recent molecular docking-based study indicated that flavonoid naringenin has capacity to bind to active site amino residues of M^{pro} protein of SARS-CoV-2 virus through H-bond interactions indicating mitigation capacity of flavonoid to SARS-CoV-2 (Khaerunnisa et al., 2020). Hesperidin, another flavonoid, has shown a binding tendency towards SARS-CoV-2 M^{pro}, the peptidase domain of ACE-2 (PD-ACE-2) and the receptor binding domain of S protein (RBD-S) (Tallei et al., 2020; Utomo et al., 2020). Molecular docking study on quercetin flavonoid also depicted inhibitory action against SARS-CoV-2 (Sekiou et al., 2020). It depicted tremendous binding towards Mpro protein. In another *in silico* study, it was revealed that food-based flavonoid bioactive compounds like cyanidin and genistein exhibited similar binding affinity to RdRp and Mpro compared with synthetic drugs like Lopinavir and Nelfinavir (Pendyala & Patras, 2020). Another computational analysis revealed the significance of numerous flavonoids like luteolin-7-glucoside, quercetin, kaempferol, naringenin, apigenin-7-glucoside, catechin, epigallocatechin to SARSCoV-2 Mpro (Khaerunnisa et al., 2020). In another study, Rutin was screened and proposed a key compound that might be active against the COVID-19 Mpro/3CLpro. Chrysin exhibited exceptional binding towards Mpro of MERS-CoV, SARS-CoV, and SARS-CoV-2 (Tables 13.1–13.4).

TABLE 13.1 Food Items Having Flavonoids

Types of flavonoids	Food item	References
Flavones	Parsely	Ayoub et al, (2016)
Anthocyanidins	Blueberries	Levaj et al., (2009)
Flavan-3-ols	Black tea	
Flavonoids include hesperidin (a glycoside of the flavanone hesperetin), quercetin, rutin (two glycosides of the flavonol quercetin), and the flavone tangeritin	citrus	Ayoub et al, (2016)
Flavonoids	Wine	Ayoub et al, (2016)
Flavonoids	Dark chocolate	Levaj et al., (2009)
Flavonoids	Peanut (red) Skin	Levaj et al., (2009)
Quercetin	Capers Buckwheat Onions	Ghidoli et al., (2021)
Kaempferol	Capers Saffron Brassicaceae	Ghidoli et al., (2021)
Naringenin hesperetin	Citrus fruits Tomatoes	Ghidoli et al., (2021)

TABLE 13.2 Flavonoid Classes Identified *In-Silico* as Potential Inhibitors to SARS-CoV-2 Targets

Flavonoid Compound	SARS-CoV-2 target	References
Apigenin, Catechin, Cyanidin Epigallocatechin, Genistein, Hesperidin, Luteolin, Myricetin Icariin, Naringenin, Quercetin	Mpro	Chikale et al (2020), Khaerunnisa et al., (2020), Ngwa et al., (2020), Pendyala and Patras (2020), Sekiou et al., (2020)
Caflanone, Hesperidin, Kaempferol, Linebacker	ACE-2	Chikale et al (2020), Khaerunnisa et al., (2020), Ngwa et al., (2020), Sekiou et al., (2020)
Cyanidin, Genistein	RdRp	Pendyala and Pantras (2020)
Kaempferol, Luteolin	S	Chikale et al (2020), Khaerunnisa et al., (2020), Sekiou et al., (2020)
Myrecetin, Icariin, Naringin	TMPRSS2	Cheng et al., (2020), Chikale et al (2020), Khaerunnisa et al., (2020), Sekiou et al., (2020)

TABLE 13.3 *In Vitro* Studies Reporting Antiviral Activity of Natural Flavonoids Against Coronaviruses

Corona Virus	Flavonoid Type	Model	Effects	References
SARS-CoV	Baicalin	fRhK4 cell line	EC_{50} = 12.5–25 µg/ml	Chen et al. (2004)
	Cinnamomi	–	IC_{50} = 7.8 µg/ml	Zhuang et al. (2009)
	Luteolin	Vero E6 cells	EC_{50} = 10.6 µM	Yi et al. (2004)
	Procyanidin A2 Procyanidin B1 Cinnamtannin B1		IC_{50} = 30–40 µM	Zhuang et al. (2009)
	Kaempferol derivatives	Heterologously expression of 3a protein of SARS-CoV in Xenopus oocyte.	IC_{50} = 2.3 µM	Schwarz et al. (2014)
	Purified chalcones	SARS-CoV proteases (3CLpro and PLpro) expressed in *E. coli* BL21	IC_{50} = 5.8–50.8 µM	Park et al. (2016)
	Purified flavonoids		IC_{50} = 30.2–233.3 µM	Park et al. (2016)
	Quercetin-β-galactoside		IC_{50} = 128.8 µM	Park et al. (2017)
	Theaflavins	SARS-CoV proteases (3CLpro) expressed in *E. coli.*	IC_{50} = 3–9.5 µM	Zu et al. (2012)

TABLE 13.3 *(Continued)*

Corona Virus	Flavonoid Type	Model	Effects	References
	Geranylated flavonoids (tomentin AE)	fluorogenic peptide Z-RLRGG-AMC	IC_{50} = 5.0–14.4 μM	Kim et al. (2014)
	Bavachinin Corylifol A Isobavachalcone 4′-O-methylbavachalcone Neobavaisoflavone	fluorogenic peptide Z-RLRGG-AMC	IC_{50} = 4.2–38.4 μM	Cho et al. (2013)
	Herbacetin Pectolinarin Rhoifolin	Recombinant protein; FRET method	IC_{50} = 33.17 IC_{50} = 27.45 IC_{50} = 37.78 μM	Jo et al. (2020)
	Amentoflavone	Recombinant protein; FRET method.	IC_{50} = 8.3 μM	Ryu et al. (2010)
	Epigallocatechin gallate gallocatechin gallate quercetin	Recombinant protein; FRET method	IC_{50} = 47–73 μM.	Nguyen et al. (2012)
	Quercetin	Recombinant protein FRET-based dsDNA unwinding assay.	IC_{50} = 8.1 μM	Lee et al. (2009)
	7-O-arylmethylquercetin derivatives	Recombinant protein FRET-based dsDNA unwinding assay.	IC_{50} = 2.7–5.2 μM	Park et al. (2012)
	Myricetin	Recombinant protein FRET-based dsDNA unwinding assay.	IC_{50} = 2.71 μM	Yu et al. (2012)
HIV/SARS pseudo-typed virus	Cinnamomi cortex extract.	Vero E6 cells	IC_{50} = 37.3 μg/ml	Zhuang et al. (2009)
SARS-CoV-2	Baicalein	hACE2 transgenic mice infected with SARS-CoV-2.	200 mg/kg	Zhan et al. (2021)
	Flavonoid compounds	SARS-CoV proteases (recombinant 3CLpro) expressed in Pichia pastoris GS115.	IC_{50} = 47–381 μM	Nguyen et al. (2014)
	Isoliquiritigenin	SARS-CoV proteases (3CLpro and PLpro) expressed in *E. coli* BL21.	IC_{50} = 61.9 μM IC_{50} = 33.9 μM	Park et al. (2017)
	Kaempferol	MERS-CoV proteases (3CLpro and PLpro) expressed in *E. coli* BL21.	IC_{50} = 116.3 μM IC_{50} = 206.6 μM,	Park et al. (2017)

TABLE 13.3 *(Continued)*

Corona Virus	Flavonoid Type	Model	Effects	References
	Naringenin	Vero E6 cells infected with HCoVOC43, HCoV229E, and SARS-CoV-2.	62.5, 250 μM	Clementi et al. (2020)
	Quercetin	ACE2h cells infected with SARS-CoV-2.	50 μM	Song et al. (2021)
Porcine epidemic diarrhea virus (PEDV)	Quercetin 7-rhamnoside	Vero cells	IC_{50} = 0.014 μg/	Choi et al. (2009)
Bovine coronavirus (BCV)	Theaflavins	HRT-18 cells	EC_{50} = 34.7 μg/ml	Clark et al. (1998)
HIV/SARS pseudotyped virus	Quercetin	Hep-G2 cells	EC_{50} = 83.4 μM	Yi et al. (2004)

TABLE 13.4 Immunomodulatory and Anti-Inflammatory Effects of Different Flavonoid Classes

Flavonoid Class	Immunomodulatory mechanism of action	References
Apigenin	Inhibits CCL5, IL-6, VCAM1 and ICAM1	Zhang et al., (2014)
Caflanone	Inhibition of 5-lipoxygenase and microsomal prostaglandin E synthase 1	Erridge et al., (2020)
Chrysin	Inhibits COX-2 and MPO activity	Shen et al., (2015)
		Sassi et al., (2017)
Fisetin	Impedes COX-2, PKCd activity and Prostaglandin E2 production	Peng et al., (2018); Lee et al., (2018)
Hesperetin	Inhibition of NF-jb and ERK pathway	Ma et al., (2015); Ye et al., (2019)
Luteolin	Increase the number of CD4, CD25 regulatory T-cells	Kuo et al., (2011); Kim et al., (2018)
Naringenin	Inhibition of ERK, decreases iNOS	Fouad et al., (2016); Ali et al., (2017)
Quercetin	Regulates Th1/Th2 balance	Michalski et al., (2000)

Wan et al. (2020) recently observed that a well-defined therapeutic strategy to combat infection of SARS-CoV-2 virus is targeting the ACE-2 receptors on host cells. S-proteins of SARS-CoV viruses are the key protein involved in this interaction. Several docking studies at the molecular level were executed to study the necessary affinity of various flavonoids like myricetin, hesperetin, caflanone, and linebacker, which revealed nice affinity towards S protein, ACE-2 receptor, and helicase and thus could block the virus entry (Ngwa et al., 2020). Naringenin flavonoid exhibited a strong affinity towards the ACE-2 receptor (Cheng et al., 2020). An *in-silico* study exposed that flavonoid baicalin exhibited tremendous binding affinity towards S protein comparable to synthetic drugs like abacavir and hydroxychloroquine (Pandey et al., 2021).

Hoffmann et al. (2020) observed that Human TMPRSS2 is a very important protease involved in virus activation via S protein cleavage. Through docking studies, it was shown that various flavonoids like quercitrin myricitrin naringin, neohesperidin, and icariin have an excellent binding affinity to TMPRSS2 (Chikhale et al., 2020). It was also cited by a comprehensive computational study that silybin is involved in binding to TMPRSS2 required for viral entry. Chrysin also mitigated the contact of S protein with ACE-2 in SARS-CoV-2 (Jha et al., 2020). Docking analysis on MERS-CoV 3CLpro protein revealed the interaction of helichrysetin flavonoid via forms of a hydrogen bond (Jo et al., 2019). Using computational methods, a recent study demonstrated that narcissoside, present in several wild plants, is an effective inhibitor of SARS-CoV-2 3-CLpro protease (Dubey et al., 2020). Docking simulation studies to find the binding affinity of flavonoids, like luteolin apigenin, daidzein, epigallocatechin, quercetin, and kaempferol, indicated potential binding towards SARS-CoV 3-CLpro (Jo et al., 2019). A similar observation on molecular interaction and inhibitory effect was observed by molecular docking study on quercetin-3-β-galactoside against SARS-CoV 3-CLpro (Chen et al., 2006). It was summarized that interaction was mediated by H-bond interactions (Chen et al., 2006).

RdRP is an important viral RNA polymer involved in virus replication. Zandi et al. (2021) revealed the *in vitro* antiviral effect of baicalein and baicalin against infection of SARS-CoV-2 in the Vero CCL-81 cell line through RdRp inhibition, with a higher potency by baicalein. Further *in silico* evaluations showed these two compounds to have a higher affinity to RdRp in comparison to remdesivir. The attachment site of baicalin and baicalein also seems to be different from that of remdesivir; thus, these flavonoids can be used as an adjuvant treatment along with remdesivir.

13.3 FLAVONOIDS-MEDIATED MITIGATION MECHANISMS OF COVID-19: *IN VIVO/VITRO* STUDIES

Flavonoids like baicalin and baicalein have recently been cited as the unique, natural product inhibitors of Mpro/3CL protease *in vitro* (Jo et al., 2020; Chen et al., 2005, 2006; Su et al., 2020; Park et al., 2016; Liu et al., 2020), and could be potential anti-COVID-19 inhibitors. The same authors screened about 64-flavonoid library against SARS-CoV 3CLpro, and It was concluded that pectolinarin, rhoifolin, and herbacetin are the utmost potent inhibitors with IC_{50} values of 37.78, 27.45, and 33.17 μM, respectively.

Using enzymatic assays in a competitive mode, Chen et al. (2006) described inhibition of 3CLpro inhibition activity by glycoside flavonoids quercetin-3-β-galactose with IC_{50} value of 42.79 ± 4.97 μM. A similar study examined the required interaction way of the compounds with virus proteases. Puerarin displayed less inhibitor activity against 3CLpro, with an IC_{50} value of 381 μM (Nguyen et al., 2012). In an additional study, through cell-free cleavage assay, Lin et al. (2005) identified numerous compounds containing the isoflavone daidzein and flavanone hesperetin showing IC_{50} values against SARS-CoV 3CLpro, of 105 and 60 μM, respectively. Four flavonoids: sciadopytiscin, amentoflavone, bilobetin, and ginkgetin, were Fractionation from leaves extract of *Torreya nucifera* that revealed 3CLpro inhibitory activity with IC_{50} values are 38.4, 32.0, 72.3, and 8.3 μM, respectively (Ryu et al., 2010).

In vitro analyses against proteases of SARS-CoV, comprising PLpro, were observed with nine alkylated chalcones sequestered from *Angelika keiskei*. IC_{50} of 1.2, 5.6, 11.7, 19.3, 11.7, 21.1, 26.0, and 46.4 μM against PLpro were observed for xanthoangelol E, xanthoangelol F, xanthoangelol, xanthoangelol D, xanthoangelol B, xanthokeistal A, 4-Hydroxyderricin and xanthoangelol G, respectively, with xanthoangelol E is the extreme active (Park et al., 2016). The same authors described isobavachalcone with IC_{50} of 13.0 μM against PLpro of SARS-CoV.

Bioactive extracts obtained from *Paulownia tomentosa* fruits directed to the separation of flavonoids and its evaluation against SARS-CoV PLpro, and it was observed that entire compounds displayed inhibitory activity. Tomentin E, tomentin D, tomentin C, tomentin B, tomentin A, 40-O-methoxydiplacol, 30-O-methyldiplacol, 40-O-methyldiplacone, 30-O-methyldiplacone, 6-geranyl-40,5,7-trihydroxy-30,50-dimethoxyflavanone, diplacone, and mimulone and showed IC_{50} values of 5.0, 12.5, 11.6, 6.1, 6.2, 9.2, 9.5, 12.7, 13.2, 13.9, 10.4, and 14.4 μM, respectively (Cho et al., 2013).

Nguyen et al. (2012) equated the activity of epigallocatechin gallate (EGCG), ampelopsin, epigallocatechin, and gallocatechin gallate (GCG) from *Pichia pastoris*. It was suggested that flavonoid compounds EGCG and ampelopsin possess stronger 3CLpro inhibitory activity with IC_{50} values of 73 and 47 μM, respectively. An additional study on components of black tea validated that theaflavin-3,30-gallate and 3-theaflavin-3-gallate were active against SARS-CoV 3CLpro, with IC_{50} values of 9.5 and 7.0 μM. It indicated that black tea possibly will avert or diminish infection of CoV (Chen et al., 2005).

Zhuang et al. (2009) used Cinnamomi cortex extracts and demonstrated inhibition in wild-type SARS-CoV. Authors isolated dimer cinnamtannin B1, procyanidin B2, and procyanidin A2, viewing complete inhibitory effects having IC_{50} values of 32.9, 161.1, and 120.7 μM, respectively (Zhuang et al., 2009). IC_{50} value of 2.5 μM was observed with the flavonoids juglanin against SARS-CoV (Schwarz et al., 2014).

In another study, authors revealed that quercetin abridged infection of bovine and human coronaviruses, NCDCV, and OC43, respectively, by 50% at 60 μg/ml concentration. Kaempferol, at a concentration of 10 μg/ml, abridged virus replication by 50% in OC43 and 65% in NCDCV (Debiaggi et al., 1990). Using the HRT-18 cell line, it was proved that theaflavins from black tea (theaflavin, theaflavin-3-mono gallate, theaflavin-3,3′ gallate, theaflavin-3′-mono gallate) were very effective against Bovine CoV, BCV, with a mean value of EC50 of 34.7 μg/ml. The other two flavonoids, quercetin and luteolin, depicted the capability to occlude the SARS-CoV entry within host cells (Yi et al., 2004). Results using infection of SARS-CoV of Vero E6 cells revealed that Luteolin inhibited in a dose-dependent manner, with EC50 value of 10.6 μM (CC50 = 155 μM), whereas quercetin mitigate HIV-luc/SARS pseudo-typed entry of virus with value of EC50 of 83.4 μM (CC50 = 3.32 mM) (Yi et al., 2004). Traditional Chinese medicine (TMC) from *Scutellaria baicalensis Georgi* (flavone glycoside baicalin) is used for the deterrence and management of SARS-CoV. This molecule was verified on fRhK4 cell lines using SARS-CoV coronavirus from 10 diverse patients. Flavone glycoside baicalin presented a value of EC50 of 12.5–25 μg/ml at 48 h, deprived of substantial cytotoxicity (Yi et al., 2004).

By using cell-based and cell-free methods by measuring SARS-CoV 3CLpro cleavage activity, aqueous flavonoid having extract from *Isatis indigotica* root depicted a dose-dependent capacity to mitigate the 3CLpro proteolytic cleavage activity with a value of IC_{50} of 191.6 and 53.8 μg/ml, respectively (Soukhova et al., 2004). In the same assays, various herb-derived flavonoids were tested (naringenin, quercetin, hesperetin). Results showed that merely hesperetin depicted dose-dependently inhibited cleavage activity of the 3CLpro in cell-based and cell-free assays with IC_{50} 8.3 and 60 μM, respectively (Soukhova et al., 2004).

Another set of flavonoids was verified in a diverse work (Nguyen et al., 2012). It was observed that gallocatechin gallate (GCG) and quercetin depicted utmost effective inhibitory activity of recombinant SARS-CoV 3CLpro with a value of IC_{50} in the range 47 to 73 μM. Others, like puerarin, daidzein, and ampelopsin, exposed to an IC_{50} greater than 350 μM. In ways, novel and infrequent geranylated flavonoids like: tomentin E, D, C, B, and A repressed SARS-CoV PLpro with IC_{50} raging between 5.0 and 14.4 μM (Cho et al., 2013). Papyriflavonol A, a prenylated quercetin derivative, presented a powerful, non-competitive inhibitory action on PLpro of COVID-19 virus with a value of IC_{50} of 3.7 μM (Park et al., 2017).

Against MERS-CoV 3CLpro, researchers established that amongst 40 flavonoids verified at 20 μM concentration, only four of them, viz. quercetin, isobavachalcone, herbacetin, and helichrysetin were the maximum active with an IC_{50} value of 35.85, 40.59, 67.04, and 37.03 μM, respectively (Jo et al., 2019). Other identified flavonoids were: rhoifolin, pectolinarin, and herbacetin, and as the very projecting inhibitors with an IC_{50} of 27.45, 37.78, and 33.17 μM, respectively (Jo et al., 2020).

Another study on chalcones indicated their capability to hinder both SARS-CoV PLpro protease activity and SARS-CoV 3CLpro using cell-free assays against SARS-CoV PLpro with IC_{50} value of 1.2–26.0 μM and against 3CLpro with IC_{50} of 11.4–39.4 μM. The very potent xanthoangelol E displayed a value of IC_{50} of 1.2 and 11.4 μM for SARS-CoV PLpro and SARSCoV 3CLpro, respectively. In the case of cell-based cleavage, this chalcone ensued to inhibit SARS-CoV 3CLpro with a CC50 of 65.6 μM and a value of IC_{50} of 7.1 μM (Park et al., 2020).

Ramalingam et al. (2018), through expression studies in *E. coli* on Isoliquiritigenin, a chalcone, showed that it could be utilized as a therapeutic or remedial agent on SARS-CoV and MERS-CoV 3CLpro and PLpro. Using Vero CCL-81 cell, a topical examination by Zandi et al. (2021) discovered the *in vitro* antiviral consequence of baicalein and baicalin against SARS-CoV-2 infections by inhibition of RdRp. Another *in vitro* study using quercetin and isorhamnetin on SARS-CoV-2 revealed that these flavonoids pose the capability to bind ACE2 receptors and thus decline viral entry by inhibition of spike protein attachment to ACE2 (Zhan et al., 2021). Another study using Vero E6 cells and hACE2 transgenic mice evaluated the result of baicalein on SARS-CoV-2 infection and observed a substantial decrease in infection of the virus (Song et al., 2021).

13.4 IMMUNOMODULATING AND ANTI-INFLAMMATORY POTENTIAL OF FLAVONOIDS

Infection of COVID-19 has been shown to be a major reason behind the emerging storm of cytokine (Mahmudpour et al., 2020). It is associated with life-threatening complications termed acute respiratory distress syndrome (ARDS), which might signify>33% of COVID-19 hospitalized patients with a mortality rate of 40% (Tzotzos et al., 2020). It was observed that in COVID-19 patients' levels of various cytokines like TNF-a, GCSF, MIP1A, MCP1, IP10, IL-10, IL-7, IL-6, and IL-2, increased, which is termed as "cytokine storm" by doctors recently (Cheng et al., 2020). Inside the human body, modulation of immune responses and Inhibition of hyper-inflammatory response is a key approach to diminishing the storm of cytokines (Mahmudpour et al., 2020). Notably, some flavonoids exhibited strong anti-inflammatory activities and can be used to ameliorate complicated COVID-19 symptoms.

Flavonoid Naringenin depicted promising immunomodulatory activity in rats when exposed to benzopyrene by demising the intensity of inflammatory responses by reducing the proinflammatory cytokines (Ali et al., 2017; Tutunchi et al., 2020). Another study cited that by administration of flavonoid naringenin, expression of TNF-a, iNOS, and NF-jB reduced in the lungs of rats with sepsis, thus concluding that naringenin can be used as an immune-modulatory drug in infection of SARS-CoV-2 (Fouad et al., 2016). In another *in-vitro* study, Yoshida et al. (2010) observed that naringenin and hesperetin downregulate the expression of inflammatory mediator TNF-a in NF-jB and pathways, resulting in the inhibition of IL-6 transcription on mouse adipocytes. In Rats suffering from an acute lung injury, it highlighted that flavonoid hesperetin alleviates the proliferator's expression of peroxisome-activated receptor gamma. Which consequently inhibited the NF-jB pathway, thus lowering the production of the inflammatory cytokines comprising TNF-a, IL-1b, and IL-6 (Ma et al., 2015; Ye et al., 2019). In murine mice asthma model, flavonoid Quercetin triggered inhibition of leukocyte and inflammation that regulated the Th1/Th2 balance (Park et al., 2009). Authors observed that quercetin-loaded micro-emulsion in the murine asthma model depicted the same results as that of the synthetic drug dexamethasone, which showed a drastic decrease in the production of mucus inside the lungs (Rogerio et al., 2010). A study made by Lu et al.

(2020) in COVID-19 patients exhibited that MUC5AC and MUC1 mucin proteins level increased. Notably, in rats and epithelial NCI-H292 cells, *in vitro* studies demonstrated that flavonoid quercetin inhibited EGFR-based tyrosine phosphorylation in NF-jB pathways, thus caused in the clampdown of mucin synthesis, which led to a reduction in the production of mucus and thus solved breathing difficulty (Yang et al., 2012). In human dendritic cells (DCs), Furthermore, quercetin exerts immunomodulatory effects by producing down expression of CD83 (Michalski et al., 2020). This, resulted in the activation of T-cells coupled with the relocation of matured DCs (Michalski et al., 2020).

In human lung epithelial cells, flavonoid Fisetin exhibited significant anti-inflammatory and immunomodulatory potential by causing inhibition of TNF-a, COX-2, MCP1, IL-6, IL8, prostaglandin E2 and CCL5 (Peng et al., 2018). It was cited that Fisetin likewise down-regulates the ERK1/2 pathway, NF-jB pathway, thus resulting in the decrease of expression of ICAM1, which is associated with the adhesion of monocytes (Peng et al., 2018). Lee et al. (2018) in human airway epithelial cell lines observed that Fisetin deleteriously controls the activity of PKC-d, which is important for the stimulation of the IKK/TNF-a/NF-jB signaling cascade.

Fisetin obstructs PKCd activity, phosphorylation of ERK1/2 pathways, inhibits NF-jB, production of prostaglandin E2 and COX-2 and declines TNF-a, MPC1, IL-8, IL-6, and CCL5 levels (Lee et al., 2018; Peng et al., 2018) Flavones Chrysin works as PPAR-c agonist, inhibits MPO, COX-2 activity and NF-jB pathway. Furthermore, it also inhibits IL-8, IL-1b, iNOS, and TNF-a levels, excites lysosomal activity of macrophage, and inhibits nitric oxide (NO) production (Shen et al., 2015; Sassi et al., 2017; Zeinali et al., 2017). Apigenin hinders CCL5, IL-6, VCAM1, and ICAM1 (Zhang et al., 2014). Luteolin proliferates CD4. CD25 number of regulatory T-cells declines the immune cells number, for example CD3-CCR3?, CD4?T, CD19?B, and CD11b.Gr-1?, inhibits NF-jB and MARK pathways, reduces TNF-a, IL-6, IL-1b levels, and inhibits MPO activity (Kim et al., 2018; Kuo et al., 2011; Liu et al., 2018). Caflanone constrains 5-lipoxygenase and microsomal prostaglandin E synthase 1 (Erridge et al., 2020).

Another study has revealed the anti-inflammatory and immunomodulatory potential of chrysin flavonoids, which suppresses the NF-jB pathway that regulates COX-2 expression and iNOS genes (Zeinali et al., 2017). In mice, it was experimentally observed by Shen et al. (2015) that when mice were exposed to smoking cigarettes in order to prompt epithelial cells inflammation, chrysin mitigated the inflammation by overpowering the discharge of TNF-a, MPO, IL-8, IL1b inside the lung tissue. Chrysin similarly down-regulates sp38 and phosphorylation of ERK (Shen et al., 2015). In additional study on rat peritoneal macrophages, it was deduced that chrysin stimulates macrophage lysosomal activity that regulates killing and assimilating the pathogenic microbes coupled with NO_2 production (Sassi et al., 2017). The same has been validated by a docking study that indicated the binding potential of chrysin to COX-2 enzymes, thus decreasing the likelihood of undesirable adverse effects of GIT (Rauf et al., 2015). Likewise, apigenin flavonoid, upon pre-treatment of pre-inflamed macrophage of humans, exhibited IL-6 mRNA stability and inhibition of IL-6 secretion (Zhang et al., 2014). It was found that Apigenin not only inhibits pro-inflammatory cytokines but also adhesion molecules (ICAM1 and VCAM1) and inflammatory chemokines (CCL5) (Zhang et al., 2014).

Another flavonoid Luteolin also considerably amplified the number of regulatory T-cells, $CD25^+$ and $CD4^+$ in murine splenic that were enthused by anti-CD3/anti-CD28 (Kim et al., 2018). In the lungs of inflamed airway mouse models, Luteolin also offered immunomodulatory potential by diminishing the immune cells like CD4 + T, CD19+ B, $CD3\text{-}CCR3^+$, $CD11b^+$ (Kim et al., 2018). Liu et al. (2018) observed that luteolin also inhibited the NF-jB pathway, thus reducing IL-6 and TNF-a levels and inhibiting MPO activity. It was shown that when mice were fed with lipo-polysaccharides (LPS), Luteolin depicted a protective effect by inhibition of the NF-jB pathway, MAPK pathways, and degradation of IKB (Kuo et al., 2011).

Erridge et al. (2020) cited that caflanone owns anti-inflammatory action via the inhibition of prostaglandin E synthase and 5-lipoxygenase. In situ study made by Li et al. (2018) studied the effect of flavonoids like: vicenin-1, vicenin-2, apigenin, C-glycosides (ACGs), shaftoside, isoshaftoside, isovitexin, vitexin, violanthin, and isoviolanthin on lung inflammation and cytokine level determination. By metabolic profiling studies, It was observed that ACGs condensed microvascular permeability and pulmonary edema by down-regulating lipo-polysaccharides induced IL-1β, IL-6, TNF-α expression involved in TLR4/TRPC signaling pathway activation.

Wei et al. (2015), in the *in-vitro* study, cited that baicalin flavonoids have shown antioxidant, antiapoptosis, and anti-inflammatory activities. An *in-vitro* analysis revealed that this molecule reduces oxidative stress coupled with endothelial dysfunction via improvement of ACE2 activity that endorsed repression of endothelial cells of human umbilical vein and endothelial-dependent vasodilation and apoptosis. Baicalin also reduced the expression of the pro-apoptotic protein Bax, and cleaved caspase-3 was involved in the increase of Bcl-2 expression. Authors also cited that Baicalin also pointedly up-regulating the AKT/PI3K//eNOS pathways and altered expression of Mas receptor mRNA (Wei et al., 2015).

In vivo and *in vitro* experiments have approved that flavonoid Astilbin, exhibited to reduce Lipo-polysaccharide induced ARDS by regulating pro-inflammatory cytokines TNF-α/IL-6 and suppression of heparinase (pro-inflammatory enzyme), MAPK phosphorylation inhibition, and lessened degradation of heparin sulfate (Kong et al., 2016).

An *in vivo* study on Rutin was investigated for anti-inflammatory effects (Guardia et al., 2001). It was shown that at 21 days, it reduced the acute phase of inflammation. A study done on licorice flavonoids liquiritin, liquiritin apioside, and liquiritigenin showed that these molecules successfully lessen Lipo-polysaccharides induced pulmonary inflammation via inhibition of inflammatory mediator release and decrease of IL-1β and TNF-α expression. Effects were similar to the standard drug dexamethasone when used at a concentration of 1 mg/kg (Xie al., 2009).

Another flavonoid, eriodictyol, depicted anti-inflammatory and anti-oxidant properties. In LPS induced ALI model, Eriodictyol established inhibition of oxidative injury and pro-inflammatory cytokine expression by triggering the Nrf2 pathway (Zhu et al., 2015).

In addition, oroxylin-A amended the increased level of the white blood cell counts, raised TNF-α and NO, thickened alveolar septa, and augmented pulmonary edema (Tseng et al., 2012). An *in vitro* mouse model of LPS-induced studies using pretreatment with flavonoid pinocembrin lessened pulmonary edema, macrophage infiltration by regulating the production of IL-6, IL-10, IL-1β, and TNF-α through inhibition of JNK, ERK1/2, phosphorylation of p38 MAPK and IκBα (Soromou et al., 2012).

Furthermore, in sepsis-induced mice, luteolin pretreatment displayed a noteworthy decrease in cytokines (pro-inflammatory) such as IL-1β and IL-6 and reduction via the decrease in NF-κB, ICAM-1, and partially iNOS pathway (Rungsung et al., 2018). It further reduced activities of catalase and superoxide, peroxidation of lipids, and thus controlled exudation of IL-8 (KC), ICAM-1, and TNF-α. Another flavonoid Sakuranetin demonstrated anti-inflammatory activity by diminishing neutrophils number. Furthermore, sakuranetin declined the macrophages count and cytokines (pro-inflammatory) like IL-8 and TNF-α in mice (Bittencourt-Mernak et al., 2017).

13.5 CONCLUSION

It was observed that foods which are in flavonoids could be of noteworthy prominence meant for the deterrence and SARS-CoV-2 treatment. Notably, flavonoids depicted strong inhibitory activities against perilous viral targets, mandatory to enable their admittance and reproduction, like RBD of the S-protein Mpro, RdRp, besides TMPRSS2 and human ACE-2 receptors. In addition, the immune-modulatory activities of flavonoids have been established by the retardation of key pathways involved in inflammatory reactions and pro-inflammatory cytokines. We expect that the discussions and hypotheses offered at this point can arouse researchers to propose apt experimentations to demonstrate that naturally available flavonoid molecules or their byproducts could amend anticoronavirus prevention and treatment.

KEYWORDS

- **antiviral**
- **COVID-19**
- **flavonoids**
- **herbal plants**
- **immune stimulating**
- ***in vivo/vitro* studies**

REFERENCES

Ahmad, A., Kaleem, M., Ahmed, Z., et al., (2015). Therapeutic potential of flavonoids and their mechanism of action against microbial and viral infections—A review. *Food Res. Int., 77*, 231–235.

Ahmadian, R., Rahimi, R., & Bahramsoltani, R. (2020). Kaempferol: An encouraging flavonoid for COVID-19. *Bol. Latinoam. Caribe Plantas Med Aromát., 19*, 492–495. doi: 10.37360/blacpma.20.19.5.33.

Ali, R., Shahid, A., Ali, N., et al., (2017). Amelioration of benzo[a]- pyrene-induced oxidative stress and pulmonary toxicity by Naringenin in Wistar rats: A plausible role of COX-2 and NF-jB. *Hum. Exp. Toxicol., 36*, 349–364.

Ayoub, M., De Camargo, A. C., & Shahidi, F. (2016). Antioxidants and bioactivities of free, esterified, and insoluble-bound phenolics from berry seed meals. *Food Chemistry, 197*(Pt A), 221–232. doi: 10.1016/j.foodchem.2015.10.107. PMID 26616944.

Bahramsoltani, R., Sodagari, H. R., Farzaei, M. H., Abdolghaffari, A. H., Gooshe, M., & Rezaei, N., (2016). The preventive and therapeutic potential of natural polyphenols on influenza. *Expert. Rev. Anti Infect. Ther., 14*, 57–80.

Bhardwaj, V. K., Singh, R., Sharma, J., Rajendran, V., Purohit, R., & Kumar, S., (2020). Identification of bioactive molecules from tea plants as SARS-CoV-2 main protease inhibitors. *Journal of Biomolecular Structure and Dynamics, 5*, 1–10.

Bittencourt-Mernak, M. I., Pinheiro, N. M., Santana, F. P. R., Guerreiro, M. P., Saraiva-Romanholo, B. M., Grecco, S. S., & Prado, C. M., (2017). Prophylactic and therapeutic treatment with the flavonone sakuranetin ameliorates LPS-induced acute lung injury. *American Journal of Physiology-Lung Cellular and Molecular Physiology, 312*, L217–L230.

Chen, F., Chan, K. H., Jiang, Y., Kao, R. Y., Lu, H. T., Fan, K. W., Cheng, V. C., et al., (2004). *In vitro* susceptibility of 10 clinical isolates of SARS coronavirus to selected antiviral compounds. *J. Clin. Virol., 31*, 69–75.

Chen, L., Gui, C., Luo, X., Yang, Q., Günther, S., Scandella, E., et al., (2005). Cinanserin is an inhibitor of the 3C-like proteinase of severe acute respiratory syndrome coronavirus and strongly reduces virus replication *in vitro. J. Viorl., 79*, 7095–7103. doi: 10.1128/JVI.79.11.7095-7103.2005.

Chen, L., Li, J., Luo, C., Liu, H., Xu, W., Chen, G., Liew, O. W., et al., (2006). Binding interaction of quercetin-3-beta-galactoside and its synthetic derivatives with SARS-CoV 3CL(pro): Structure-activity relationship studies reveal salient pharmacophore features. *Bioorg. Med. Chem., 14*, 8295–8306.

Cheng, L., Zheng, W., Li, M., et al., (2020). *Citrus Fruits are Rich in Flavonoids for Immunoregulation and Potential Targeting ACE2*. Preprints 2020:2020020313.

Chikhale, R. V., Gupta, V. K., Eldesoky, G. E., et al., (2020). Identification of potential anti-TMPRSS2 natural products through homology modeling, virtual screening, and molecular dynamics simulation studies. *J. Biomol. Struct. Dyn.* https://doi.org/10.1080/07391102.2020.1798813.

Cho, J. K., Curtis-Long, M. J., Lee, K. H., Kim, D. W., Ryu, H. W., Yuk, H. J., & Park, K. H., (2013). Geranylated flavonoids displaying SARS-CoV papain-like protease inhibition from the fruits of *Paulownia tomentosa. Bioorg. Med. Chem., 21*, 3051–3057.

Choi, H. J., Kim, J. H., Lee, C. H., Ahn, Y. J., Song, J. H., Baek, S. H., & Kwon, D. H., (2009). Antiviral activity of quercetin 7-rhamnoside against porcine epidemic Diarrhea virus. *Antiviral Res., 81*, 77–81.

Clark, K. J., Grant, P. G., Sarr, A. B., Belakere, J. R., Swaggerty, C. L., Phillips, T. D., & Woode, G. N., (1998). An *in vitro* study of theaflavins extracted from black tea to neutralize bovine rotavirus and bovine coronavirus infections. *Vet. Microbiol., 63*, 147–157.

Clementi, N., Scagnolari, C., D'Amore, A., Palombi, F., Criscuolo, E., Frasca, F., Pierangeli, A., et al., (2020). Naringenin is a powerful inhibitor of SARS-CoV-2 infection *in vitro. Pharmacol. Res., 163*, 105255.

Debiaggi, M., Tateo, F., Pagani, L., Luini, M., & Romero, E., (1990). Effects of propolis flavonoids on virus infectivity and replication. *Microbiol., 13*, 207–213.

Dubey, K., & Dubey, R., (2020). Computation screening of narcissoside a glycosyloxyflavone for potential novel coronavirus 2019 (COVID-19) inhibitor. *Biomed. J.*

Erridge, S., Mangal, N., Salazar, O., et al. (2020). Cannflavins—From plant to patient: A scoping review. *Fitoterapia, 146*, 104712.

Fouad, A. A., Albuali, W. H., & Jresat, I., (2016). Protective effect of naringenin against lipopolysaccharide-induced acute lung injury in rats. *Pharmacology, 97*, 224–232.

Gansukh, E., Kazibwe, Z., Pandurangan, M., Judy, G., & Kim, D. H., (2016). Probing the impact of quercetin-7-O-glucoside on influenza virus replication influence. *Phytomedicine, 23*, 958–967.

Ghidoli, M., Colombo, F., Sangiorgio, S., Landoni, M., Giupponi, L., Nielsen, E., & Pilu, R., (2021). Food containing bioactive flavonoids and other phenolic or sulfur phytochemicals with antiviral effect: Can we design a promising diet against COVID-19? *Front. Nutr., 8*, 661331.

Guardia, T., Rotelli, A. E., Juarez, A. O., & Pelzer, L. E., (2001). Anti-inflammatory properties of plant flavonoids. Effects of rutin, quercetin and hesperidin on adjuvant arthritis in rat. *Il. Farmaco, 56*, 683–687.

Hoffmann, M., Kleine-Weber, H., Schroeder, S., et al., (2020). SARSCoV-2 cell entry depends on ACE2 and TMPRSS2 and is blocked by a clinically proven protease inhibitor. *Cell, 181*, 271–280.e278.

Huang, C., Wang, Y., Li, X., et al., (2020). Clinical features of patients infected with 2019 novel coronavirus in Wuhan, China. *Lancet, 395*, 497–506.

Jha, N. K., Jeyaraman, M., Rachamalla, M., et al., (2021). Current understanding of novel coronavirus: Molecular pathogenesis, diagnosis, and treatment approaches. *Immuno., 1*, 30–66.

Jin, J., Chen, S., Wang, D. C., Chen, Y. J., Wang, Y. X., Guo, M., Zhou, C. L., & Dou, J., (2018). Oroxylin A suppresses influenza A virus replication correlating with neuraminidase inhibition and induction of IFNs. *Biomed. Pharmacother., 97*, 385–394.

Jo, S., Kim, H., Kim, S., Shin, D. H., & Kim, M. S., (2019). Characteristics of flavonoids as potent MERS-CoV 3C-like protease inhibitors. *Chem. Biol. Drug Des., 94*, 2023–2030.

Jo, S., Kim, S., Shin, D. H., & Kim, M. S., (2020). Inhibition of SARS-CoV 3CL protease by flavonoids. *J. Enzyme Inhib. Med. Chem., 35*, 145–151.

Khaerunnisa, S., Kurniawan, H., Awaluddin, R., et al., (2020). *Potential Inhibitor of COVID-19 Main Protease (Mpro) from Several Medicinal Plant Compounds by Molecular Docking Study*. Preprints 2020, 2020030226.

Kim, D. W., Seo, K. H., Curtis-Long, M. J., Oh, K. Y., Oh, J. W., Cho, J. K., Lee, K. H., & Park, K. H., (2014). Phenolic phytochemical displaying SARS-CoV papain-like protease inhibition from the seeds of *Psoralea corylifolia. J. Enzyme Inhib. Med. Chem., 29*, 59–63.

Kim, S. H., Saba, E., Kim, B. K., et al., (2018). Luteolin attenuates airway inflammation by inducing the transition of CD4(?)CD25(-) to CD4(?)CD25(?) regulatory T cells. *Eur. J. Pharmacol., 820*, 53–64.

Kong, G., Huang, X., Wang, L., Li, Y., Sun, T., Han, S., & Wang, X., (2016). Astilbin alleviates LPS-induced ARDS by suppressing the MAPK signaling pathway and protecting pulmonary endothelial glycocalyx. *International Immunopharmacology, 36*, 51–58.

Krych, J., & Gebicka, L., (2013). Catalase is inhibited by flavonoids. *Int. J. Biol. Macromol., 58*, 148–153.

Kumar, S., & Pandey, A. K., (2013). Chemistry and biological activities of flavonoids: An overview. *Sci. World J., 2013*, 162750.

Kuo, M. Y., Liao, M. F., Chen, F. L., et al., (2011). Luteolin attenuates the pulmonary inflammatory response and involves the abilities of antioxidation and inhibition of MAPK and NFjB pathways in mice with endotoxin-induced acute lung injury. *Food Chem. Toxicol., 49*, 2660–2666.

Lee, C., Lee, J. M., Lee, N. R., Kim, D. E., Jeong, Y. J., & Chong, Y., (2009). Investigation of the pharmacophore space of severe acute respiratory syndrome coronavirus (SARSCoV)

NTPase/helicase by dihydroxychromone derivatives. *Bioorg. Med. Chem. Lett., 19*, 4538–4541.

Lee, S., Ro, H., In, H. J., et al., (2018). Fisetin inhibits TNF-alpha/NFkappaB-induced IL-8 expression by targeting PKC delta in human airway epithelial cells. *Cytokine, 108*, 247–254.

Levaj, B., et al., (2009). Determination of flavonoids in pulp and peel of mandarin fruits (table 1) (PDF). *Agriculturae Conspectus Scientificus, 74*(3), 223.

Lin, C. W., Tsai, F. J., Tsai, C. H., Lai, C. C., Wan, L., Ho, T. Y., & Chao, P. D. L., (2005). Anti-SARS coronavirus 3C-like protease effects of *Isatis* indigotica root and plant-derived phenolic compounds. *Antiviral Research, 68*, 36–42.

Lipinski, C. A., (2004). Lead- and drug-like compounds: The rule-of-five revolution. *Drug Discovery Today. Technologies, 1*, 337–341.

Liu, X., & Wang, X. J., (2020). Potential inhibitors against 2019-nCoV coronavirus M protease from clinically 13 of 14 approved medicine. *J. Genet. Genomics.* doi: 10.1016/j.jgg.2020.02.001.

Lu, H. (2020). Drug treatment options for the 2019-new coronavirus (2019-nCoV). *Biosci. Trends*. doi: 10.5582/bst.01020.

Ma, H., Feng, X., & Ding, S. (2015). Hesperetin attenuates ventilator-induced acute lung injury through the inhibition of NF-jB-mediated inflammation. *Eur. J. Pharmacol., 769*, 333–341.

Mahmoud, I. S., Jarrar, Y. B., Alshaer, W., et al., (2020). SARS-CoV-2 entry in host cells-multiple targets for treatment and prevention. *Biochimie, 175*, 93–98.

Mahmudpour, M., Roozbeh, J., Keshavarz, M., et al., (2020). COVID-19 cytokine storm: The anger of inflammation. *Cytokine, 133*, 155151.

Michalski, J., Deinzer, A., Stich, L., et al., (2020). Quercetin induces an immunoregulatory phenotype in mature human dendritic cells. *Immunobiology, 225*, 151929.

Millet, J. K., Kien, F., Cheung, C. Y., et al., (2012). Ezrin interacts with the SARS coronavirus spike protein and restrains infection at the entry stage. *Plos One, 7*, e49566.

Nguyen, T. T. H., Kang, H. K., Kim, Y. M., Jang, T. S., & Kim, D., (2014). Inhibition effect of flavonoid compounds against neuraminidase expressed in *Pichia pastoris. Biotechnol. Bioprocess Eng., 19*, 70–75.

Nguyen, T. T., Woo, H. J., Kang, H. K., Nguyen, V. D., Kim, Y. M., Kim, D. W., Ahn, S. A., et al., (2012). Flavonoid-mediated inhibition of SARS coronavirus 3C-like protease expressed in *Pichia pastoris. Biotechnol. Lett., 34*, 831–838.

Ngwa, W., Kumar, R., Thompson, D., et al. (2020). Potential of flavonoid-inspired phytomedicines against COVID-19. *Molecules, 25*, 2707.

Pandey, P., Khan, F., Rana, A. K., et al., (2021). A drug repurposing approach towards elucidating the potential of flavonoids as COVID-19 spike protein inhibitors. *Biointerface Res. Appl. Chem., 11*, 8482–8501.

Park, H. R., Yoon, H., Kim, M. K., Lee, S. D., & Chong, Y., (2012). Synthesis and antiviral evaluation of 7-O-arylmethylquercetin derivatives against SARS-associated coronavirus (SCV) and hepatitis C virus (HCV). *Arch. Pharm. Res., 35*, 77–85.

Park, J. Y., Ko, J. A., Kim, D. W., Kim, Y. M., Kwon, H. J., Jeong, H. J., Kim, C. Y., Park, K. H., Lee, W. S., & Ryu, Y. B., (2016). Chalcones isolated from *Angelica keiskei* inhibit cysteine proteases of SARS-CoV. *J. Enzyme Inhib. Med. Chem., 31*, 23–30.

Park, J. Y., Yuk, H. J., Ryu, H. W., Lim, S. H., Kim, K. S., Park, K. H., Ryu, Y. B., & Lee, W. S., (2017). Evaluation of polyphenols from *Broussonetia papyrifera* as coronavirus protease inhibitors. *J. Enzym. Inhib. Med. Chem., 32*, 504–515.

Pendyala, B., & Patras, A., (2020). *In Silico Screening of Food Bioactive Compounds to Predict Potential Inhibitors of COVID-19 Main Protease (Mpro) and RNA-Dependent RNA Polymerase (RdRp).* ChemRxiv. Preprint 12051927.

Peng, H. L., Huang, W. C., Cheng, S. C., et al., (2018). Fisetin inhibits the generation of inflammatory mediators in interleukin-1betainduced human lung epithelial cells by suppressing the NFjB and ERK1/2 pathways. *Int. Immunopharmacol., 60*, 202–210.

Ragab, F. A., Yahya, T. A., El-Naa, M. M., et al., (2014). Design, synthesis, and structure-activity relationship of novel semisynthetic flavonoids as antiproliferative agents. *Eur. J. Med. Chem., 82*, 506–520.

Ramalingam, M., Kim, H., Lee, Y., & Lee, Y. I., (2018). Phytochemical and pharmacological role of liquiritigenin and isoliquiritigenin from radix glycyrrhiza in human health and disease models. *Front. Aging Neurosci., 10*, 348.

Rauf, A., Khan, R., Raza, M., et al., (2015). Suppression of inflammatory response by chrysin, a flavone isolated from *Potentilla evestita* Th. Wolf. In silico predictive study on its mechanistic effect. *Fitoterapia, 103*, 129–135.

Rogerio, A. P., Dora, C. L., Andrade, E. L., et al., (2010). Anti-inflammatory effect of quercetin-loaded microemulsion in the airways allergic inflammatory model in mice. *Pharmacol. Res., 61,* 288–297.

Rungsung, S., Singh, T. U., Rabha, D. J., Kumar, T., Cholenahalli, L. M., Parida, S., & Kumar, D., (2018). Luteolin attenuates acute lung injury in an experimental mouse model of sepsis. *Cytokine, 110*, 333–343.

Ryu, Y. B., Jeong, H. J., Kim, J. H., Kim, Y. M., Park, J. Y., Kim, D., Nguyen, T. T., et al., (2010). Bioflavonoids from *Torreya nucifera* are displaying SARS-CoV 3CL(pro) inhibition. *Bioorg. Med. Chem., 18*, 7940–7947.

Sadati, S. M., Gheibi, N., Ranjbar, S., & Hashemzadeh, M. S., (2019). Docking study of flavonoid derivatives as potent inhibitors of influenza H1N1 virus neuraminidase. *Biomed. Rep., 10*, 33–38.

Sassi, A., Mokdad, B. I., Mustapha, N., et al., (2017). Immunomodulatory potential of hesperetin and chrysin through the cellular and humoral response. *Eur. J. Pharmacol., 812*, 91–96.

Schwarz, S., Sauter, D., Wang, K., Zhang, R., Sun, B., Karioti, A., Bilia, A. R., Efferth, T., & Schwarz, W., (2014). Kaempferol derivatives as antiviral drugs against the 3a channel protein of coronavirus. *Planta Med., 80*, 177.

Sekiou, O., Bouziane, I., Frissou, N., et al., (2020). In-silico identification of potent inhibitors of COVID-19 main protease (Mpro) from natural products. *International Journal of Biochemistry & Physiology (IJBP), 5*, 000189.

Sharma, A. D., & Inderjeet, K., (2020a). Eucalyptol (1,8 cineole) from eucalyptus essential oil is a potential inhibitor of COVID-19 Coronavirus infection by molecular docking studies. *Notulae Scientia Biologicae, 12*, 536–545.

Sharma, A. D., & Inderjeet, K., (2020b). Bioactive molecules from eucalyptus essential oil as potential inhibitors of COVID-19 coronavirus infection by molecular docking studies. *Kragujevac J. Sci., 42*, 29–43.

Sharma, A. D., & Inderjeet, K., (2020c). Jensenone from eucalyptus essential oil is a potential inhibitor of COVID-19 coronavirus infection. *Research and Reviews in Biotechnology and Biosciences, 7*, 60–67.

Shen, Y., Tian, P., Li, D., et al., (2015). Chrysin suppresses cigarette smoke-induced airway inflammation in mice. *Int. J. Clin. Exp. Med., 8*, 2001–2008.

Sithisarn, P., Michaelis, M., Schubert-Zsilavecz, M., & Cinatl, J. Jr., (2013). Differential antiviral and anti-inflammatory mechanisms of the flavonoids biochanin A and baicalein in H5N1 influenza A virus-infected cells. *Antivir. Res., 97*, 41–48.

Solnier, J., & Fladerer, J. P., (2020). Flavonoids: A complementary approach to conventional therapy of COVID-19? *Phytochem. Rev*. https://doi.org/10.1007/s11101-020-09720-6.

Song, J., Zhang, L., Xu, Y., Yang, D., Yang, S., Zhang, W., Wang, J., Tian, S., Yang, S., & Yuan, T., (2021). The comprehensive study on the therapeutic effects of baicalein for the treatment of COVID-19 *in vivo* and *in vitro*. *Biochem. Pharmacol., 183*, 114302.

Soromou, L. W., Chu, X., Jiang, L., Wei, M., Huo, M., Chen, N., & Deng, X., (2012). *In vitro* and *in vivo* protection provided by pinocembrin against lipopolysaccharide-induced inflammatory responses. *International Immunopharmacology, 14*, 66–74.

Soukhova, N., Soldin, O. P., & Soldin, S. J., (2004). Isotope dilution tandem mass spectrometric method for T4/T3. *Clin. Chim. Acta, 343*, 185–190.

Su, H., Yao, S., Zhao, W., Li, M., Liu, J., Shang, W., et al. (2020). Discovery of baicalin and baicalein as novel, natural product inhibitors of SARS-CoV-2 3CL protease *in vitro*. *BioRxiv*. doi: 10.1101/2020.04.13.038687.

Tallei, T., Tumilaar, S., Niode, N., et al., (2020). Potential of plant bioactive compounds as SARS-CoV-2 main protease (Mpro) and spike (S) glycoprotein inhibitors: A molecular docking study. *Scientifica*. https://doi.org/10.1155/2020/ 6307457.

Tseng, T. L., Chen, M. F., Tsai, M. J., Hsu, Y. H., Chen, C. P., & Lee, T. J. F., (2012). Oroxylin-A rescues LPS-induced acute lung injury via regulation of NF-κB signaling pathway in rodents. *PLoS One, 7*, e47403–e47413.

Tutunchi, H., Naeini, F., Ostadrahimi, A., & Hosseinzadeh-Attar, M. J., (2020). Naringenin, a flavanone with antiviral and anti-inflammatory effects: A promising treatment strategy against COVID-19. *Phytother. Res., 34*, 3137–3147.

Tzotzos, S. J., Fischer, B., Fischer, H., et al., (2020). Incidence of ARDS and outcomes in hospitalized patients with COVID-19: A global literature survey. *Crit. Care, 24*, 516.

Utomo, R. Y., Ikawati, M., & Meiyanto, E., (2020). *Revealing the Potency of Citrus and Galangal Constituents to Halt SARSCoV-2 Infection*. Preprints 2020:2020030214.

Wan, Y., Shang, J., Graham, R., et al., (2020). Receptor recognition by the novel coronavirus from Wuhan: An analysis based on decade-long structural studies of SARS coronavirus. *J. Vitro, 94*, e00120–e127.

Wei, X., Zhu, X., Hu, N., Zhang, X., Sun, T., Xu, J., & Bian, X., (2015). Baicalin attenuates angiotensin II-induced endothelial dysfunction. *Biochemical and Biophysical Research Communications, 465*, 101–107.

Wu, C., Liu, Y., Yang, Y., Zhang, P., Zhong, W., Wang, Y., Wang, Q., et al., (2020). Analysis of therapeutic targets for SARS-CoV-2 and discovery of potential drugs by computational methods. *Acta Pharmaceutica Sinica B, 10*, 766–788.

Xie, Y. C., Dong, X. W., Wu, X. M., Yan, X. F., & Xie, Q. M., (2009). Inhibitory effects of flavonoids extracted from licorice on lipopolysaccharide-induced acute pulmonary inflammation in mice. *International Immunopharmacology, 9*, 194–200.

Yahia, E. M., Garcı´a-Solı´s, P., & Celis, M. E. M., (2019). Contribution of fruits and vegetables to human nutrition and health. *Postharvest Physiology and Biochemistry of Fruits and Vegetables* (pp. 19–45). Woodhead Publishing, Cambridge.

Yang, T., Luo, F., Shen, Y., et al., (2012). Quercetin attenuates airway inflammation and mucus production induced by cigarette smoke in rats. *Int. Immunopharmacol., 13*, 73–81.

Ye, J., Guan, M., Lu, Y., et al., (2019). Protective effects of hesperetin on lipopolysaccharide-induced acute lung injury by targeting MD2. *Eur. J. Pharmacol., 852*, 151–158.

Yi, L., Li, Z., Yuan, K., Qu, X., Chen, J., Wang, G., Zhang, H., et al., (2004). Small molecules blocking the entry of severe acute respiratory syndrome coronavirus into host cells. *J. Virol., 78,* 11334–11339.

Yoshida, H., Takamura, N., Shuto, T., et al., (2010). The citrus flavonoids hesperetin and naringenin block the lipolytic actions of TNF-alpha in mouse adipocytes. *Biochem Biophys Res. Commun., 394*, 728–732.

Yu, M. S., Lee, J., Lee, J. M., Kim, Y., Chin, Y. W., Jee, J. G., Keum, Y. S., & Jeong, Y. J., (2012). Identification of myricetin and scutellarein as novel chemical inhibitors of the SARS coronavirus helicase, nsP13. *Bioorg. Med. Chem. Lett., 22*, 4049–4054.

Zandi, K., Musall, K., Oo, A., Cao, D., Liang, B., Hassandarvish, P., Lan, S., Slack, R. L., Kirby, K. A., & Bassit, L., (2021). Baicalein and baicalin inhibit SARS-CoV-2 RNA-dependent-RNA polymerase. *Microorganisms, 9*, 893.

Zeinali, M., Rezaee, S. A., & Hosseinzadeh, H., (2017). An overview of immunoregulatory and anti-inflammatory properties of chrysin and flavonoid substances. *Biomed. Pharmacother., 92*, 998–1009.

Zhan, Y., Ta, W., Tang, W., Hua, R., Wang, J., Wang, C., & Lu, W., (2021). Potential antiviral activity of isorhamnetin against SARS-CoV-2 spike pseudotyped virus *in vitro*. *Drug. Dev. Res.*

Zhang, R., Wang, X., Ni, L., Di, X., Ma, B., Niu, S., Liu, C., & Reiter, R. J., (2020). COVID-19: Melatonin as a potential adjuvant treatment. *Life Sci., 250*, 117583.

Zhang, X., Wang, G., Gurley, E. C., & Zhou, H., (2014). Flavonoid apigenin inhibits lipopolysaccharide-induced inflammatory response through multiple mechanisms in macrophages. *PLoS One, 9*(9), e107072.

Zhu, G. F., Guo, H. J., Huang, Y., Wu, C. T., & Zhang, X. F. (2015). Eriodictyol, a plant flavonoid, attenuates LPS-induced acute lung injury through its antioxidative and anti-inflammatory activity. *Experimental and Therapeutic Medicine, 10*, 2259–2266.

Zhuang, M., Jiang, H., Suzuki, Y., Li, X., Xiao, P., Tanaka, T., Ling, H., et al., (2009). Procyanidins and butanol extract of cinnamomi cortex inhibit SARS-CoV infection. *Antiviral Res., 82*, 73–81.

Zu, M., Yang, F., Zhou, W. L., Liu, A. L., Du, G. H., & Zheng, L. S. (2012). *In vitro* anti-influenza virus and anti-inflammatory activities of theaflavin derivatives. *Antivir. Res., 94*, 217–224.

INDEX

B

D

E

F

G

H

Q

R